T0256011

Security and Privacy Vision in 6G

Security and Privacy Vision in 6G

A Comprehensive Guide

Pawani Porambage
VTT Technical Research Centre of Finland
University of Oulu

Madhusanka Liyanage
University College Dublin
University of Oulu

IEEE PRESS
WILEY

For general information on our other products and services or for technical support, please contact our Customer Care Department within the United States at (800) 762-2974, outside the United States at (317) 572-3993 or fax (317) 572-4002.

Wiley also publishes its books in a variety of electronic formats. Some content that appears in print may not be available in electronic formats. For more information about Wiley products, visit our web site at www.wiley.com.

Library of Congress Cataloging-in-Publication Data

Names: Porambage, Pawani, author. | Liyanage, Madhusanka, author.
Title: Security and privacy vision in 6G : a comprehensive guide / Pawani Porambage, Madhusanka Liyanage.
Description: Hoboken, New Jersey : Wiley-IEEE Press, [2023] | Includes index.
Identifiers: LCCN 2023013357 (print) | LCCN 2023013358 (ebook) | ISBN 9781119875406 (cloth) | ISBN 9781119875413 (adobe pdf) | ISBN 9781119875420 (epub)
Subjects: LCSH: 6G mobile communication systems–Security measures. | Data privacy.
Classification: LCC TK5103.252 .P67 2023 (print) | LCC TK5103.252 (ebook) | DDC 621.3845/6–dc23/eng/20230417
LC record available at https://lccn.loc.gov/2023013357
LC ebook record available at https://lccn.loc.gov/2023013358

Cover Design: Wiley
Cover Image: © KanawatTH/Shutterstock

Set in 9.5/12.5pt STIXTwoText by Straive, Chennai, India

To 6G Seekers

Contents

Acronyms

5G	fifth generation
6G	sixth generation of wireless networks
3D	three-dimensional
ABS	aerial base station
AES	advanced encryption standard
AF	amplify-and-forward
AI	artificial intelligence
AN	artificial noise
APs	access points
BS	base station
CFR	channel frequency response
CIR	channel impulse response
CJ	cooperative jamming
CSI	channel state information
CPU	central processing unit
DF	decode-and-forward
EMI	electro-magnetic interference
FEC	forward error correction
FD	full-duplex
FSO	free space optics
HDBN	hierarchical dynamic Bayesian network
IRS	intelligent reflecting surface
IoT	Internet of Things
IoV	Internet of Vehicles
KPIs	key performance indicators
JCS	joint communications and sensing
LoS	line of sight
MEC	mobile-edge computing
MIMO	multiple-input multiple-output

MISO	multiple-input single-output
MC	molecular communications
ML	machine learning
mMIMO	massive MIMO
mmWave	millimeter-wave
mMTC	massive machine-type communications
PLS	physical layer security
PLA	physical layer authentication
PSD	power spectral density
QoS	quality-of-service
RF	radio-frequency
RIS	reconfigurable intelligent surfaces
RSS	received signal strength
SKG	secret key generation
SNR	signal-to-noise ratio
SOP	secrecy outage probability
TDD	time-division duplexing
THz	Terahertz
UAVs	unmanned aerial vehicles
UEs	user equipment
URLLC	ultrareliable low-latency communications
VLC	visible-light communication
V2X	vehicular-to-everything
WBAN	wireless body-area networks

About the Authors

Pawani Porambage, D.Sc., is Senior Scientist at the VTT Technical Research Centre of Finland and a Docent at University of Oulu. She has over eleven years experience in network security research and has authored or co-authored more than 70 publications.

Madhusanka Liyanage is an Assistant Professor/Ad Astra Fellow and Director of Graduate Research at the School of Computer Science, University College Dublin, Ireland. He is also an Adjunct Professor at the University of Oulu, Finland, the University of Ruhuna, Sri Lanka, and the University of Sri Jarawardhenepura, Sri Lanka. He is also an expert consultant at the European Union Agency for Cybersecurity (ENISA) and a Funded Investigator of the Science Foundation Ireland CONNECT Research Centre, Ireland. He has over a decade of research experience in telecommunication network security. He has co-authored over 150 publications, including three authored books, four edited books with Wiley, and two patents.

Foreword

With each generation of communication networks, digital technologies progress such that they transform the way we live and work. Improvements in network speed and latency make new applications possible and change what we are able to do with existing applications. As we look toward 6th-generation (6G) networks, we see potential for new classes of real-time, immersive, and tactile applications. The Internet will not only be a source of information but also a tool to actuate the world around us. Edge clouds will make the power of cloud computing local, an integral and activating part of the built environment of our homes, communities, and personal spaces. Higher performance edge connectivity is expected to usher in a new generation of connected vehicles, drones, and robotic devices, overcoming many of the limitations we are seeing in fully autonomous constructs. As part of this, networks will increasingly become distributed sensors and distributed artificial intelligence (AI) platforms, collecting and processing data relating to virtually everything that affects our lives.

While the power of data and AI that 6G will bring promises many benefits, it also carries significant risks that cannot be ignored. Security and privacy truly become paramount in a 6G-empowered world. Historically, these issues have been treated as add-on features or afterthought technologies in communication networks. If we want to properly center security and privacy for 6G, as we must, then a good starting point is a vision for security and privacy in 6G, which is exactly what this book accomplishes. For those who want to contribute to or understand the formation of 6G and 6G applications, this book is an essential first stop on their journey. Indeed, it provides a comprehensive overview of the technologies, architectures, and applications that will underpin and form 6G. Here the reader will gain a full picture of what is emerging for 6G from a pair of authors who are well situated within the maelstrom of the 6G technology R&D community. The reader is guided through the evolution of mobile technologies and the key enabling technologies of 6G that are vital to achieving the desired performance and functional enhancements. However, it is all done from the vantage point of security and privacy. The deployment-focused perspectives of device security, Open-RAN security (i.e. access networks), edge intelligence security (i.e. the mobile edge expanding to the core), and network slicing security (i.e. network management) cover all the architectural requirements of the 6G vision. In addition, the reader is introduced to major application trends including Industry 5.0, Metaverse, Society 5.0, Internet-of-Vehicles, and Smart Grid 2.0, and learns about the corresponding security implications. Importantly, the security requirements are aligned with the privacy obligations associated with the use of these systems and applications. Furthermore, the assimilated requisites for each are presented against the prevailing security standardization and legal frameworks for better clarity on their global compliance.

The latter half of the book delves more deeply into specific expert topics. Blockchain is one of the developing pillars for 6G due to its adaptability as a scalable and distributed security and privacy solution. It is widely discussed throughout, and a chapter is dedicated specifically to distributed ledger technologies and blockchain. Similar treatment is given to AI and machine learning (ML), explainable artificial intelligence (XAI), zero-touch network and service management (ZSM), physical layer security (PLS), and quantum security. These chapters serve as succinct introductions and reference material for the respective topics and further examine the security and privacy aspects.

Whether seeking a first introduction to 6G or searching for specific details on emerging 6G technologies and applications, this book will prove an essential resource and one that properly brings focus to the critical issues of security and privacy. From this perspective, readers will find this a reliable reference for information across the entire breadth of what we refer to as 6G – one that will get heavy wear whether on the digital or physical bookshelf.

Dan Kilper
Director and Principal Investigator
SFI CONNECT Center
Ireland

Preface

The evolution of wireless telecommunication networks started with the first-generation cellular networks in the 1980s. Since then, many advancements have been introduced in 2G, 3G, and 4G cellular networks. Currently, we are experiencing the fifth-generation (5G) wireless networks which are yet to evolve mostly on software-based till the 2025 providing the full coverage. Even though 5G coverage is not yet being fully implemented, the vision for sixth generation (6G) mobile communication is already projected. It is envisioned that the 6G standardization will start somewhere in 2026 with a great touch of intelligent network orchestration and management. The most significant driving force of 6G vision is the added intelligence in the telecommunication networks. Similar to the network softwarization and cloudification which pave the way to 5G, artificial intelligence (AI) and machine learning (ML) techniques will lead the journey of 6G toward the intelligent telecommunication networks. 6G vision may include many novelties and advancements in terms of applications, architecture, technologies, policies, and standardizations. These novelties and added intelligence may also have a close fusion with the security and privacy aspects of 6G. On the other hand, the adversaries also become more powerful and intelligent and capable of creating new forms of security threats. As an example, detecting zero-day attacks is always challenging, whereas prevention from their propagation is the most achievable mechanism. Therefore, 6G security needs to be architected to not only protect from the threats in the foreseen 6G networks but also to address the increased and evolving threat landscape. Adequate security should include prediction, detection, mitigation, and prevention mechanisms, and the ability to limit the propagation of such vulnerabilities, with greater intelligence, visibility, and real-time protection. It is also equally significant to ensure privacy and trust in the respective domains and among the stakeholders. Security and privacy are two closely coupled topics where security relates to the safeguarding of the actual data and privacy ensures the covering up of the identities related to that data.

The security and privacy considerations in the envisioned 6G networks need to be addressed with respect to many areas. There are specific security issues that may arise with the novel 6G architectural framework as stated above. In addition to that, there are many hypes on blending novel technologies such as blockchain, visible light communication (VLC), TeraHertz (THz), and quantum computing features in 6G intelligent networking paradigms in such a way to tackle the security and privacy issues. Therefore, 6G security considerations need to be also discussed with respect to the physical layer security (PLS), network information security, application security, and deep learning related security. This leads to a need to strengthen certain security functional areas. Attack resistance needs to be a design consideration when defining new 6G protocols and key performance indicators. Security and privacy are cornerstones for 6G to become a platform for the Networked Society. Cellular systems pioneered the creation of security solutions for public communication, providing a vast, trustworthy ecosystem – 6G will drive new requirements due to new business and trust models, new service delivery models, an evolved threat landscape and an increased concern for privacy.

Primary market: This book will be of key interest for the following:

- **Telecommunication researchers**: 6G security and privacy are two key areas of interest for telecommunications researchers as security challenges outpaces the traditional tools available to market and to introduce ground-breaking solutions. This book will offer a single source of all the security-related topics for 6G researchers and provide leads for basics of 6G security and privacy vision.

- **Academics**: Mobile network security has already been an area of research and study for major educational institutions across the world. At the very initial stage of 6G evolution as the future of mobile networks, there is no such reference and book available that academics can use for teaching this critical area of interest.

- **Mobile network operators (MNOs)** will be looking to embrace 6G technology to offer new and state-of-the-art secure services to their customers. This book will offer the required guidelines, methods, tools, and mechanisms to secure their network while getting ready to reach the next generation of 6G.

- **Mobile network virtual operators (MVNOs)** would like to equally reach the extremely large customer base who is going to switch to 6G networks. Security is key requirement while connecting MVNOs with the core networks of large operators.

- **Technology architects**: 6G is going to surpass the traditional mobility borders and is going to have an equal impact to enterprises and organizations who are planning to transform into digital businesses. It would be critical for architects to start aligning their technology and security architectures to the anticipated future needs of 6G standards. This book offers a roadmap and the tentative resources to design and build a security architecture and maintain it.

- **IoT service providers**: The trend toward 6G is moving from Internet of Things (IoT) to Internet of Everything (IoE). Most of the daily consumable devices are becoming more advanced with a greater connectivity and performance levels. The devices may encounter new security threats and users will be more privacy aware. Therefore, the IoT service providers in this domain should be well aware of the novel technological advancements and 6G trends together with the security and privacy considerations to provide consistent services.

1. Book Organization

This book provides a comprehensive overview about the security and privacy aspects related to 6G vision and it is composed of six parts. The first part of the book is the introduction with three chapters that describe evolution of mobile networks, high-level overview about key 6G technologies, and 6G security vision. Part II of the book includes the security and privacy considerations with respect to the architectural phases of 6G mobile network. The chapters are allocated for 6G device security, open-RAN and RAN-Core convergence, edge intelligence and specialized 6G networks, and network slicing. Part III of the book describes the most compelling 6G application areas and their security concerns. The applications are discussed in three main areas such as Society 5.0, Internet of Vehicles, smart grid 2.0. Part IV is devoted for the privacy considerations of 6G including detailed analysis on challenges, issues, and potential solutions. Moreover, Part IV includes one chapter for legal aspects and security standardization. In Part V, the chapters are aligned with the security considerations in 6G technologies in terms of Distributed Ledger Technologies (DLTs) and Blockchain, AI/ML, Explainable AI, Zero Touch Network and Service Management, Physical Layer Security, and quantum security. Finally, we end the book with the concluding remarks in Part VI.

6 April 2023

Pawani Porambage
Oulu Finland

Acknowledgments

The book is focused on security and privacy vision of 6G and formed with the valuable inputs received from many people. We would like to thank all those who contributed to selected chapters of the book. In particular we very much appreciate the help of Dr. Onel Alcaraz Lopez, Dr. Diana Moya Osorio, Proff. Kimmo Halunen, Sara Nikula, Jose Vega Sanchez, Edgar Olivo, Dr. Andre N. Barreto, Saeid Sheikhi, Chamara Sandeepa, Yushan Senavirathne, Charithri Yapa, Thulitha Senavirathna, Tharaka Hewa, Dr. Pasika Ranaweera, Dr. Anshuman Kalla and Dr. Chamitha de Alwis for additional contribution to some chapters.

The initial idea for this book originated during our joint research work in 6G Flagship and the research articles about 6G security and privacy published in IEEE Open Journal of Communications and EuCNC (European Conference on Networks and Communications) and 6G Summit. The concept of publishing this book to facilitate 6G Security related studies, research, development, and standardization came into light during our research work in projects such as the European Union funded SPATIAL project, Academy of Finland funded 6Genesis project, and Science Foundation Ireland funded CONNECT phase 2 project. We would also like to acknowledge all the partners of those projects. Moreover, we thank the anonymous reviewers who evaluated the proposal and gave plenty of useful suggestions for improving it. We also thank Sandra Grayson, Juliet Booker and Becky Cowan from John Wiley and Sons for help and support in getting the book published.

Also, the Authors are grateful to VTT Technical Research Centre of Finland, School of Computer Science at University College Dublin, Centre for Wireless Communication (CWC) at University of Oulu, for hosting the 6G Security related research projects which helped us to gain the fundamental knowledge for this book. Last but not least, we would like to thank our core and extended families and our friends for their love and support in getting the book completed.

Pawani Porambage
Madhusanka Liyanage

Part I

Introduction

1

Evolution of Mobile Networks

This chapter has focused on the evolution, driving trends, and key requirements of future 6G wireless systems. After reading this chapter, you should be able to:

- Understand the evolution of mobile networks from 0G to 6G.
- Understand the present context of 6G development.

1.1 Introduction

While 5G mobile communication networks are deployed worldwide, multitude of new applications and use-cases driven by current trends are already being conceived, which challenges the capabilities of 5G. This has motivated researchers to rethink and work toward the next-generation mobile communication networks "hereafter 6G" [1, 2]. The 6G mobile communication networks are expected to mark a disruptive transformation to the mobile networking paradigm by reaching extreme network capabilities to cater to the demands of the future data-driven society.

So far, mobile networks have evolved through five generations during the last four decades. A new generation of mobile networks emerges every 10 years, packing more technologies and capabilities to empower humans to enhance their work and lifestyle. The precellphone era before the 1980s is marked as the zeroth-generation (0G) of mobile communication networks that provided simple radio communication functionality with devices such as walkie-talkies [3]. The first-generation (1G) introduced publicly and commercially available cellular networks in the 1980s. These networks provided voice communication using analog mobile technology [4]. The second generation (2G) of mobile communication networks marked the transition of mobile networks from analog to digital. It supported basic data services such as short message services in addition to

Security and Privacy Vision in 6G: A Comprehensive Guide, First Edition.
Pawani Porambage and Madhusanka Liyanage.
© 2023 The Institute of Electrical and Electronics Engineers, Inc. Published 2023 by John Wiley & Sons, Inc.

voice communication [5]. The third-generation (3G) introduced improved mobile broadband services and enabled new applications such as multimedia message services, video calls, and mobile TV [6]. Further improved mobile broadband services, all-IP communication, Voice Over IP (VoIP), ultrahigh definition video streaming, and online gaming were introduced in the fourth-generation (4G) [7].

5G mobile communication networks are already being deployed worldwide. 5G supports enhanced Mobile Broadband (eMBB) to deliver peak data rates up to 10 Gbps. Furthermore, ultra Reliable Low Latency Communication (uRLLC) minimizes the delays up to 1 ms, while massive Machine-Type Communication (mMTC) supports over 100× more devices per unit area compared to 4G. The expected network reliability and availability are over 99.999% [8]. Network softwarization is a prominent 5G technology that enables dynamicity, programmability, and abstraction of networks [9]. Capabilities of 5G have enabled novel applications such as virtual reality (VR), augmented reality (AR), mixed reality (MR), autonomous vehicles, Internet of Things (IoT), and Industry 4.0 [10, 11].

Recent developments in communications have introduced many new concepts such as edge intelligence (EI), beyond sub 6 GHz to THz communication, non-orthogonal multiple access (NOMA), large intelligent surfaces (LIS), swarm networks, and self-sustaining networks (SSN) [12, 13]. These concepts are evolving to become full-fledged technologies that can power future generations of communication networks. On the other hand, applications such as holographic telepresence (HT), unmanned aerial vehicles (UAV), extended reality (XR), smart grid 2.0, Industry 5.0, space, and deep-sea tourism are expected to emerge as mainstream applications of future communication networks. However, requirements of these applications such as ultra-high data rates, real-time access to powerful computing resources, extremely low-latency, precision localization and sensing, and extremely high reliability and availability surpass the network capabilities promised by 5G [14, 15]. IoT, which is enabled by 5G, is even growing to become Internet of Everything (IoE) that intends to connect massive numbers of sensors, devices, and cyber-physical systems (CPS) beyond the capabilities of 5G. This has inspired the research community to envision 6G mobile communication networks. 6G is expected to harness the developments of new communication technologies, fully support emerging applications, connect a massive number of devices, and provide real-time access to powerful computational and storage resources.

1.2 6G Mobile Communication Networks

6G networks are expected to be more capable, intelligent, reliable, scalable, and power-efficient to satisfy all the expectations that cannot be realized with 5G. 6G is also required to meet any new requirements, such as support for new technologies, applications, and regulations, raised in the coming decade. Figure 1.1 illustrates

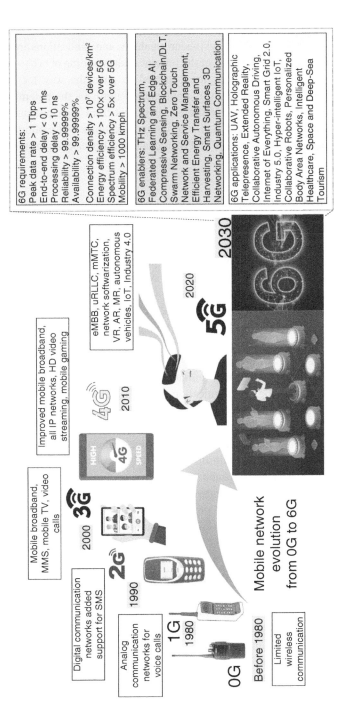

Figure 1.1 Evolution of mobile networks from 0G to 6G. Source: vectorplus/Adobe Stock; hakule/Adobe Stock; ostapenko/123RF; limpix/Shutterstock.

the evolution of mobile networks, elaborating key features of each mobile network generation. Envisaged 6G requirements, vision, enablers, and applications are also highlighted to formulate an overview of the present understanding of 6G.

Summary: THz communications are expected to pave the way for Tbps data rate to meet the demands of future applications and have the potential to strengthen backhaul networks. Nevertheless, it suffers from high propagation losses and demands Line of Sight (LoS) for communications. More efforts are required to understand the behavior of THz signals and better channel models are required.

1.2.1 6G as Envisioned Today

6G mobile communication networks, as envisioned today, are expected to provide extreme peak data rates over 1 Tbps. The end-to-end delays will be imperceptible and lie even beneath 0.1 ms. 6G networks will provide access to powerful edge intelligence that has processing delays falling below 10 ns. Network availability and reliability are expected to go beyond 99.99999%. An extremely high connection density of over 10^7 devices/km^2 is expected to be supported to facilitate IoE. The spectrum efficiency of 6G will be over 5× than 5G, while support for extreme mobility up to 1000 kmph is expected [12].

It is envisioned that the evolution of 6G will focus around a myriad of new requirements such as Further enhanced Mobile Broadband (FeMBB), ultra-massive Machine-Type Communication (umMTC), Mobile BroadBand and Low-Latency (MBBLL), and massive Low-Latency Machine Type communication (mLLMT). These requirements will be enabled through emerging technologies such as THz spectrum, federated learning (FL), edge artificial intelligence (AI), compressive sensing (CS), blockchain/distributed ledger technologies (DLT), and 3D networking. Moreover, 6G will facilitate emerging applications such as UAVs, HT, IoE, Industry 5.0, and collaborative autonomous driving. In light of this vision, many new research work and projects are themed toward developing 6G vision, technologies, use cases, applications, and standards [1, 2].

1.3 Key Driving Trends Toward 6G

A new generation of mobile communication has emerged every 10 years over the last four decades to cater to society's growing technological and societal needs. This trend is expected to continue, and 6G is seen on the horizon to meet the requirements of the 2030 society [16, 17]. The technologies, trends, requirements, and expectations that force the shift from 5G toward the next generation of

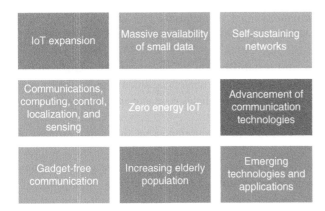

Figure 1.2 6G driving trends. Source: [18, 19]/IEEE.

networks are identified as 6G driving trends. These driving trends will shape 6G into the key enabler of a more connected and capable 2030 society.

This chapter discusses the key 6G driving trends elaborating why and how each trend demands a new generation of communication networks. Figure 1.2 illustrates the 6G driving trends that are discussed in this section.

- **Expansion of IoTs**: It is expected that the number of IoT devices in the world will grow up to 24 billion by 2030. Moreover, the revenue related to IoT will hit the market capitalization of US$ 1.5 trillion by 2030 [20].
- **Massive availability of small data**: Due to the anticipated popularity of 6G-based IoT devices and new 6G-IoT services, 6G networks will trend to generate an increasingly high volume of data. Most of such data will be small, dynamic, and heterogeneous in nature [12, 21].
- **Availability of self-sustained networks**: 6G mobile systems need to be energy self-sustainable, both at the infrastructure side and at the device side, to provide uninterrupted connectivity in every corner of the world. The development of energy harvesting capabilities will extend the life cycle of both network infrastructure devices and end devices such as IoE devices [22, 23].
- **Convergence of communication, sensing, control, localization, and computing**: Development of sensor technologies and direct integration of them with mobile networks accompanied by low-energy communication capabilities will lead to advanced 6G networks [12, 24]. Such a network will be able to provide sensing and localization services in addition to the exciting communication and computing features [12, 24, 25].
- **Zero energy IoT**: Generally, IoT devices will consume significantly more energy for communication than sensing and processing [26]. The development

of ultra-low-power communication mechanisms and efficient energy harvesting mechanisms will lead to self-energy sustainable or zero energy IoT devices [26].

- **More bits, spectrum, and reliability**: The advancement of wireless communication technologies, including coding schemes and antenna technologies, will allow to utilize new spectrum as well as reliably send more information bits over existing wireless channels [12, 16].
- **Gadget-free communication**: The integration of an increasing number of smart and intelligent devices and digital interfaces in the environment will lead to a change from gadget-centric to user-centric or gadget-free communication model. The hyperconnected digital surroundings will form an "omnipotential" atmosphere around the user, providing all the information, tools, and services that a user needs in his or her everyday life [27–29].
- **Increasing elderly population**: Due to factors such as advanced healthcare facilities and the development of new medicines, the world's older population continues to grow at an unprecedented rate. According to the "An Aging World: 2015" report, nearly 17% (1.6 billion) of the world's population will be aged 65 and over by 2050 [19].
- **Emergence of new technologies**: By 2030, the world will experience new technological advancements such as stand-alone cars, AI-powered automated devices, smart clothes, printed bodies in 3D, humanoid robots, smart grid 2.0, industry 5.0, and space travel [12, 16]. 6G will be the main underline communication infrastructure to realize these technologies.

1.4 6G Requirements/Vision

To realize new applications, 6G networks have to provide extended network capabilities beyond 5G networks. Figure 1.3 depicts such requirements which need to be satisfied by 6G networks to enable future applications.

As adopted from various studies [30–34], 6G networking requirements can be divided into different categories as follows:

- **Further enhanced Mobile Broadband (FeMBB)**: The mobile broadband speed has to be further improved beyond the limits of 5G and provide the peak data rate at Terabits per second (Tbps) level. Moreover, the user-experienced data rate should also be improved up to Gigabits per second (Gbps) level [35].
- **Ultra-massive Machine-Type Communication (umMTC)**: Connection density will further increase in 6G due to the popularity of IoT devices and the

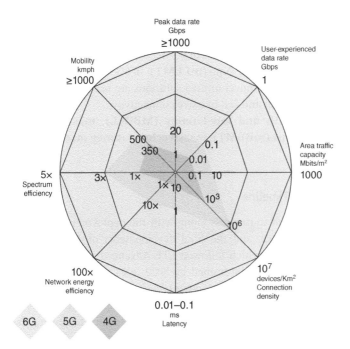

Figure 1.3   6G requirements.

novel concept of IoE. These devices communicate with each other and offer collaborative services in an autonomous manner [36, 37].

- **Enhanced Ultra-Reliable, Low-Latency Communication (ERLLC/ eURLLC)**: The E2E latency in 6G should be further reduced up to µs level to enable new high-end real-time 6G applications [17].
- **Extremely Low-Power Communications (ELPC)**: The network energy efficiency of 6G will be improved by 10× than 5G and 100× than 4G. It will enable extremely low-power communication channels for resource constrained IoT devices [17, 38].
- **Long-Distance High-Mobility Communications (LDHMC)**: With the support of fully integrated satellite technologies, 6G will provide communication for extreme places such as space and the deep sea. Moreover, AI-based automated mobility management systems and proactive migration systems will be able to support seamless mobility at speed beyond 1000 kmph [35].
- **High-Spectrum Efficiency**: The spectrum efficiency will be further improved in 6G up to 5 times as in 4G and nearly two times as in 5G networks [17].

- **High-Area Traffic Capacity**: The exponential growth of IoT will demand the improvement of the area traffic capacity by 100 times than 5G networks. It will lead up to 1 Gbps traffic per square meter in 6G networks.
- **Massive Low-Latency Machine Type (mLLMT)**: In 6G, URLLC and mMTC services should be linked, and novel unified solutions are needed to meet the challenge of offering efficient and fast massive connectivity.
- **Mobile broad bandwidth and low-latency (MBBLL), massive broad bandwidth machine type (mBBMT), massive, low-latency machine type (mLLMT)**

1.4.1   6G Development Timeline

G developments are expected to progress along with the deployment and commercialization of 5G networks, and the final developments of 4G long-term evolution (LTE), being LTE-C, which followed LTE-Advanced and LTE-B [39]. The vision for 6G is envisaged to be framed by 2022–2023 to set forth the 6G requirements and evaluate the 6G development, technologies, standards, etc. Standardization bodies such as the International Telecommunication Union (ITU) and 3rd Generation Partnership Project (3GPP) are expected to develop the specifications to develop 6G by 2026–2027 [39]. Network operators will start 6G research and development (R&D) work by this time to do 6G network trials by 2028–2029, to launch 6G communication networks by 2030 [14, 39–41]. Global 6G development initiatives are illustrated in Figure 1.5, while the expected timeline for 6G development, standardization, and launch is presented in Figure 1.4.

		2020	2021	2022	2023	2024	2025	2026	2027	2028	2029	2030
4G	4G network operators	LTE-C										
5G	5G network operators	R&D Trial										
						5G commercialization						
	5G research					5G evolution (beyond 5G)						
6G	ITU standardization			6G vision			6G requirements			6G evaluation		
	3GPP standardization					6G study		6G specifications		6G products		
	6G research		Structuring and framing				Research projects		Standardization		Evolution	
	6G network operators										R&D Trial	Launch

Figure 1.4   Expected timeline of 6G development, standardization, and launch [14, 39–41].

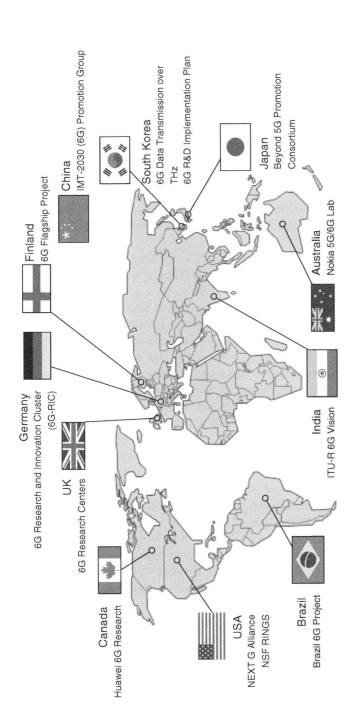

Figure 1.5 Global 6G development initiatives.

References

1 "Focus Group on Technologies for Network 2030," International Telecommunication Union, 2019, [Accessed on 29.03.2021]. [Online]. Available: https://www.itu.int/en/ITU-T/focusgroups/net2030/Pages/default.aspx.

2 "6G Flagship," University of Oulu, 2020, [Accessed on 29.03.2021]. [Online]. Available: https://www.oulu.fi/6gflagship/.

3 M. R. Bhalla and A. V. Bhalla, "Generations of mobile wireless technology: A survey," *International Journal of Computer Applications*, vol. 5, no. 4, pp. 26–32, 2010.

4 V. Pereira and T. Sousa, "Evolution of Mobile Communications: From 1G to 4G," *Department of Informatics Engineering of the University of Coimbra, Portugal*, 2004.

5 J. R. Churi, T. S. Surendran, S. A. Tigdi, and S. Yewale, "Evolution of networks (2G-5G)," in *International Conference on Advances in Communication and Computing Technologies (ICACACT)*, vol. 51, no. 4. Citeseer, 2012, pp. 8–13.

6 S. Won and S. W. Choi, "Three decades of 3GPP target cell search through 3G, 4G, and 5G," *IEEE Access*, vol. 8, pp. 116 914–116 960, 2020.

7 P. Datta and S. Kaushal, "Exploration and comparison of different 4G technologies implementations: A survey," in *2014 Recent Advances in Engineering and Computational Sciences (RAECS)*. IEEE, 2014, pp. 1–6.

8 P. Popovski, K. F. Trillingsgaard, O. Simeone, and G. Durisi, "5G wireless network slicing for eMBB, URLLC, and mMTC: A communication-theoretic view," *IEEE Access*, vol. 6, pp. 55 765–55 779, 2018.

9 M. Liyanage, A. Gurtov, and M. Ylianttila, *Software Defined Mobile Networks (SDMN): Beyond LTE Network Architecture*. John Wiley & Sons, 2015.

10 S. Wijethilaka and M. Liyanage, "Realizing Internet of Things with network slicing: Opportunities and challenges," in *2021 IEEE 18th Annual Consumer Communications & Networking Conference (CCNC)*. IEEE, 2021, pp. 1–6.

11 Y. Siriwardhana, C. De Alwis, G. Gür, M. Ylianttila, and M. Liyanage, "The fight against the COVID-19 pandemic with 5G technologies," *IEEE Engineering Management Review*, vol. 48, no. 3, pp. 72–84, 2020.

12 W. Saad, M. Bennis, and M. Chen, "A vision of 6G wireless systems: Applications, trends, technologies, and open research problems," *IEEE Network*, vol. 34, no. 3, pp. 134–142, 2019.

13 F. Fang, Y. Xu, Q.-V. Pham, and Z. Ding, "Energy-efficient design of IRS-NOMA networks," *IEEE Transactions on Vehicular Technology*, vol. 69, no. 11, pp. 14 088–14 092, 2020.

14 Y. Lu and X. Zheng, "6G: A survey on technologies, scenarios, challenges, and the related issues," *Journal of Industrial Information Integration*, vol. 19, p. 100158, 2020.

15 Y. Liu, X. Yuan, Z. Xiong, J. Kang, X. Wang, and D. Niyato, "Federated Learning for 6G Communications: Challenges, Methods, and Future Directions," *arXiv preprint arXiv:2006.02931*, 2020.

16 M. Latva-Aho and K. Leppänen, "Key drivers and research challenges for 6G ubiquitous wireless intelligence (white paper)," *Oulu, Finland: 6G Flagship*, 2019.

17 Z. Zhang, Y. Xiao, Z. Ma, M. Xiao, Z. Ding, X. Lei, G. K. Karagiannidis, and P. Fan, "6G wireless networks: Vision, requirements, architecture, and key technologies," *IEEE Vehicular Technology Magazine*, vol. 14, no. 3, pp. 28–41, 2019.

18 L. Mucchi, S. Jayousi, S. Caputo, E. Paoletti, P. Zoppi, S. Geli, and P. Dioniso, "How 6G technology can change the future wireless healthcare," in *2020 2nd 6G Wireless Summit (6G SUMMIT)*. IEEE, 2020, pp. 1–6.

19 W. He, D. Goodkind, and P. R. Kowal, "An Aging World: 2015," 2016.

20 "Global IoT Market will Grow to 24.1 Billion Devices in 2030, Generating $1.5 Trillion Annual Revenue," Transforma Insights research, May, 2020, [Accessed on 29.03.2021]. [Online]. Available: https://transformainsights.com/news/iot-market-24-billion-usd15-trillion-revenue-2030.

21 T.-Y. Chan, Y. Ren, Y.-C. Tseng, and J.-C. Chen, "Multi-slot allocation protocols for massive IoT devices with small-size uploading data," *IEEE Wireless Communications Letters*, vol. 8, no. 2, pp. 448–451, 2018.

22 M. Z. Chowdhury, M. Shahjalal, S. Ahmed, and Y. M. Jang, "6G wireless communication systems: Applications, requirements, technologies, challenges, and research directions," *IEEE Open Journal of the Communications Society*, vol. 1, pp. 957–975, 2020.

23 J. Hu, Q. Wang, and K. Yang, "Energy self-sustainability in full-spectrum 6G," *IEEE Wireless Communications*, vol. 28, no. 1, pp. 104–111, 2020.

24 S. Wang, X. Zhang, Y. Zhang, L. Wang, J. Yang, and W. Wang, "A survey on mobile edge networks: Convergence of computing, caching and communications," *IEEE Access*, vol. 5, pp. 6757–6779, 2017.

25 A. Bourdoux, A. N. Barreto, B. van Liempd, C. de Lima, D. Dardari, D. Belot, E.-S. Lohan, G. Seco-Granados, H. Sarieddeen, H. Wymeersch et al., "6G White Paper on Localization and Sensing," *arXiv preprint arXiv:2006.01779*, 2020.

26 T. Higashino, A. Uchiyama, S. Saruwatari, H. Yamaguchi, and T. Watanabe, "Context recognition of humans and objects by distributed zero-energy IoT devices," in *2019 IEEE 39th International Conference on Distributed Computing Systems (ICDCS)*. IEEE, 2019, pp. 1787–1796.

27 T. Kumar, P. Porambage, I. Ahmad, M. Liyanage, E. Harjula, and M. Ylianttila, "Securing gadget-free digital services," *Computer*, vol. 51, no. 11, pp. 66–77, 2018.

28 I. Ahmad, T. Kumar, M. Liyanage, M. Ylianttila, T. Koskela, T. Braysy, A. Anttonen, V. Pentikinen, J.-P. Soininen, and J. Huusko, "Towards gadget-free internet services: A roadmap of the naked world," *Telematics and Informatics*, vol. 35, no. 1, pp. 82–92, 2018.

29 M. Liyanage, A. Braeken, and M. Ylianttila, "Gadget free authentication," *IoT Security: Advances in Authentication*, vol. 1, pp. 143–157, 2020.

30 E. Markoval, D. Moltchanov, R. Pirmagomedov, D. Ivanova, Y. Koucheryavy, and K. Samouylov, "Priority-based coexistence of eMBB and URLLC traffic in industrial 5G NR deployments," in *2020 12th International Congress on Ultra Modern Telecommunications and Control Systems and Workshops (ICUMT)*. IEEE, 2020, pp. 1–6.

31 Z. Na, Y. Liu, J. Shi, C. Liu, and Z. Gao, "UAV-supported clustered NOMA for 6G-enabled Internet of Things: Trajectory planning and resource allocation," *IEEE Internet of Things Journal*, vol. 8, no. 20, pp. 15 041–15 048, 2021.

32 H. E. Melcherts, *The Internet of Everything and Beyond*. Wiley Online Library, 2017.

33 M. A. Siddiqi, H. Yu, and J. Joung, "5G ultra-reliable low-latency communication implementation challenges and operational issues with IoT devices," *Electronics*, vol. 8, no. 9, p. 981, 2019.

34 J. Zhao, "A Survey of Intelligent Reflecting Surfaces (IRSs): Towards 6G Wireless Communication Networks with Massive MIMO 2.0," *arXiv preprint arXiv:1907.04789*, 2019.

35 S. Nayak and R. Patgiri, "6G Communication Technology: A Vision on Intelligent Healthcare," *arXiv preprint arXiv:2005.07532*, 2020.

36 Z. Bojkovic, D. Milovanovic, T. P. Fowdur, and M. A. Hosany, "What 5G has been and what should 5G+ be?" *Athens Journal of Technology & Engineering*, vol. 8, no. 1, pp. 27–38, 2021.

37 F. Jameel, U. Javaid, B. Sikdar, I. Khan, G. Mastorakis, and C. X. Mavromoustakis, "Optimizing blockchain networks with artificial intelligence: Towards efficient and reliable IoT applications," in *Convergence of Artificial Intelligence and the Internet of Things*. Springer, 2020, pp. 299–321.

38 M. Naresh, "Towards 6G: Wireless communication," *Tathapi with ISSN 2320-0693 is an UGC CARE Journal*, vol. 19, no. 9, pp. 335–341, 2020.

39 K. B. Letaief, W. Chen, Y. Shi, J. Zhang, and Y.-J. A. Zhang, "The roadmap to 6G: AI empowered wireless networks," *IEEE Communications Magazine*, vol. 57, no. 8, pp. 84–90, 2019.

40 M. Giordani, M. Polese, M. Mezzavilla, S. Rangan, and M. Zorzi, "Toward 6G networks: Use cases and technologies," *IEEE Communications Magazine*, vol. 58, no. 3, pp. 55–61, 2020.

41 S. Dang, O. Amin, B. Shihada, and M.-S. Alouini, "What should 6G be?" *Nature Electronics*, vol. 3, no. 1, pp. 20–29, 2020.

2

Key 6G Technologies

The Chapter 1 has focused on the evolution, driving trends, and key requirements of future 6G wireless systems. Several key technologies have been proposed to realize 6G, and these technologies will be discussed in this chapter. After reading this chapter, you should be able to:

- Gain an overview of key technologies in future 6G wireless systems.
- Explore each of the key technologies with a preliminary, the role in 6G, and a review of representative studies.

2.1   Radio Network Technologies

In this section, we present important 6G radio network technologies, including THz communications and nonterrestrial networks toward 3D Networking.

2.1.1   Beyond Sub 6 GHz Toward THz Communication

The rapid increase in wireless data traffic is estimated to have a sevenfold increase in mobile data traffic from 2016 to 2021 [1]. Wide radio bands such as millimeter-waves (up to 300 GHz) are expected to fulfill the demand for data in 5G networks. However, applications such as holographic telepresence (HT), brain–computer interfaces (BCIs), and extended reality (XR) are expected to require data rates in the range of Tbps, which would be difficult with mmWave systems [2]. This requires exploring the Terahertz (THz) frequency band (0.1–10 THz). This type of communication will especially be useful for ultrahigh data rate communication with zero error rates within short distances.

6G is expected to deliver over a 1000× increase in the data rates compared to 5G to meet the target requirement of 1 Tbps. More spectrum resources beyond sub

Security and Privacy Vision in 6G: A Comprehensive Guide, First Edition.
Pawani Porambage and Madhusanka Liyanage.
© 2023 The Institute of Electrical and Electronics Engineers, Inc. Published 2023 by John Wiley & Sons, Inc.

6 GHz are explored by researchers to cater to this significant increase in data rates. Early 6G systems are expected to bank on sub 6 GHz mmWave wireless networks. However, 6G is expected to progress by exploiting frequencies beyond mmWave, at the THz band [3]. The size of 6G cells is expected to shrink further from small cells in 5G toward tiny cells that will have a radius of only a few tens of meters. Thus, 6G networks will require to have a new architectural design and mobility management techniques that can meet denser network deployments than 5G [4]. 6G transceivers will also be required to support integrated frequency bands ranging from microwave to THz spectra. The applications of THz for 6G networks are illustrated in Figure 2.1, where THz communication is used for high-speed transmissions between radio towers and mobile devices, integrated access and backhaul networks, and high-speed satellite communication links.

THz waves are located between mmWave and optical frequency bands. This allows the usage of electronics-based and photonics-based technologies in future communication networks. As for electronic devices, nanofabrication technologies can facilitate the progress of semiconductor devices that operates in the THz

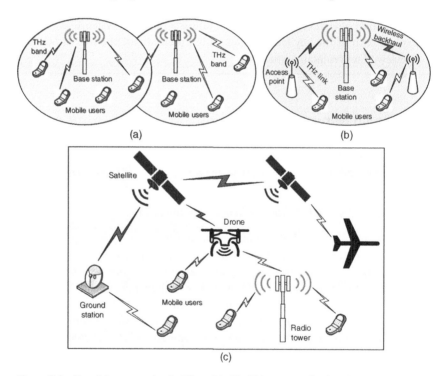

Figure 2.1 Promising scenarios in 6G enabled by THz communication: (a) high-speed transmission, (b) integrated access and backhaul networks, and (c) high-speed satellite communication links.

frequency band. The electronics in these devices are made from indium, gallium, arsenide, phosphide, and various silicon-based technologies [5]. A scalable silicon architecture allows synthesis and shaping of THz wave signals in a single microchip [6]. The feeding mechanism of optical fibers to THz circuits is prominent to achieve higher data rates in terms of photonics devices. Conventional materials used at lower frequencies in the microwave and mmWave ranges are not efficient enough for high-frequency wireless communication. Devices made from such materials exhibit large losses at the THz frequency range. THz waves require electromagnetically reconfigurable materials. In this context, graphene is identified as a suitable candidate to reform THz electromagnetic waves by using thin graphene layers [7, 8]. Graphene-based THz wireless communication components have exhibited promising results in terms of generating, modulating, and detecting THz waves [9]. THz wireless communication allows small antenna sizes to achieve both diversity gain and antenna directivity gain using multiple-input multiple-output (MIMO). For example, 1024×1024 ultra-Massive MIMO was introduced in [10] as an approach to increase the communication distance in THz wireless communication systems.

THz band channel is considered highly frequency-selective [11]. These channels suffer from high atmospheric absorption, atmospheric attenuation, and free-space path loss. This requires the development of new channel models to mimic the behavior of THz communication channels [1]. The work in [12] proposed the first statistical model for THz channels. This model depends on performing extensive ray-tracing simulations to obtain statistical parameters of the channel. Recent studies in [13, 16] provided more accurate channel models. Various research works have also focused on applications of THz communication. A hybrid radio frequency and free-space optical system is presented in [17], where a THz/optical link is envisaged as a suitable method for future wireless communication. In addition, Mollahasani and Onur [18] present using THz links in data centers to improve performance while achieving massive savings in minimizing the cable usage.

Summary: THz communications are expected to pave the way for Tbps data rate to meet the demands of future applications and have the potential to strengthen backhaul networks. Nevertheless, it suffers from high propagation losses and demands Line of Sight (LoS) for communications. More efforts are required to understand the behavior of THz signals, and better channel models are required.

2.1.2 Nonterrestrial Networks Toward 3D Networking

In conventional ground-centric mobile networks, the functioning of base stations is optimized to primarily cater to the needs of ground uses. Moreover, the elevation angle provided to the antennas at ground base stations focuses on the ground user for better directivity and hence cannot support aerial users [19]. Such a mobile network allows marginal vertical movement (i.e. above and below the ground surface), thus predominantly offering two-dimensional (2D) connectivity. Nonterrestrial networks expand the 2D connectivity by adding altitude as the third dimension [20–22]. Nonterrestrial networks are capable of providing coverage, trunking, backhauling, and supporting high-speed mobility in unserved or underserved areas through the integration of unmanned aerial vehicles (UAVs), satellites (in particular very low Earth orbit [VLEO]), tethered balloons, and high altitude platform (HAP) stations [4, 20, 23]. The development of protocols and architectural solutions for new radio (NR) operations in nonterrestrial networks is promoted in 3GPP Rel-17 and is expected to continue in Rel-18 and Rel-19 [20]. 3D networking further extends the nonterrestrial network paradigm allowing 6G to emerge as a global communication system by extending its coverage from ground to air toward space, underground, and underwater [24]. Interestingly, aerial base stations powered by UAV technology can offer on-demand, broadband, and reliable wireless coverage in a cost-effective and agile way. Some of the promising characteristics of UAV-enabled aerial base stations [19, 24] are as follows:

- Intelligent 3D mobility and ease of maneuvering.
- Varying capabilities in terms of computation, storage, power backup, etc., to meet heterogeneous demands.
- LoS communication links that allows effective beamforming in 3D.
- High flexibility in terms of the number of antenna elements when UAVs are used to create antenna arrays for 3D MIMO.

There has been an exponential increase in the number of connected devices, and the trend will continue with a higher rate of increase in the future. In particular, the future is expected to see a significant increase in aerial users or aerial-connected devices. Technological advancements in various fields, such as electronics and sensor technology, high-speed links, data communication networking, aviation technology, provide a necessary ecosystem for the robust growth of UAVs (aka drones), which have, in turn, extended the horizon of UAV's applications. By 2022, the fleet of small model UAVs (primarily used for recreational purposes by hobbyists) is expected to reach a mark of 1.38 million units, whereas small nonmodel UAVs (primarily used for commercial purposes) are forecast to be 789,000 million units as per the Federal Aviation Administration FAA's report [25]. Moreover, by the same year, i.e. 2022, the global market of UAVs is estimated to value at US$ 68.6 billion [26]. Hence, 6G mobile networks are expected to provide the required connectivity to such an increasing number of aerial users. To fulfill this expectation, the 3D networking paradigm is going to play a key enabling role in 6G.

A framework for UAV-based 3D cellular network for beyond 5G provides solutions for placement of UAV-enabled aerial base stations in 3D (using the truncated octahedron approach) as well as latency-sensitive association of UAV-based users to UAV-enabled aerial base stations [19]. In [27], a 3D nonstationary geometry-based stochastic model (GBSM) was investigated for UAV-to-ground channels that are envisioned in UAV-integrated 6G mobile networks. GBSM is proposed to work well for mmWave and massive MIMO configurations. The work in [28] proposed to extend the intelligence to edge 3D networks (i.e. beyond the premises of 2D networks) by leveraging edge computing. In particular, the work envisioned to integrate flying base stations with terrestrial stations by using emerging technologies, such as mobile edge computing, software-defined networking, and artificial intelligence (AI).

Summary: Nonterrestrial networks are evolved toward 3D networking to enable global radio coverage and capacity in 3D for future 6G networks. Nonterrestrial networks represent a gamut of technologies such as UAVs, HAPs, satellites, and other flying gadgets that are anticipated to work in harmony to offer seamless coverage over space, air, ground, underwater, and underground. AI/machine learning (ML)-based solutions are expected to play an important role in overcoming the limitations posed by the physical absence of human beings.

2.2 AI/ML/FL

Thanks to distinctive features and remarkable abilities, AI has various applications in wireless and mobile networking. Massive data generated by massive IoT devices can be exploited by AI approaches to extract valuable information, thus improving the network operation and performance. Recently, federated learning (FL) has emerged as a new AI concept that leverages on-device processing power and improves user data privacy [29–31]. The rationale is to collaboratively train a shared model such that participating devices train the local models and only share the updates (instead of data) with the centralized parameter server [32, 33]. Mobile and IoT devices, such as mobile phones, IoT devices, and autonomous are getting increasingly powerful in terms of storage and computational capabilities and this has paved the growing interest in FL. According to [34], FL can be classified into horizontal FL, vertical FL, and federated transfer learning:

- **Horizontal FL**: The overall feature spaces of various datasets are the same, but FL users have different sample distributions, as shown in Figure 2.2a.
- **Vertical FL**: The feature spaces of various datasets are different, but FL users share the same sample distribution, as illustrated in Figure 2.2b.
- **Federated transfer learning**: FL users have datasets with different feature spaces and sample space distributions, as shown in Figure 2.2c.

With 6G being envisioned to have AI/ML at its core, the role of AI/FL becomes important to 6G. The use of conventional centralized ML approaches is suitable

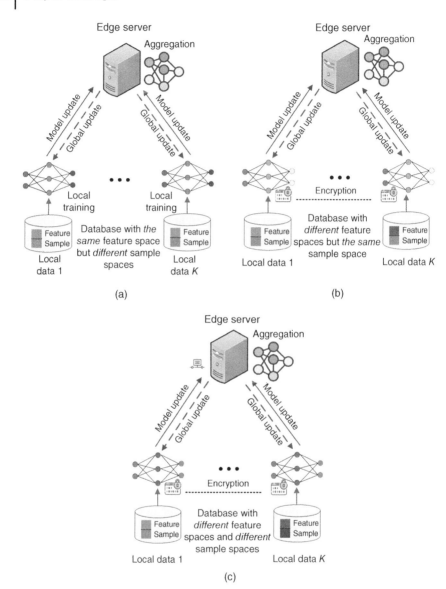

Figure 2.2 Three categories of FL: (a) horizontal FL, (b) vertical FL, and (c) federated transfer learning.

for network scenarios, where centralized data collection and processing are available. Numerous problems in future 6G networks related to the rapid increase in the amount of mobile data and services, advancements in computing hardware performance, can be effectively solved by AI approaches such as modulation classification, waveform detection, signal processing, and physical layer design [35–37]. To overcome limitations of centralized AI such as privacy concerns and huge communication overhead [32], and moreover, due to the centralized nature of such ML-based systems, they suffer from single-point-of-failure as well as security vulnerabilities. Thus, FL is gaining popularity and is emerging as a viable distributed AI solution that enables "ubiquitous AI" vision of 6G communications [38]. FL offers a multitude of benefits to 6G, as summarized by authors in [39], communication-efficient distributed AI, support for heterogeneous data originating from different devices pertaining to different services that can lead to nonidentically distributed (non-IID) dataset, privacy-protection since data remain locally and is not uploaded anywhere, and enable large-scale deployment.

The last few years have witnessed the use of different AI techniques for numerous problems in wireless networks. For example, the work in [40] reviews applications of deep reinforcement learning (DRL) for three important topics, including network access and rate control, data caching and computation offloading, security and connectivity preservation, and for a number of miscellaneous issues such as resource allocation, traffic routing, signal detection, and load balancing. The applications of ML for wireless networks are reviewed in [41], where AI-enabled resource management, networking, mobility, and localization solutions are discussed. The use of transfer learning, deep learning, and swarm intelligence for future wireless networks can be found in the studies of [42–44]. Some of the related work pertain to use of FL in wireless networks. The work in [45] proposed a Stackelberg game-based approach to incentivize the interaction between the global server and devices participating in training model. The work in [38] proposed an FL framework for IoT networks with the aim to simultaneously maximize the utilization efficiency of edge resources and minimize the cost for IoT networks. An incentive FL mechanism using contract theory is proposed in [46] to allow highly reputed devices to engage in the training task. Moreover, a reputation scheme based on a multiweight model was proposed for selection devices and blockchain was used for managing this reputation system. Despite the hype, there are numerous challenges that need to be addressed to gain the maximum benefits of FL in the 6G realm. Some challenges of FL are highlighted in [29, 32], including the cost of communications, significant hardware heterogeneity, high device churn, privacy leakage through model update, and security issues.

Summary: AI and FL techniques enable the design of intelligent mechanisms for future 6G networks via exploiting massive mobile data and increasingly computing resources available at the network edge. Participating devices in FL locally train ML models (in use) by leveraging their on-board resources. By doing so, these devices will not have to share the raw data, which may be private and prone to security attacks. Therefore, what is shared is the updated model at the centralized servers, thereby making the learning to be federated. Nevertheless, in order to effectively realize AI, the very first challenge is the existence of high-quality training datasets. Another challenge is the inclusion of AI in beyond 5G and future 6G networks [47]. Moreover, the bottlenecks of AI caused by 6G wireless networks with many new advanced technologies, device heterogeneity, and emerging intelligent applications should be further studied. Regarding FL, the devices need to be incentivized to participate in the training process, and also, the rogue devices need to be detected as early as possible. Moreover, the challenges like privacy leakage via model updates and hardware heterogeneity also need to be met [48].

2.3 DLT/Blockchain

The last couple of years have witnessed the rise of distributed ledger technologies (DLT), in particular, blockchain technology. DLT is envisioned to unlock the doors to the decentralized future by overcoming the well-known impediments of centralized systems [49]. Blockchain is a type of DLT, which maintains a digital ledger in a secure and distributed way. This ledger holds all the transactions in a chronological order and is cryptographically sealed [50]. The illustration a blockchain is depicted in Figure 2.3. If a hacker intends to edit the transaction in a blockchain, he has to modify hash of not only that block but also every other hash, which is nearly impossible. This security property of blockchain makes it an ideal choice for usage in many sectors like banking, insurance, government services, and supply chain management. Moreover, blockchain offers numerous advantages like disintermediation, immutability, nonrepudiation, proof of provenance, integrity, and pseudonymity; therefore, blockchain has received all-around attention equally by industry and academia [31].

Many sectors have already acknowledged the pragmatic use of blockchain technology and its efficacy as they are running/offering blockchain-based technological solutions [51]. Examples of some of these business sectors are finance and banking, industrial supply chain and manufacturing, shipping and transportation, medical health care and patient records, and educational processes and credentialing. The blockchain technology can play a cardinal role in improving the following aspects of future 6G wireless systems.

- Management and orchestration in terms of interference mitigation, resource allocation, spectrum, and mobility management [52, 53].

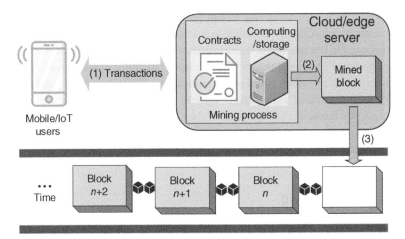

Figure 2.3 Illustration of the general transaction process of blockchain: (1) a mobile user or an IoT device sends a transaction to smart contracts to request data; (2) a new block is created to represent the verified transaction; and (3) a mined block is appended to the blockchain.

- Operations in terms of cell-free communications and 3D-networking.
- Business models in terms of decentralized and trustless digital markets involving various stakeholders, such as Infrastructure Providers (InPs), network tenants, industry verticals, over-the-top (OTT) providers, and edge providers [54, 55].

Further, blockchain has an immense potential to strengthen the existing service arena of mobile networks as well as set the floor for futuristic applications and use cases of 6G.

Blockchain has been identified as one of the key enabling technologies for 6G. Numerous efforts are being made to leverage its potential to improve both the technical aspects of 6G as well as use cases of the 6G ecosystem. For instance, the work in [52] presented the use of blockchain for decentralized network management of 6G. In particular, the work showed the blockchain plus smart contract for spectrum trading. In [56], blockchain-based radio access network was proposed (B-RAN) to allow small subnetworks to collaborate in a trustless environment to create larger cooperative network. The work in [53] presented how blockchain can be leveraged to remove the intermediate layer and develop a distributed Mobility-as-a-Service that is hosted as an application on edge computing facility. Improved transparency and trust among all the stakeholders are the manifested advantages of this work. The work in [57] advocated the use of blockchain for ensuring data security AI-powered applications for 6G. The two main applications are indoor positioning and autonomous vehicles.

Summary: Blockchain being one of the prominent types of DLT has turned out to be a very promising key enabling technology because of its built-in strong security nature. On the one hand, it can enhance the technical aspects of 6G like dynamic spectrum sharing, resource management, mobility management, and on the other hand, it enables unforeseen applications like holographic telepresence, XR, fully connected autonomous vehicle, Industry 5.0, and many more. Nevertheless, to harness the best use of blockchain for 6G challenges like computational overheads, lightweight consensus algorithms, high transaction throughput, quantum resistance, and storage scalability need to be mitigated.

2.4 Edge Computing

One of the very first edge-computing concepts, the so-called "cloudlet," was proposed in 2009 [58]. Conceptually, a cloudlet is defined as a trusted and resource-rich computer or a cluster of computers located in a strategic location at the edge and well connected to the Internet. The main purpose of cloudlet is to extend cloud computing to the network edge and support resource-poor mobile users in running resource-intensive and interactive applications. Although the storage and computing capabilities of a cloudlet are relatively smaller than those of a data center in cloud computing, cloudlets have the advantages of low deployment cost and high scalability [59]. Mobile users exploit virtual machine (VM)-based virtualization on customizing service software on proximate cloudlets and then using those cloudlet services over a wireless local area network to offload intensive computations [58, 60]. There are two different VM approaches for computation offloading [58], including VM migration and VM synthesis. Since VM provisioning is used in cloudlets, cloudlets can operate in standalone mode without the intervention of the cloud [60, 61]. Mobile users can access cloudlet services through Wi-Fi. The idea of cloudlet to distribute the cloud computing in close proximity to mobile users is similar to the Wi-Fi concept in providing Internet access [62]. The Wi-Fi connection between users and cloudlets can be a serious drawback. In this way, mobile users are unable to access cloudlets in the long distance and use both Wi-Fi and cellular connection simultaneously [59], i.e. users have to switch between the mobile network and Wi-Fi if they use cloudlet services.

To address the inherent drawbacks of cloud computing, fog computing has emerged as a promising solution. Fog computing, a term put forward by Cisco in 2012, refers to the extension of the cloud computing from the core to the network edge, thus bringing computing resources closer to end-users [63]. The main purpose of fog computing is to bring computational resources to end-users as well as reduce the amount of data needed to transfer to the cloud for processing and storage. Therefore, instead of being transferred to the cloud, most intensive-computations from mobile users and data collected by end-users (e.g. sensors and IoT devices) can be processed and analyzed by fog nodes at the network edge, thus reducing the execution latency and network congestion [64]. As a complement to cloud computing, fog computing stands out in the following features [61, 65]:

- Edge location, low latency, and location awareness,
- Widespread geographic distribution,
- Support for mobility and real-time applications,
- A very large number of nodes as a consequence of geographic distribution,
- Heterogeneity of fog nodes and predominance of wireless access.

Due to its characteristics, fog computing plays an important role in many use cases and applications [65]. In terms of the node type, a fog node can be built from heterogeneous elements (e.g. routers, switches, and IoT gateways) and a cloudlet can be referred to as a cloud data center in a box [66]. In terms of node location, a fog node can be deployed at a strategic location ranging from end devices to centralized clouds and a cloudlet can be deployed indoors as well as outdoors. The close similarity between cloudlet and fog computing is that cloudlets and fog nodes are not integrated into the mobile network architecture; thus, fog nodes and cloudlets are commonly implemented and owned by private enterprises, and it is not easy to provide mobile users with the quality of service (QoS) and quality of experience (QoE) guarantees [61, 62].

The mobile edge computing (MEC) concept was initiated in late 2014. As a complement of the C-RAN architecture, MEC aims to unite the telecommunication and IT cloud services to provide the cloud-computing capabilities within radio access networks in the close vicinity of mobile users [67]. The general architecture of MEC can be illustrated in Figure 2.4. The main purposes of MEC are [68]: optimization of mobile resources by hosting compute-intensive applications, optimization of the large data before sending to the cloud, enabling cloud services within the close proximity of mobile subscribers, and providing context-aware services with the help of radio access network information. From the illustration in Figure2.4, MEC enables a wide variety of applications, where the real-time response is strictly required, e.g. driverless vehicles, virtual reality (VR), augmented reality (AR), robotics, and immersive media. To reap additional benefits of MEC with heterogeneous access technologies, e.g. 4G/5G/6G, Wi-Fi, and fixed connection, the name of mobile edge computing is changed to mean multiaccess edge computing in 2017 [69]. After this scope expansion, MEC servers can be deployed by the network operators at various locations within RAN and/or collocated with different elements that establish the network edge. While the BSs in traditional cellular networks are mainly used for communication purposes,

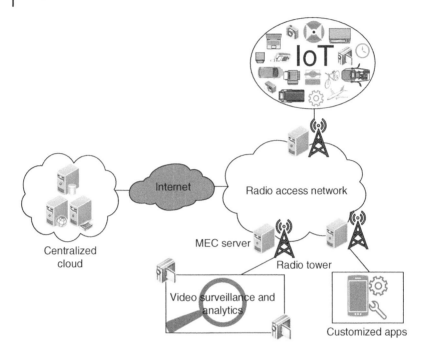

Figure 2.4 A general architecture of MEC.

MEC enables them to collocate with MEC servers for facilitating mobile users with additional services. This transformation pushes intelligence toward the traditional BSs so that they can be used for not only communication purposes but also computation, caching, and control services [70].

Edge computing has been a key technology in 5G networks and will still be a key technology in beyond 5G and future 6G networks. The implementation of edge computing enables many new services and applications, for example, autonomous vehicles, smart IoT, smart cities, smart healthcare, and more. As a result, there exists a strong need for the close integration of edge computing with enabling technologies, such as artificial intelligence, big data, intelligent surfaces, blockchain, IoT, and VR/AR/metaverse. Significantly, there will be more applications and services requiring the deployment of AI at the edge of the network and the deployment of intelligent edge, creating the concept of edge AI, as presented in Chapter 6. Moreover, there are several new concepts thanks to the combination of edge computing with emerging technologies. For example, in-network computing can further enhance conventional edge computing concepts. In particular, in-network computing enables in-network devices, such as routers and switches, to perform computing functions instead of forwarding the data transmission [71]. The amalgamation of aerial access networks and edge computing introduces a novel concept, referred to as aerial computing [72]. Aerial computing is expected to provide advanced services, e.g. communication, computing, caching, sensing, navigation, and control, at a global scale. Facilitated by advantages of high mobility, fast deployment, global availability, scalability, and flexibility, aerial computing complements conventional computing paradigms (e.g. cloud computing, fog computing, and MEC) and is thus considered a pillar of the comprehensive computing infrastructure in future 6G networks.

Summary: Different edge computing paradigms are proposed to provide applications and services at the edge networks. The availability of edge computing resources and the explosion of IoT data enable the deployment of many new services with low latency and intelligent capabilities.

2.5 Quantum Communication

Quantum computing-enabled communications has derived a lot of research and development interests recently. Slowly but beyond mere science fiction, quantum computing (QC) will be a transformative reality in the 6G paradigm in the next decade or so. As 6G must meet stringent requirements, such as massive data rates, fast computing, and strong security, QC will be a potential enabler. The driving force of QC is that it exploits traditional concepts of physics, i.e. photons are being used to process the computation to the quantum qubits. These qubits are then sent from a sender (or emitter) machine to a receiver machine. Using flying qubits in communications brings enormous advantages, such as weak interaction with an environment, faster computations and communications, quantum teleportation, communication security, and low transmission losses in communication.

As per [73], the role of quantum systems for telecommunication and networking fall under two different categories: quantum communication and quantum computing. Quantum communication is a way to transfer a quantum state from a sender to a receiver [74]. It can enable the execution of tasks that either cannot be performed or are inefficiently performed using classical techniques [73, 74]. Some of the interesting offerings of quantum communication are quantum key distribution (QKD), quantum secure direct communication (QSDC), quantum secret sharing (QSS), quantum teleportation, quantum network (quantum channel, quantum repeaters, quantum memory, and quantum server) [75]. One of the promising uses of quantum communication is the secure distribution of cryptographic keys referred to QKD. The techniques that use quantum entanglement for this purpose are called entanglement-based QKD [76]. Any kind of man-in-middle attack while using QKD can be detected easily as the attacker disturbs the quantum (shared joint) state and this disturbance can be known by examining the correlations between the communicating entities [76].

Quantum communication is foreseen to play a crucial role in realizing secure 6G communications. In particular, the underlying principles of quantum entanglement and its nonlocality, superposition, inalienable law, and noncloning theorem pave the way for strong security. The next generation of services that are going to be increasingly supported by quantum communication is HT, tactile Internet, BCI, extremely massive, and intelligent communications [73, 77]. These QKD protocols have shown the most progress and numerous practical implementations of such protocols have been shown which reflects their potential applicability in future 6G wireless networks [73, 78]. Yet another interesting use of quantum communication is its applicability for secure long-distance communication [78]. This, in particular, would be interesting since it is envisioned that 6G will have a special focus on LDHMC that would deal with extremely long-distance communications.

The work in [79] surveyed various paradigms like quantum communication, quantum computing-assisted communication, and quantum assisted ML-based communication that utilizes ML and quantum computing in a synergetic manner. The advent of quantum computing facilities poses a significant treat to traditional cryptographic techniques that are used to encrypt data in current wireless communication systems. Thus, the work in [80] exploited the QKD mechanism for key generation and management in 5G IoT scenarios, namely Quantum Key GRID for Authentication and Key Agreement (QKG-AKA), and analytically showed that security cannot be broken in polynomial time. The work in [81] proposed two hash functions (for 5G applications) that utilize quantum walks (QW), namely, QWHF-1 and QWHF-2 (Quantum Walk Hash Function 1 and 2). Further, QWHF-1 is in turn used to develop Authentication Key Distribution (AKD) protocol and QWHF-2 is used to develop Authenticated Quantum Direction Communication (AQDC) protocol for Device-to-Device (D2D) communications. The work in [82] studied the applicability of QKD for securing fronthaul that offers low-latency connectivity to a myriad of 5G terminals. In particular, a fronthaul link integrates BB84 and QKD and supports advanced encryption standard (AES) encryption. Their work paves the way toward quantum secure fronthaul infrastructure for 5G/B5G networks.

Summary: Quantum communication is a very promising application of the principles of quantum physics in the world of communication. Some of the techniques provided by quantum communication are quantum teleportation, quantum network, QKD, QSDC, and QSS. Various upcoming applications of quantum communication are quantum optical twin, holographic telepresence, tactile Internet, brain–computer interface, and long-distance intelligent communication.

In spite of the hype, the establishing synergy between quantum communication and classical communication will be challenging. So far, most advancements are limited to QKD.

2.6 Other New Technologies

2.6.1 Visible Light Communications

Visible light communication (VLC), which uses visible light for short-range communication, is one of the promising optical wireless communication (OWC) technologies [83]. The frequency spectrum for VLC is between 430 and 790 THz. Further, in VLC, the most common devices used for transmission are light emitting diode (LED) and light amplification by stimulated emission of radiation (LASER) diodes, whereas for reception photodetectors such as silicon photodiode, PIN photo-diode (PD), and PIN avalanche photo-diode (APD) are used [83–85]. Some of the advantages of using VLC technology based on [83, 85, 86] are as follows:

- The visible light spectrum is free to use since it falls under the unlicensed band,
- Very high bandwidth as compared to radio frequency (RF) signals (visible light spectrum is 10^4 times higher than radio waves [85]),
- High spatial reusability since visible light is blocked by objects like walls,
- Precise estimation of direction-of-arrival,
- Very high data rate, e.g. 10 Gbps using LED and 100 Gbps using LASER diodes [87],
- Low-energy consumption with the use of LEDs,
- Inherently secure due to unidirectional propagation, signal isolation, and non-penetrating nature of visible light,
- Less costly as compared to radio communication especially in mmWave and THz range,
- Safe to be used for communication since it meets the eye and skin regulations [85].

Moreover, the existing radio frequency band and the visible light band are well separated; thus, there is no electromagnetic interference [88].

VLC came into existence with the emergence of white LEDs and has matured over the last two decades for it to be considered as an enabling technology for 6G [89]. The technology has been successfully used for various applications scenarios, such as vehicular communications, underwater communications, indoor scenarios, visible light identification systems, and wireless local area networks, underground mines. Since VLC operates in the THz range, thus it offers ultra-high bandwidth, which means it can well satisfy the capacity and data rate demands of 6G [90]. From the 6G point-of-view, a hybrid communication infrastructure can be developed leveraging the best of VLC and other conventional communications like RF, Wi-Fi, infrared (IR), and power line communication (PLC) [89, 91, 92]. For instance, to establish RF-VLC hybrid system, use of reconfigurable intelligent surfaces (RIS) has been suggested. Here, RIS can control the propagation environment and ensure the LoS communication between base stations and mobile devices equipped with a photodetector [93].

2.6.2 Large Intelligent Surfaces

The emerging paradigm for controlling the propagation environment via smart and intelligent surfaces has been referred to with numerous names like large intelligent surface (LIS), intelligent reflecting surface (IRS), RIS, software-defined surface (SDS), and many more [94]. LIS plays a vital role when direct LoS communication is not feasible or degrades in quality such that it hampers sensible communications. LIS provides a way to transform man-made structures (e.g. building, roads, indoor walls/ceilings) into an intelligent and electromagnetically active wireless environment [95]. This transformation is performed by augmenting these structures either with large array of small, low-cost, and passive antennas or with meta-material (aka meta-surfaces), such that they help in controlling the characteristics like reflection, scattering, and refraction of propagation environment [94]. Although changes can be made to different characteristics of the incident electromagnetic signals by LIS, most of the work focuses on change in the phase of the incident signal.

LIS has been identified as a key technological enabler for 6G since it is going to operate at higher frequencies and is expected to go beyond the massive MIMO [4, 96]. Various advantages of LIS over conventional massive MIMO, as highlighted in [95], include reduced noise, lower inter-user interference, and reliable communication. Interestingly, LIS would allow the use of holographic radio frequency as well as holographic MIMO [4]. Moreover, in the 6G realm, LIS can be used to better sense wireless environments by capturing channel state information (CSI) [97].

2.6.3 Compressive Sensing

Sampling is an integral part of modern digital signal processing and stands at the interface between analog (physical) and digital worlds. Traditionally, for efficient transmission, flexible processing, noise immunity, security inclusion (using encryption and decryption), low cost, etc., the Nyquist sampling theorem has been used. According to this, for a band-limited signal, if the samples are taken at a rate greater than equal to twice the highest frequency of that signal, then the exact replica of the signal can be reconstructed using these samples. Sampling is usually followed by a compression process, where the sampled data are compressed to maintain some acceptable level of quality [98]. As pointed out in [99], with the increase in the transmission bandwidth with 5G and future 6G mobile networks, the continued use of the Nyquist sampling technique will result in numerous challenges like significant overheads, large complexity, and higher-power consumption. In this context, compressive sensing, also known as compressive sampling or sparse sampling, has been proposed as an intriguing solution that has the potential to overcome the limits imposed by traditional sampling. compressive sensing is basically a sub-Nyquist sampling framework. Compressive sensing states that provided a signal is characterized by sparsity and incoherence, it can be sampled at a rate lower than the Nyquist rate, and the resulting (smaller set of) samples are sufficient to reconstruct the original signal [99]. This is achieved in a computationally efficient manner and by finding a solution to the underdetermined linear systems. Moreover, in compressive sensing both the sampling and compression are carried out at the same time. In nutshell, the sampling rate in compressive sensing depends on sparsity and incoherence characteristic of the signal being sampled and does not depend on the bandwidth of the signal [100]. This property of compressive sensing opens the door for its applicability to 6G networks. In general, compressive sensing is proposed to be used for reducing the data generated by IoT devices for mMTC [101]. Another proposed use of compressive sensing is to enable non-orthogonal multiple access (NOMA) at the transmitter in the landscape of mMTC scenario [102]. This is carried out by assigning nonorthogonal spreading codes to devices and applying compressive sensing-based multiuser detection technique since very less percentage of total devices are active at any given time. The advantage is this scheme incurs no control signaling overhead. Yet another use of compressive sensing along with deep learning techniques is to overcome the issues pertaining to the expected intensive usage of LISs in the next-generation networks [103].

2.6.4 Zero-Touch Network and Service Management

Zero-touch network and service management (ZSM) is an evolving concept that aims to provide a framework for building a fully automated network management, primarily driven by the initiative of ETSI. The idea of ZSM is to empower the network so that they can perform self-configuration to carry out autonomous configuration without the need for explicit human intervention, self-optimization to better adapt as per the prevailing situation, self-healing to ensure correct functioning, self-monitoring to track its functioning, and self-scaling to dynamically engage or disengage resources as per need [104]. To stress the importance of the ZSM framework, ETSI in [105] identifies a list of scenarios that are grouped into seven different broad categories as follows:

- The end-to-end network and service management category talks about the automation of operational and functional tasks involved in the end-to-end lifecycle management of different types of network resources and services that are part of core network, transport network, and radio access network.
- Network-as-a-Service (NaaS) presents the requirement of exposing some of the service capabilities from all the parts of the network to enable zero-touch automation.
- Analytics and ML scenarios emphasize the need for integration of ML and AI capabilities for realizing ZSM.
- Collaborative service management category emphasizes the need for collaborative management spanning domains of multiple operators.
- Security highlights the need for strong security and privacy mechanisms for ZSM framework.
- Testing scenario points out the need for automated testing of resources as well as services.
- Tracing scenario needs are driven by requirements for automated troubleshooting and root cause analysis.

In this direction, a framework, named Self-Evolving Networks (SENs) [106], is proposed that aims to automate the network management with self-efficient resource utilization, coordination and conflict management, inherent security and trust, reduce cost, and high QoE.

Future 6G wireless networks will be heterogeneous with multitenancy, multioperator, and multi(micro)-services features. To make such networks work at their best and at a low cost, they are envisioned to be fully automated. Thus, ZSM becomes highly important in 6G. The use of AI/ML capabilities within the framework of ZSM has indeed the potential to add many new capabilities (as mentioned above) and set the floor for AI-enabled autonomous networks. However, security remains a great concern. This is because ML techniques are vulnerable to attacks like poisoning attacks or evasion attacks [107]. Here, the use of blockchain as a common communication channel can do the required. Further, enabling automated service updates without affecting service interoperability as well as end-user experience is another challenge when using ZSM [73].

2.6.5 Efficient Energy Transfer and Harvesting

Energy harvesting has been the much-sought area of research when it comes to a future sustainable way of energizing the growing number of connected devices.

The aim of energy harvesting is to replace the conventional ways of powering devices and sensors by tapping the energy from ambient environments. The two broad classes of sources for energy harvesting are natural sources and man-made sources [108]. Natural sources include renewable energy sources like solar, mechanical vibrations, wind, thermal, microbial fuel cell, and human activity powered [108]. Man-made energy harvesting happens through wireless energy transfer (WET), where a dedicated power beacon is used to transfer energy from source to destination [108]. Since the natural sources of energy harvesting suffer unpredictability and periodicity, they fail to offer guaranteed QoS; thus, WET is the hot research area [109].

With the vision of a universal communication system and providing solid underpinning to IoE, the future 6G networks will be proliferated with a large number of connected devices. The conventional way of powering these devices with rechargeable or replaceable batteries might not efficiently scale in the 6G era. The reason being such solutions are, in general, costly, inconvenient, risky, and have adverse effects when the devices are operating inside the body [109]. Thus, energy harvesting technologies are considered to be an efficient alternative solution for next-generation mobile networks [108]. In this context, much of the excitement revolves around the idea that radio signals can simultaneously transfer energy as well as information [110]. This is referred to as radio frequency energy harvesting (RF-EH) [111]. Nevertheless, to have the pragmatic use of such techniques and to gain the maximum benefit, the challenge lies in efficient integration of both wireless information transfer and WET provided that their hardware and operational requirements are different [108]. The other open issues are high mobility, multiuser energy and information scheduling, resource allocation and interference management, health issues, and security issues [109, 111].

References

1 H. Elayan, O. Amin, R. M. Shubair, and M.-S. Alouini, "Terahertz communication: The opportunities of wireless technology beyond 5G," in *2018 International Conference on Advanced Communication Technologies and Networking (CommNet)*. IEEE, 2018, pp. 1–5.

2 C. Han, Y. Wu, Z. Chen, and X. Wang, "Terahertz Communications (TeraCom): Challenges and Impact on 6G Wireless Systems," *arXiv preprint arXiv:1912.06040*, 2019.

3 Y. Xing and T. S. Rappaport, "Propagation measurement system and approach at 140 GHz-moving to 6G and above 100 GHz," in *2018 IEEE Global Communications Conference (GLOBECOM)*. IEEE, 2018, pp. 1–6.

4 W. Saad, M. Bennis, and M. Chen, "A vision of 6G wireless systems: Applications, trends, technologies, and open research problems," *IEEE Network*, vol. 34, no. 3, pp. 134–142, 2019.

5 T. Nagatsuma, "Terahertz technologies: Present and future," *IEICE Electronics Express*, vol. 8, no. 14, pp. 1127–1142, 2011.

6 X. Wu and K. Sengupta, "Dynamic waveform shaping with picosecond time widths," *IEEE Journal of Solid-State Circuits*, vol. 52, no. 2, pp. 389–405, 2016.

7 J. M. Jornet and I. F. Akyildiz, "Graphene-based plasmonic nano-transceiver for terahertz band communication," in *The 8th European Conference on Antennas and Propagation (EuCAP 2014)*. IEEE, 2014, pp. 492–496.

8 C. J. Docherty and M. B. Johnston, "Terahertz properties of graphene," *Journal of Infrared, Millimeter, and Terahertz Waves*, vol. 33, no. 8, pp. 797–815, 2012.

9 M. Hasan, S. Arezoomandan, H. Condori, and B. Sensale-Rodriguez, "Graphene terahertz devices for communications applications," *Nano Communication Networks*, vol. 10, pp. 68–78, 2016.

10 I. F. Akyildiz and J. M. Jornet, "Realizing ultra-massive MIMO (1024× 1024) communication in the (0.06–10) terahertz band," *Nano Communication Networks*, vol. 8, pp. 46–54, 2016.

11 J. M. Jornet and I. F. Akyildiz, "Channel modeling and capacity analysis for electromagnetic wireless nanonetworks in the terahertz band," *IEEE Transactions on Wireless Communications*, vol. 10, no. 10, pp. 3211–3221, 2011.

12 S. Priebe and T. Kurner, "Stochastic modeling of THz indoor radio channels," *IEEE Transactions on Wireless Communications*, vol. 12, no. 9, pp. 4445–4455, 2013.

13 S. Kim and A. Zajić, "Statistical modeling of THz scatter channels," in *2015 9th European Conference on Antennas and Propagation (EuCAP)*. IEEE, 2015, pp. 1–5.

14 S. Kim and A. Zajić, "Statistical modeling and simulation of short-range device-to-device communication channels at sub-THz frequencies," *IEEE Transactions on Wireless Communications*, vol. 15, no. 9, pp. 6423–6433, 2016.

15 D. He, K. Guan, A. Fricke, B. Ai, R. He, Z. Zhong, A. Kasamatsu, I. Hosako, and T. Kürner, "Stochastic channel modeling for Kiosk applications in the terahertz band," *IEEE Transactions on Terahertz Science and Technology*, vol. 7, no. 5, pp. 502–513, 2017.

16 A. R. Ekti, A. Boyaci, A. Alparslan, İ. Ünal, S. Yarkan, A. Görçin, H. Arslan, and M. Uysal, "Statistical modeling of propagation channels for terahertz band," in *2017 IEEE Conference on Standards for Communications and Networking (CSCN)*. IEEE, 2017, pp. 275–280.

17 M. A. Khalighi and M. Uysal, "Survey on free space optical communication: A communication theory perspective," *IEEE Communications Surveys & Tutorials*, vol. 16, no. 4, pp. 2231–2258, 2014.

18 S. Mollahasani and E. Onur, "Evaluation of terahertz channel in data centers," in *NOMS 2016-2016 IEEE/IFIP Network Operations and Management Symposium*. IEEE, 2016, pp. 727–730.

19 M. Mozaffari, A. T. Z. Kasgari, W. Saad, M. Bennis, and M. Debbah, "Beyond 5G with UAVs: Foundations of a 3D wireless cellular network," *IEEE Transactions on Wireless Communications*, vol. 18, no. 1, pp. 357–372, 2018.

20 M. Giordani and M. Zorzi, "Non-terrestrial networks in the 6G era: Challenges and opportunities," *IEEE Network*, vol. 35, no. 2, pp. 244–251, 2020.

21 C. Liu, W. Feng, X. Tao, and N. Ge, "MEC-Empowered Non-terrestrial Network for 6G Wide-Area Time-Sensitive Internet of Things," *arXiv preprint arXiv:2103.11907*, 2021.

22 D. Wang, M. Giordani, M.-S. Alouini, and M. Zorzi, "The Potential of Multi-Layered Hierarchical Non-Terrestrial Networks for 6G," *arXiv preprint arXiv:2011.08608*, 2020.

23 M. Marchese, A. Moheddine, and F. Patrone, "IoT and UAV integration in 5G hybrid terrestrial-satellite networks," *Sensors*, vol. 19, no. 17, p. 3704, 2019.

24 N.-N. Dao, Q.-V. Pham, N. H. Tu, T. T. Thanh, V. N. Q. Bao, D. S. Lakew, and S. Cho, "Survey on aerial radio access networks: Toward a comprehensive 6G access infrastructure," *IEEE Communications Surveys & Tutorials*, vol. 23, no. 2, pp. 1193–1225, 2021.

25 "FAA Aerospace Forecast Fiscal Year 201–2038," Federal Aviation Administration, https://www.faa.gov/data_research/aviation/aerospace_forecasts/media/FY2019-39_FAA_Aerospace_Forecast.pdf. accessed on 2021-03-29.

26 S. D. Intelligence, "The Global UAV Payload Market 2012–2022," *Strategic Defence Intelligence: White Papers*, 2013.

27 H. Chang, C.-X. Wang, Y. Liu, J. Huang, J. Sun, W. Zhang, and X. Gao, "A novel non-stationary 6G UAV-to-ground wireless channel model with 3D arbitrary trajectory changes," *IEEE Internet of Things Journal*, vol. 8, no. 12, pp. 9865–9877, 2020.

28 E. C. Strinati, S. Barbarossa, T. Choi, A. Pietrabissa, A. Giuseppi, E. De Santis, J. Vidal, Z. Becvar, T. Haustein, N. Cassiau et al., "6G in the sky: On-demand intelligence at the edge of 3D networks," *ETRI Journal*, vol. 42, no. 5, pp. 643–657, 2020.

29 T. Li, A. K. Sahu, A. Talwalkar, and V. Smith, "Federated learning: Challenges, methods, and future directions," *IEEE Signal Processing Magazine*, vol. 37, no. 3, pp. 50–60, 2020.

30 S. Niknam, H. S. Dhillon, and J. H. Reed, "Federated learning for wireless communications: Motivation, opportunities, and challenges," *IEEE Communications Magazine*, vol. 58, no. 6, pp. 46–51, 2020.

31 D. C. Nguyen, M. Ding, Q.-V. Pham, P. N. Pathirana, L. B. Le, A. Seneviratne, J. Li, D. Niyato, and H. V. Poor, "Federated learning meets blockchain in edge computing: Opportunities and challenges," *IEEE Internet of Things Journal*, vol. 8, no. 16, pp. 12 806–12 825, 2021.

32 Y. Liu, X. Yuan, Z. Xiong, J. Kang, X. Wang, and D. Niyato, "Federated Learning for 6G Communications: Challenges, Methods, and Future Directions," *arXiv preprint arXiv:2006.02931*, 2020.

33 M. Parimala, R. M. Swarna Priya, Q.-V. Pham, K. Dev, P. K. R. Maddikunta, T. R. Gadekallu, and T. Huynh-The, "Fusion of Federated Learning and Industrial Internet of Things: A Survey," *arXiv preprint arXiv:2101.00798*, 2021.

34 Q. Yang, Y. Liu, T. Chen, and Y. Tong, "Federated machine learning: Concept and applications," *ACM Transactions on Intelligent Systems and Technology*, vol. 10, no. 2, pp. 1–19, 2019.

35 T. Huynh-The, C.-H. Hua, Q.-V. Pham, and D.-S. Kim, "MCNet: An efficient CNN architecture for robust automatic modulation classification," *IEEE Communications Letters*, vol. 24, no. 4, pp. 811–815, 2020.

36 H. Li, K. Ota, and M. Dong, "Learning IoT in edge: Deep learning for the Internet of Things with edge computing," *IEEE Network*, vol. 32, no. 1, pp. 96–101, 2018.

37 M. Alsenwi, N. H. Tran, M. Bennis, S. R. Pandey, A. K. Bairagi, and C. S. Hong, "Intelligent resource slicing for eMBB and URLLC coexistence in 5G and beyond: A deep reinforcement learning based approach," *IEEE Transactions on Wireless Communications*, vol. 20, no. 7, pp. 4585–4600, 2021.

38 Y. Xiao, Y. Li, G. Shi, and H. V. Poor, "Optimizing resource-efficiency for federated edge intelligence in IoT networks," in *2020 International Conference on Wireless Communications and Signal Processing (WCSP)*. IEEE, 2020, pp. 86–92.

39 Y. Xiao, G. Shi, and M. Krunz, "Towards Ubiquitous AI in 6G with Federated Learning," *arXiv preprint arXiv:2004.13563*, 2020.

40 N. C. Luong, D. T. Hoang, S. Gong, D. Niyato, P. Wang, Y.-C. Liang, and D. I. Kim, "Applications of deep reinforcement learning in communications and networking: A survey," *IEEE Communications Surveys & Tutorials*, vol. 21, no. 4, pp. 3133–3174, 2019.

41 Y. Sun, M. Peng, Y. Zhou, Y. Huang, and S. Mao, "Application of machine learning in wireless networks: Key techniques and open issues," *IEEE Communications Surveys & Tutorials*, vol. 21, no. 4, pp. 3072–3108, 2019.

42 C. T. Nguyen, N. Van Huynh, N. H. Chu, Y. M. Saputra, D. T. Hoang, D. N. Nguyen, Q.-V. Pham, D. Niyato, E. Dutkiewicz, and W.-J. Hwang, "Transfer Learning for Future Wireless Networks: A Comprehensive Survey," *arXiv preprint arXiv:2102.07572*, 2021.

43 C. Zhang, P. Patras, and H. Haddadi, "Deep learning in mobile and wireless networking: A survey," *IEEE Communications Surveys & Tutorials*, vol. 21, no. 3, pp. 2224–2287, 2019.

44 Q.-V. Pham, D. C. Nguyen, S. Mirjalili, D. T. Hoang, D. N. Nguyen, P. N. Pathirana, and W.-J. Hwang, "Swarm Intelligence for Next-Generation Wireless Networks: Recent Advances and Applications," *arXiv preprint arXiv:2007.15221*, 2020.

45 L. U. Khan, S. R. Pandey, N. H. Tran, W. Saad, Z. Han, M. N. Nguyen, and C. S. Hong, "Federated learning for edge networks: Resource optimization and incentive mechanism," *IEEE Communications Magazine*, vol. 58, no. 10, pp. 88–93, 2020.

46 J. Kang, Z. Xiong, D. Niyato, S. Xie, and J. Zhang, "Incentive mechanism for reliable federated learning: A joint optimization approach to combining reputation and contract theory," *IEEE Internet of Things Journal*, vol. 6, no. 6, pp. 10 700–10 714, 2019.

47 K. B. Letaief, W. Chen, Y. Shi, J. Zhang, and Y.-J. A. Zhang, "The roadmap to 6G: AI empowered wireless networks," *IEEE Communications Magazine*, vol. 57, no. 8, pp. 84–90, 2019.

48 W. Y. B. Lim, N. C. Luong, D. T. Hoang, Y. Jiao, Y.-C. Liang, Q. Yang, D. Niyato, and C. Miao, "Federated learning in mobile edge networks: A comprehensive survey," *IEEE Communications Surveys & Tutorials*, vol. 22, no. 3, pp. 2031–2063, 2020.

49 H. F. Atlam and G. B. Wills, "Intersections between IoT and distributed ledger," in *Advances in Computers*. Elsevier, 2019, vol. 115, pp. 73–113.

50 Y. Lu, "The blockchain: State-of-the-art and research challenges," *Journal of Industrial Information Integration*, vol. 15, pp. 80–90, 2019.

51 S. Daley, "25 Blockchain Applications & Real-World Use Cases Disrupting the Status Quo," 2020, accessed on 29.03.2021. [Online]. Available: https://builtin .com/blockchain/blockchain-applications.

52 T. Maksymyuk, J. Gazda, M. Volosin, G. Bugar, D. Horvath, M. Klymash, and M. Dohler, "Blockchain-empowered framework for decentralized network managementin 6G," *IEEE Communications Magazine*, vol. 58, no. 9, pp. 86–92, 2020.

53 D. C. Nguyen, P. N. Pathirana, M. Ding, and A. Seneviratne, "Blockchain for 5G and beyond networks: A state of the art survey," *Journal of Network and Computer Applications*, vol. 166, p. 102693, 2020.

54 S. Yrjölä, "How could blockchain transform 6G towards open ecosystemic business models?" in *2020 IEEE International Conference on Communications Workshops (ICC Workshops)*. IEEE, 2020, pp. 1–6.

55 T. Hewa, G. Gür, A. Kalla, M. Ylianttila, A. Bracken, and M. Liyanage, "The role of blockchain in 6G: Challenges, opportunities and research directions," in *2020 2nd 6G Wireless Summit (6G SUMMIT)*. IEEE, 2020, pp. 1–5.

56 X. Ling, J. Wang, Y. Le, Z. Ding, and X. Gao, "Blockchain radio access network beyond 5G," *IEEE Wireless Communications*, vol. 27, no. 6, pp. 160–168, 2020.

57 W. Li, Z. Su, R. Li, K. Zhang, and Y. Wang, "Blockchain-based data security for artificial intelligence applications in 6G networks," *IEEE Network*, vol. 34, no. 6, pp. 31–37, 2020.

58 M. Satyanarayanan, P. Bahl, R. Caceres, and N. Davies, "The case for VM-based cloudlets in mobile computing," *IEEE Pervasive Computing*, vol. 8, no. 4, pp. 14–23, 2009.

59 S. Barbarossa, S. Sardellitti, and P. D. Lorenzo, "Communicating while computing: Distributed mobile cloud computing over 5G heterogeneous networks," *IEEE Signal Processing Magazine*, vol. 31, no. 6, pp. 45–55, Nov. 2014.

60 C. Mouradian, D. Naboulsi, S. Yangui, R. H. Glitho, M. J. Morrow, and P. A. Polakos, "A comprehensive survey on fog computing: State-of-the-art and research challenges," *IEEE Communications Surveys & Tutorials*, vol. 20, no. 1, pp. 416–464, 2018.

61 M. Mukherjee, L. Shu, and D. Wang, "Survey of fog computing: Fundamental, network applications, and research challenges," *IEEE Communications Surveys $ Tutorials*, vol. 20, no. 3, pp. 1826–1857, 2018.

62 P. Mach and Z. Becvar, "Mobile edge computing: A survey on architecture and computation offloading," *IEEE Communications Surveys & Tutorials*, vol. 19, no. 3, pp. 1628–1656, 2017.

63 M. Chiang and T. Zhang, "Fog and IoT: An overview of research opportunities," *IEEE Internet of Things Journal*, vol. 3, no. 6, pp. 854–864, 2016.

64 H. F. Atlam, R. J. Walters, and G. B. Wills, "Fog computing and the Internet of Things: A review," *Big Data and Cognitive Computing*, vol. 2, no. 2, p. 10, 2018.

65 F. Bonomi, R. Milito, J. Zhu, and S. Addepalli, "Fog computing and its role in the Internet of Things," in *Proceedings of the First Edition of the MCC Workshop on Mobile Cloud Computing*, 2012, pp. 13–16.

66 K. Dolui and S. K. Datta, "Comparison of edge computing implementations: Fog computing, cloudlet and mobile edge computing," in *2017 Global Internet of Things Summit (GIoTS)*, 2017, pp. 1–6.

67 M. Patel, B. Naughton, C. Chan, N. Sprecher, S. Abeta, and A. Neal, "Mobile-edge computing introductory technical white paper," *White Paper, Mobile-edge Computing (MEC) industry initiative*, September 2014.

68 A. Ahmed and E. Ahmed, "A survey on mobile edge computing," in *2016 10th International Conference on Intelligent Systems and Control (ISCO)*, 2016, pp. 1–8.

69 S. Kekki et al., "MEC in 5G networks," *ETSI white paper*, no. 28, pp. 1–28, 2018.

70 Q.-V. Pham, F. Fang, V. N. Ha, M. J. Piran, M. Le, L. B. Le, W.-J. Hwang, and Z. Ding, "A survey of multi-access edge computing in 5G and beyond: Fundamentals, technology integration, and state-of-the-art," *IEEE Access*, vol. 8, pp. 116 974–117 017, 2020.

71 Y. Qiao, M. Zhang, Y. Zhou, X. Kong, H. Zhang, J. Bi, M. Xu, and J. Wang, "NetEC: Accelerating erasure coding reconstruction with in-network aggregation," *IEEE Transactions on Parallel and Distributed Systems*, vol. 33, no. 10, pp. 2571–2583, 2022.

72 Q.-V. Pham, R. Ruby, F. Fang, D. C. Nguyen, Z. Yang, M. Le, Z. Ding, and W.-J. Hwang, "Aerial computing: A new computing paradigm, applications, and challenges," *IEEE Internet of Things Journal*, vol. 9, no. 11, pp. 8339–8363, 2022.

73 R. L. Aguiar, I. F. Akyldiz, A. Autenrieth, A. Azcorra, G. Bianchi, S. Bigo, N. Blefari-Melazzi, H. Bock, A. Bourdoux, G. Caire et al., "Smart Networks in the Context of NGI," *White paper, European Technology Platform NetWorld2020*, May 2020. [Online]. Available: https://www.networld2020.eu/wp-content/uploads/2018/11/networld2020-5gia-sria-version-2.0.pdf.

74 N. Gisin and R. Thew, "Quantum communication," *Nature Photonics*, vol. 1, no. 3, pp. 165–171, 2007.

75 X. Su, M. Wang, Z. Yan, X. Jia, C. Xie, and K. Peng, "Quantum network based on non-classical light," *Science China Information Sciences*, vol. 63, no. 8, pp. 1–12, 2020.

76 T. Brougham, S. M. Barnett, K. T. McCusker, P. G. Kwiat, and D. J. Gauthier, "Security of high-dimensional quantum key distribution protocols using Franson interferometers," *Journal of Physics B: Atomic, Molecular and Optical Physics*, vol. 46, no. 10, p. 104010, 2013.

77 A. Manzalini, "Quantum communications in future networks and services," *Quantum Reports*, vol. 2, no. 1, pp. 221–232, 2020.

78 F. Tariq, M. R. Khandaker, K.-K. Wong, M. A. Imran, M. Bennis, and M. Debbah, "A speculative study on 6G," *IEEE Wireless Communications*, vol. 27, no. 4, pp. 118–125, 2020.

79 S. J. Nawaz, S. K. Sharma, S. Wyne, M. N. Patwary, and M. Asaduzzaman, "Quantum machine learning for 6G communication networks: State-of-the-art and vision for the future," *IEEE Access*, vol. 7, pp. 46 317–46 350, 2019.

80 R. Arul, G. Raja, A. O. Almagrabi, M. S. Alkatheiri, S. H. Chauhdary, and A. K. Bashir, "A quantum-safe key hierarchy and dynamic security association for LTE/SAE in 5G scenario," *IEEE Transactions on Industrial Informatics*, vol. 16, no. 1, pp. 681–690, 2019.

81 A. A. Abd EL-Latif, B. Abd-El-Atty, S. E. Venegas-Andraca, and W. Mazurczyk, "Efficient quantum-based security protocols for information sharing and data protection in 5G networks," *Future Generation Computer Systems*, vol. 100, pp. 893–906, 2019.

82 D. Zavitsanos, A. Ntanos, G. Giannoulis, and H. Avramopoulos, "On the QKD integration in converged fiber/wireless topologies for secured, low-latency 5G/B5G fronthaul," *Applied Sciences*, vol. 10, no. 15, p. 5193, 2020.

83 M. Z. Chowdhury, M. Shahjalal, M. Hasan, and Y. M. Jang, "The role of optical wireless communication technologies in 5G/6G and IoT solutions: Prospects, directions, and challenges," *Applied Sciences*, vol. 9, no. 20, p. 4367, 2019.

84 L. U. Khan, "Visible light communication: Applications, architecture, standardization and research challenges," *Digital Communications and Networks*, vol. 3, no. 2, pp. 78–88, 2017.

85 D. Karunatilaka, F. Zafar, V. Kalavally, and R. Parthiban, "Led based indoor visible light communications: State of the art," *IEEE Communications Surveys & Tutorials*, vol. 17, no. 3, pp. 1649–1678, 2015.

86 A. Jovicic, J. Li, and T. Richardson, "Visible light communication: Opportunities, challenges and the path to market," *IEEE Communications Magazine*, vol. 51, no. 12, pp. 26–32, 2013.

87 D. Tsonev, S. Videv, and H. Haas, "Towards a 100 Gb/s visible light wireless access network," *Optics Express*, vol. 23, no. 2, pp. 1627–1637, 2015.

88 S. Soderi, "Enhancing security in 6G visible light communications," in *2020 2nd 6G Wireless Summit (6G SUMMIT)*. IEEE, 2020, pp. 1–5.

89 M. Katz and I. Ahmed, "Opportunities and challenges for visible light communications in 6G," in *2020 2nd 6G Wireless Summit (6G SUMMIT)*. IEEE, 2020, pp. 1–5.

90 S. Ariyanti and M. Suryanegara, "Visible light communication (VLC) for 6G technology: The potency and research challenges," in *2020 Fourth World Conference on Smart Trends in Systems, Security and Sustainability (WorldS4)*. IEEE, 2020, pp. 490–493.

91 S. U. Rehman, S. Ullah, P. H. J. Chong, S. Yongchareon, and D. Komosny, "Visible light communication: A system perspectiveoverview and challenges," *Sensors*, vol. 19, no. 5, p. 1153, 2019.

92 J. Chen and Z. Wang, "Topology control in hybrid VLC/RF vehicular ad-hoc network," *IEEE Transactions on Wireless Communications*, vol. 19, no. 3, pp. 1965–1976, 2019.

93 R. Alghamdi, R. Alhadrami, D. Alhothali, H. Almorad, A. Faisal, S. Helal, R. Shalabi, R. Asfour, N. Hammad, A. Shams et al., "Intelligent surfaces for 6G wireless networks: A survey of optimization and performance analysis techniques," *IEEE Access*, vol. 8, pp. 202795–202818, 2020.

94 M. Di Renzo, K. Ntontin, J. Song, F. H. Danufane, X. Qian, F. Lazarakis, J. De Rosny, D.-T. Phan-Huy, O. Simeone, R. Zhang et al., "Reconfigurable intelligent surfaces vs. relaying: Differences, similarities, and performance comparison," *IEEE Open Journal of the Communications Society*, vol. 1, pp. 798–807, 2020.

95 M. Jung, W. Saad, and G. Kong, "Performance Analysis of Large Intelligent Surfaces (LISs): Uplink Spectral Efficiency and Pilot Training," *arXiv preprint arXiv:1904.00453*, 2019.

96 E. Basar, "Reconfigurable intelligent surface-based index modulation: A new beyond MIMO paradigm for 6G," *IEEE Transactions on Communications*, vol. 68, no. 5, pp. 3187–3196, 2020.

97 C. J. Vaca-Rubio, P. Ramirez-Espinosa, K. Kansanen, Z.-H. Tan, E. de Carvalho, and P. Popovski, "Assessing Wireless Sensing Potential with Large Intelligent Surfaces," *arXiv preprint arXiv:2011.08465*, 2020.

98 S. Stanković, "Compressive sensing: Theory, algorithms and applications," in *2015 4th Mediterranean Conference on Embedded Computing (MECO)*. IEEE, 2015, pp. 4–6.

99 Z. Gao, L. Dai, S. Han, I. Chih-Lin, Z. Wang, and L. Hanzo, "Compressive sensing techniques for next-generation wireless communications," *IEEE Wireless Communications*, vol. 25, no. 3, pp. 144–153, 2018.

100 Y. Lu and X. Zheng, "6G: A survey on technologies, scenarios, challenges, and the related issues," *Journal of Industrial Information Integration*, vol. 19, p. 100158, 2020.

101 Y. Huo, X. Dong, W. Xu, and M. Yuen, "Enabling multi-functional 5G and beyond user equipment: A survey and tutorial," *IEEE Access*, vol. 7, pp. 116 975–117 008, 2019.

102 M. B. Shahab, R. Abbas, M. Shirvanimoghaddam, and S. J. Johnson, "Grant-free non-orthogonal multiple access for IoT: A survey," *IEEE Communications Surveys & Tutorials*, vol. 22, no. 3, pp. 1805–1838, 2020.

103 A. Taha, M. Alrabeiah, and A. Alkhateeb, "Enabling large intelligent surfaces with compressive sensing and deep learning," *IEEE Access*, vol. 9, pp. 44 304–44 321, 2021.

104 "Zero-touch network and service management (ZSM); terminology for concepts in ZSM," August 2019, [Accessed on 29.03.2021]. [Online]. Available: https://www.etsi.org/deliver/etsi_gs/ZSM/001_099/007/01.01.01_60/gs_ZSM007v010101p.pdf.

105 "Zero-touch network and Service Management (ZSM); Requirements based on documented scenarios," October 2019, [Accessed on 29.03.2021]. [Online]. Available: https://www.etsi.org/deliver/etsi_gs/ZSM/001_099/001/01.01.01_60/gs_ZSM001v010101p.pdf.

106 T. Darwish, G. K. Kurt, H. Yanikomeroglu, G. Senarath, and P. Zhu, "A Vision of Self-Evolving Network Management for Future Intelligent Vertical HetNet," *arXiv preprint arXiv:2009.02771*, 2020.

107 C. Benzaid and T. Taleb, "ZSM security: Threat surface and best practices," *IEEE Network*, vol. 34, no. 3, pp. 124–133, 2020.

108 N. H. Mahmood, H. Alves, O. A. López, M. Shehab, D. P. M. Osorio, and M. Latva-aho, "Six Key Enablers for Machine Type Communication in 6G," *arXiv preprint arXiv:1903.05406*, 2019.

109 S. Hu, X. Chen, W. Ni, X. Wang, and E. Hossain, "Modeling and analysis of energy harvesting and smart grid-powered wireless communication networks: A contemporary survey," *IEEE Transactions on Green Communications and Networking*, vol. 4, no. 2, pp. 461–496, 2020.

110 A. A. Nasir, X. Zhou, S. Durrani, and R. A. Kennedy, "Relaying protocols for wireless energy harvesting and information processing," *IEEE Transactions on Wireless Communications*, vol. 12, no. 7, pp. 3622–3636, 2013.

111 M. A. Hossain, R. M. Noor, K.-L. A. Yau, I. Ahmedy, and S. S. Anjum, "A survey on simultaneous wireless information and power transfer with cooperative relay and future challenges," *IEEE Access*, vol. 7, pp. 19 166–19 198, 2019.

3

6G Security Vision

This chapter has the focus on the 6G security vision and the relevant topics. After reading this chapter, you should be able to:

- Gain an overview of key technologies with a tendency of being used in 6G networks.
- Identify the major security considerations, threat landscape, and counter measures in future 6G devices.

3.1 Overview of 6G Security Vision

In this section, we first provide an overview about novel 6G requirements in general. Then we discuss the security considerations, 6G security vision, and the potential security key performance indicators (KPIs). In Section 3.2.1 describe the security landscape for the envisioned 6G architecture which is classified into four key areas such as functional architecture (i.e. intelligent radio and radio-core convergence), edge intelligence and cloudification, specialized subnetworks, and network management and orchestration.

3.1.1 New 6G Requirements

Future 6G applications will pose stringent requirements and require extended network capabilities compared with the currently developed 5G networks. These requirements are summarized in Figure 3.1. They are established to enable the wide range of key 6G use cases and thus can be categorized accordingly. For **Further enhanced Mobile Broadband (FeMBB)**, the mobile connection speed has to reach the peak data rate at Tbps level [2]. With **ultra-massive Machine-Type Communication (umMTC)**, the connection density will further

Security and Privacy Vision in 6G: A Comprehensive Guide, First Edition.
Pawani Porambage and Madhusanka Liyanage.
© 2023 The Institute of Electrical and Electronics Engineers, Inc. Published 2023 by John Wiley & Sons, Inc.

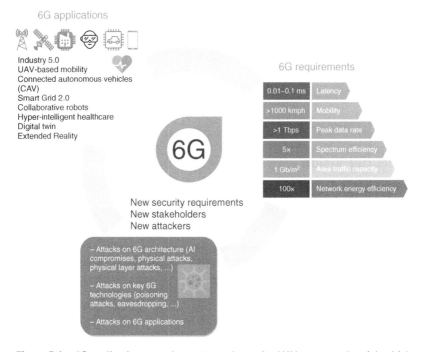

Figure 3.1 6G applications, requirements, and security. UAV, unmanned aerial vehicles. Source: [1]/IEEE/CC BY 4.0.

increase in 6G due to the novel concept of Internet of Everything (IoE) as the next phase of Internet of Things (IoT).

These devices will have to communicate with each other and the infrastructure, and provide collaborative services in an autonomous and self-driven manner [3]. For new latency extremely sensitive 6G applications in the **enhanced ultra-reliable, low-latency communication (ERLLC/eURLLC)** use case, the E2E latency in 6G should be reduced down to μs level [4]. 6G will require the network energy efficiency to be improved by 10× than 5G and 100× than 4G. It is also expected to enable extremely low-power communications for the resource-constrained devices [4]. Moreover, intelligent and proactive mobility management systems will support seamless and instant mobility beyond 1000 kmph speeds [2].

For ERLLC, the latency impact of security workflows will be considered to ensure service quality. Similarly, high reliability requirements call for very efficient security solutions protecting availability of services and resources. With FeMBB, extreme data rates will pose challenges regarding traffic processing for security such as attack detection, artificial intelligence (AI)/machine learning

(ML) pipelines, traffic analysis, and pervasive encryption. That issue can be alleviated with distributed security solutions since traffic should be processed locally and on-the-fly in different segments of the network, ranging from the edge to the core service cloud [5]. At this point, distributed ledger technologies (DLT) will be instrumental with transparency, security, and redundancy attributes. umMTC will serve critical use-cases which impose much more stringent security requirements compared to 5G. In particular, IoE with very diverse capabilities will challenge the deployment and operation of security solutions such as distributed AI/ML and privacy concerns. An important aspect is how to integrate novel security enablers in an abundance of resource-constrained devices. Nevertheless, the security enforcement will be more complex since network entities will be much more mobile, changing their edge networks frequently and getting services in different administrative domains.

3.2 6G Security Vision and KPIs

The vision of 6G networks is formed with many novelties and advancements in terms of architecture, applications, technologies, policies, and standardization. Similar to the generic 6G vision which has the added intelligence on top of the cloudified and softwarized 5G networks, 6G security vision also has a close fusion with AI which leads to security automation (Figure 3.2). At the same time, the

Figure 3.2 6G security vision. Source: [1]/IEEE/CC BY 4.0.

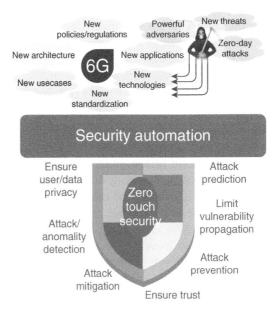

adversaries also become more powerful and intelligent and capable of creating new forms of security threats. For instance, the detecting zero-day attacks is always challenging, whereas prevention from their propagation is the most achievable mechanism. Therefore, the necessity will become more important than ever to incorporate intelligent and flexible security mechanisms for predicting, detecting, mitigating, and preventing security attacks and limiting the propagation of such vulnerabilities in the 6G networks. It is also equally significant to ensure privacy and trust in the respective domains and among the stakeholders. Especially, security and privacy are two closely coupled topics where security relates the safeguarding of the actual data and privacy ensures the covering up of the identities related to those data. While security on its own is exclusive from privacy, the vice versa is not valid: Essentially, to assure privacy, there should be always security mechanisms that protect data. Chapters 13 and 14, we discuss how security and privacy complement each other for different aspects of 6G.

To set the scope of 6G, we also think that KPIs and key value indicators (KVI) will help to take the dimensions of impact that go beyond the scope of deterministic performance measures into full account [6]. It is expected that 6G systems will incorporate novel aspects, such as integrated sensing, artificial intelligence, local compute-and-storage, and embedded devices [7]. These aspects will both lead to enhancements to existing KPIs, as well as require a whole new set of KPIs and KVIs which have not traditionally been associated with mobile networks, such as sensing accuracy, computational round-trip-time, and AI model convergence time. The KVIs will quantify the value of the new 6G-related technologies from the perspective of sustainability, security, inclusiveness, and trustworthiness stemming from the UN sustainable development goals [8, 9].

Therefore, we believe that the new aspects will have a significant impact on how security KPIs are designed and measured. Various aspects should be considered for characterizing security, such as physical layer security (PLS), network information security, and AI/ML-related security [10]. The KPIs are described with 6G impact, as follows:

Protection level: The guaranteed level of protection against certain threats and attacks; more stringent due to the pervasive utility of 6G and growing risk level.

Time to respond (mean, max, …): Time for security functions to counteract in case of malicious activity; much smaller due to compressed timescale of 6G networks, e.g. an attack can cause havoc at an order or faster.

Coverage: The coverage of security functions over the 6G service elements and functions; more challenging due to diverse 6G technologies and ultra-distributed functions.

Autonomicity level: A measure of how autonomic security controls can act; expected to be easier to implement with pervasive AI, but also may be counter-beneficial due to AI security issues.

AI robustness: The robustness of AI algorithms in the network hardened for security; more difficult to maintain consistently system-wide but more critical due to AI's role in 6G.

Security AI model convergence time: Time for learning models working for security to converge; although more advanced AI/ML models are emerging and hardware capabilities are improving, the data availability and complexity are challenging factors for this KPI.

Security function chain round-trip-time: Time for chained security functions to process for ingest, analyze, decide, and act (related to "Time to respond" KPI); security architecture in 6G supposed to be more distributed, leading to challenges. But at the same time, device-centric and edge-centric solutions will help.

Cost to deploy security functions (mean, max, …): Various cost metrics for measuring the cost of deployment; substantially increases due to complexity, thus harder to meet target KPI values.

3.2.1   Security Threat Landscape for 6G Architecture

Undoubtedly, the massive emergence of connections in the future 6g networks will increase the security and privacy vulnerabilities. Considering the foreseen technological, architectural, and application-specific aspects and their advancements in the future 6G networks, the threat landscape of 6G security is summarized in Figure 3.3. Since the attacks can be generalized based on the architecture rather than the technologies or the applications, we are taking this step forward to give the reader an insight about the security threat landscape on top of the envisioned 6G architecture. Among various visionary 6G architectures proposed by the industrial and academic research community, we have identified the vision from Nokia Bell Labs as a realistic yet ambitious proposal to facilitate our security landscape analysis for 6G architecture [11]. As stated by Ziegler et al. in [11], after investigating the potential 6G architectural innovation, they decompose the data and information architecture into four segments, namely, *platform*, *functions*, *orchestration*, and *specialization*. In the infrastructure "platform" of 6G architecture, heterogeneous clouds need to create agnostic, open, and scalable run-time environment to accelerate the hardware and improve data flow centrality. The "functional" architecture component includes topics such as RAN core convergence and intelligent radio. The "specialized" part represents the architectural enablers of flexible off-load, subnetworks, and extreme slicing. The "orchestration" component includes the intelligent network management and the cognitive closed loop and automation of 6G networks. In the rest of the section, we discuss the security considerations of these four 6G architectural components and how they are related at the consumer end.

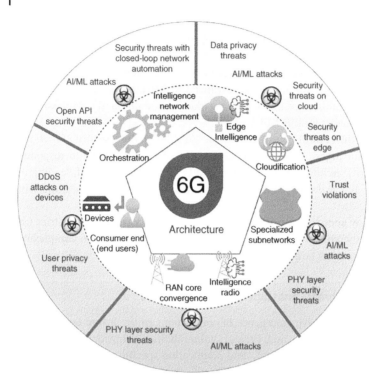

Figure 3.3 6G security threat landscape. Source: [1]/IEEE/CC BY 4.0.

However, in addition to the 6G architectural evolution, the advent and advancements of technologies may also pave the way to generate more powerful attackers who can create sophisticated attacks. For instance, while detecting AI-based malicious activities, distributed learning-based attack prediction methods give promising potential solutions within the constantly changing environments [10].

References

1 P. Porambage, G. Gür, D. P. M. Osorio, M. Liyanage, A. Gurtov, and M. Ylianttila, "The roadmap to 6G security and privacy," *IEEE Open Journal of the Communications Society*, vol. 2, pp. 1094–1122, 2021.

2 S. Nayak and R. Patgiri, "6G Communication Technology: A Vision on Intelligent Healthcare," *arXiv preprint arXiv:2005.07532*, 2020.

3 F. Jameel, U. Javaid, B. Sikdar, I. Khan, G. Mastorakis, and C. X. Mavromoustakis, "Optimizing blockchain networks with artificial intelligence:

Towards efficient and reliable IoT applications," in *Convergence of Artificial Intelligence and the Internet of Things*. Springer, 2020, pp. 299–321.

4 Z. Zhang, Y. Xiao, Z. Ma, M. Xiao, Z. Ding, X. Lei, G. K. Karagiannidis, and P. Fan, "6G wireless networks: Vision, requirements, architecture, and key technologies," *IEEE Vehicular Technology Magazine*, vol. 14, no. 3, pp. 28–41, 2019.

5 P. Ranaweera, A. D. Jurcut, and M. Liyanage, "Survey on multi-access edge computing security and privacy," *IEEE Communications Surveys & Tutorials*, vol. 23, no. 2, pp. 1078–1124, 2021.

6 M. Latva-Aho and K. Leppänen, "Key drivers and research challenges for 6G ubiquitous wireless intelligence (white paper)," *Oulu, Finland: 6G Flagship*, 2019.

7 A. Pouttu, F. Burkhardt, C. Patachia, L. Mendes, G. R. Brazil, S. Pirttikangas, E. Jou, P. Kuvaja, F. T. Finland, M. Heikkilä et al., "6G white paper on validation and trials for verticals towards 2030's." [Online]. Available: https://www.6gchannel.com/wp-content/uploads/2020/04/6g-white-paper-validation-trials.pdf.

8 United Nations (UN), "United Nations (UN) #Envision2030 Sustainable Development Goals." [Online]. Available: https://sdgs.un.org/goals.

9 V. Ziegler and S. Yrjola, "6G indicators of value and performance," in *2020 2nd 6G Wireless Summit (6G SUMMIT)*. IEEE, 2020, pp. 1–5.

10 G. Gui, M. Liu, F. Tang, N. Kato, and F. Adachi, "6G: Opening new horizons for integration of comfort, security and intelligence," *IEEE Wireless Communications*, vol. 27, no. 5, pp. 126–132, 2020.

11 V. Ziegler, H. Viswanathan, H. Flinck, M. Hoffmann, V. Räisänen, and K. Hätönen, "6G architecture to connect the worlds," *IEEE Access*, vol. 8, pp. 173 508–173 520, 2020.

Part II

Security in 6G Architecture

4

6G Device Security

This chapter has the focus on the technological evolution of networking and user devices and their new trends in the 6G wireless systems in accordance with the security aspects. After reading this chapter, you should be able to:

- Gain an overview of the key technologies with a tendency of being used in 6G networks.
- Identify the major security considerations, threat landscape, and counter measures in future 6G devices.

4.1 Overview of 6G Devices

The expectation of the scientific community is to bring major breakthroughs with 6G in many different field in terms of network and architecture, access network, and the devices and components. Among them, in this chapter, we mainly discuss the evolution of devices and components with 6G and their respective security considerations. Although it is still challenging to identify exact 6G requirements, characteristics, and the networking elements at this stage, we will bring a more futuristic overview about the expected 6G devices that will be available in the future.

Augmented reality (AR) devices will be unavoidable among the next-generation consumables. Most people will be wearing some sort of glasses in their day-to-day life in future. Moreover, the majority of the devices in our surrounding may have some kind of intelligence with a connectivity which may easily interact with you. The devices will be more intelligent and pervasive in such a way to be aware of your emotions. As a downside, this may also impose sensitive and critical privacy issues. The devices may communicate with each other which may bring lack of controllability of the humans at some point.

Security and Privacy Vision in 6G: A Comprehensive Guide, First Edition.
Pawani Porambage and Madhusanka Liyanage.

With the evolution of 6G communication technologies, the connectivity type has to be independent of the device. The devices will have multiple connectivity options which may increase their popularity and usability. Usually, the components are added to the devices which will evolve independent of the connectivity part. For instance autonomous cars are designed with multiple sensors. When the connectivity is provided, the functionality of those sensors can be upgraded. However, the connectivity is not a fundamental part for the operation of autonomous cars. In addition to that, more applications will be available to the devices.

Over the years, the devices have evolved enormously (as shown in Figure 4.1), for instance, the smart phones have evolved with the sized, thickness, camera, and battery life. This evolution may continue to the future as well. At the same time, there is a big rising wave of companion devices which are designed to work with a main or parent device which generally needs to accompany those, for instance, it can be a smartwatch or a bluetooth headset or any other wearable which is working together with a smartphone or a tablet which is the main device. This may go until smart clothing, electronic skin patches, or hearables in the near future. It is also predicted that there will be the convergence of smart phone, smart watch, and AR glasses into single extended reality (XR) device with the advancements such as immersive XR and Internet of senses. Appending such things into everyday life component such as normal eye glasses will bring more usability of the device. All miscellaneous devices that are benefited from connectivity of 6G may deploy in a wide range of applications including connected drones, security robots,

Figure 4.1 Evolution of devices and components from wired telephones to 5G wireless devices and then toward 6G vision. Source: evgtin/Adobe Stock; boyzzzzz/Adobe Stock; Verisakeet/Adobe Stock; PixieMe/Adobe Stock; Kaspars Grinvalds/Adobe Stock; Lukas Gojda/Adobe Stock; MR/Adobe Stock; khunkorn/Adobe Stock; metamorworks/Adobe Stocks; metamorworks/Adobe Stocks; Elnur/Adobe Stock; Kitreel/Adobe Stock; Kitreel/Adobe Stock.

smart cities. 6G devices will have more-advanced applications using artificial intelligence (AI) with extreme location awareness and absolute coverage.

The whole world is eagerly looking for energy-efficient technologies. New trends are emerging toward zero-energy or extreme low-power devices which never need to be charged. This is where the energy harvesting is required. Therefore, extremely energy-efficient protocols need to be developed while giving more attention to the most energy efficient security configuration. Moreover, it is necessary to minimize the amount of information that could be transmitted while moving toward radio-based technologies which are more appealing.

4.2   6G Device Security Challenges

Over a decade ago, it was hard to find smart household devices. However, the current situation is that, it is hard to find a thing which is not smart. All the things around us and our personal items are surprisingly getting smarter day by day. When we move beyond the current 5G technologies, to the 6G world, these devices may become more intelligent and operate with a higher autonomy and a expand their connectivity [1]. Although this will enhance the humans' standard of life, it may implicitly endanger the societal operations and the right of freedom of human species into a greater threat more than ever. As illustrated in Figure 4.2, some device security related challenges can be encountered in 6G.

4.2.1   Growth of Data Collection

Due to the highly connected nature of the devices, there will be a time where anyone can connect to anything from anywhere in the world. On the one hand, although the highest security protection is provided to the devices in our network, as long as the attacker or the malicious user is technologically capable to break

6G device security challenges

| Growth of data collection | Cloud connectivity | Device capacity | Ultra saturated devices |

Figure 4.2   Some device security-related challenges that may encounter in 6G vision.

the system, there is a high risk of revealing our privacy-sensitive information and reveal the most confidential content to the rest of the world.

On the other hand, the development of the devices will evolve with the collection of data. When the term "Internet of Things" (IoT) is moving toward Internet of Everything (IoE), the data collection features may also evolve by putting the customer privacy into additional risks with such enormous data collections. For instance, their home networks, localizing data, or health-related data can be hacked and that may even create life threats.

4.2.2 Cloud Connectivity

Cloud connectivity is getting popular day by day. Although it provides higher security measurements and allow offloading the resource-consuming computations, the great dependency on the cloud server operations may render the proper functionality of user devices, especially when the connectivity is lost or cloud server is down. In addition to that the connectivity between the devices and cloud servers can be exploited as a means of hacking the devices and get their control for harmful or unethical purposes.

Instead of using multifactor authentication procedures, the usual practice of using long-term authentication credentials may also endanger the IoT/IoE systems. In addition to that, the IoT/IoE threat landscape may evolve toward targeting higher value cloud-enabled devices that can create a noteworthy impact to the entire system.

4.2.3 Device Capacity

The number of connected devices will vehemently increase with the long-overdue transition to IPv6 addresses. With that, the Internet service providers (ISPs) may even support the connectivity of the very restricted devices connected to the Internet rather than operating in the local area networks. In such cases, it will be challenging to install firewalls, antivirus, or malware software in the resource-constrained devices. The attacker may directly access those devices from the Internet.

4.2.4 Ultrasaturated Devices

It is envisioned that, 6G will accommodate a more than 10-fold increase of devices over 5G. Especially, 6G will introduce more and more edge- and end-user devices to the world that may not only act as data producers and consumers but also as the repeaters themselves. With the extremely high number of devices connected to the networks, it will create an ultrasaturated device and data world with 6G, where the degree of connectivity is still unknown. Therefore, it will be like a mesh network where more complicated security attacks can occur and be delegated by giving the security professional an extremely hard time to identify the source of attack or the truncation of its mitigation. It will be really difficult to keep pace with treated actors' speed, scope, and complexity because the attackers are also more powerful and intelligent and therefore to understand and respond to the defenders tactics such as how he reacts, his moves, and repackaging techniques against them.

4.3 Addressing Device Security in 6G

From the beginning of the advanced portable communication in early generations of wireless systems, they are dependent on a physical placing of symmetric keys in a Subscriber Identity Module, which is also known as SIM card. Although the encryption computations are moved from undisclosed to universal guidelines, the alternative cryptographic instruments are introduced for the shared verification process [2]. In accordance with the general standards, 5G security model is still dependent on the SIM cards [3]. Although the SIM cards are getting smaller into nanoscale, they still need to be inserted into the device/gadgets. This may limit the appropriateness of foreseen IoE paradigm in 6G. In a way, this challenge can be tackled with using eSIMs, however, introducing some issues with physical measures. Another solution will be iSIMs which will be a part of System-on-Chip in the future gadgets. This will also face challenges due to the possible resistance coming from the telecom operators due to conceivable loss of control.

Typically, SIM cards rely on proven symmetric key encryption, which scaled well up to millions to billions of users. However, it has some serious issues with user privacy, IoT, network authentication, and fake base stations. Therefore, 6G need to consider a significant shift from symmetric crypto to asymmetric public/private keys and even to the postquantum keying mechanisms. Already 5G plans to support authentication through a public-key infrastructure (PKI) and a set of microservices communicating over HTTPS. The authentication, confidentiality, and integrity for such communications are provided by transport layer security (TLS) using elliptic curve cryptography (ECC). Experiences that come from the use of these technologies in 5G will shape the user and device authentication approaches in 6G.

In order to ensure the privacy preservation and anonymization of data, different technologies are expected to be widely incorporated with the device data management in the 6G paradigm. These include homomorphic encryption, secure multiparty communication, federated learning, and differential privacy. These topics are widely discussed in Chapters 14, 17, 18 of this book. In addition to that, the trust technologies such as Trusted Platform Modules and Trusted Execution

Environments (TEEs) will be extensively incorporated with the devices to protect data integrity and claim the proof of data ownership anchored in hardware and silicon.

Another trend in 6G networks is the incorporation of collaborative zero trust architecture (ZTA) that require sophisticated access control to protect the internal network entities [4]. The research is underway to defend against the spread of malware, Distributed Denial of Service (DDoS) attacks, and zero-day attacks through dynamic access control to the user equipment (UE). The access control of the UE can be performed at the border switches of the network to protect the internal treats from external threat.

References

1 V. Ziegler and S. Yrjölä, "How to make 6G a general purpose technology: Prerequisites and value creation paradigm shift," in *2021 Joint European Conference on Networks and Communications & 6G Summit (EuCNC/6G Summit)*. IEEE, 2021, pp. 586–591.

2 M. Ylianttila, R. Kantola, A. Gurtov, L. Mucchi, I. Oppermann, Z. Yan, T. H. Nguyen, F. Liu, T. Hewa, M. Liyanage et al., "6G White Paper: Research Challenges for Trust, Security and Privacy," *arXiv preprint arXiv:2004.11665*, 2020.

3 R. Yasmin, J. Petäjäjärvi, K. Mikhaylov, and A. Pouttu, "On the integration of LoRaWAN with the 5G test network," in *2017 IEEE 28th Annual International Symposium on Personal, Indoor, and Mobile Radio Communications (PIMRC)*. IEEE, 2017, pp. 1–6.

4 S. Rose, O. Borchert, S. Mitchell, and S. Connelly, "Zero trust architecture," National Institute of Standards and Technology, Tech. Rep., 2020.

5

Open RAN and RAN-Core Convergence

This chapter has the focus on the security and privacy challenges and opportunities related to the Open radio access network (O-RAN) which is a novel industry-level standard for radio access network (RAN). After reading this chapter, you should be able to:

- Understand the Open RAN architecture and its components
- Understand the security and privacy risks and challenges associated with Open RAN architecture
- Understand the possible security and privacy benefits of adapting Open RAN architecture for future networks
- Understand relevant security standardization efforts relevant to Open RAN security

5.1 Introduction

Mobile network communications are becoming one of the critical enablers of the current digital economy and interconnecting national critical infrastructure-based services [1]. The number of mobile subscribers and different mobile-based services is increasing in a rapid phase all across the globe [2]. However, the radio spectrum is still a scarce resource, and optimal utilization of radio resources is critical for developing a telecommunication network [3]. Thus, the orchestration and management of radio resources or the RAN are also evolved with each mobile generation. The early mobile generation mobile networks architectures such as 2G and 3G had controllers responsible for orchestration and management of RAN and its resources [4]. The flat network architecture in 4G enables a new interface (i.e. X2) to support base station-level communication to handle RAN resource allocation [5]. However, the RAN of existing mobile network generations is still

Security and Privacy Vision in 6G: A Comprehensive Guide, First Edition.
Pawani Porambage and Madhusanka Liyanage.

based on monolithic building blocks. Thus, RAN functions of existing networks, including most of the 5G network, are still contained with the proprietary vendor-specific devices called baseband units (BBUs) at the base stations [6]. However, this approach leads to the proverbial vendor lock-in RAN since different RAN vendors can design their flavor of RAN equipment. This has eliminated the possibility for mobile network operators (MNOs) to get mix-and-match services from other RAN vendors.

The introduction of the network softwarization concept in 5G [7–9] and added intelligence in beyond-5G networks have opened up a promising solution called Open RAN to mitigate this issue [9, 10]. The Open RAN Alliance [11] went back to the controller concept to enable best-of-breed Open RAN. Open RANs, also known as ORANs or O-RANs, have been considered as one of the most exciting RAN concepts, designed for 5G and beyond wireless systems. Open RAN promotes openness and added intelligence for RAN network elements that could overcome the limitations of existing RAN technologies, [12, 13]. The feature of openness allows smaller and new players in the RAN market to deploy their customized services, while the feature of intelligence is to increase automation and performance by optimizing the RAN elements and network resources. Moreover, Open RAN offers many RAN solutions and elements to the network operators to be more open and flexible. Further, the network operators can shorten the time-to-market of new applications and services to maximize the overall revenue because of the virtualization feature. Thus, the added intelligence in Open RAN could offer superior benefits even to the existing network softwarization-based virtual RAN and cloud RAN concepts.

There are two major Open RAN organizations, i.e. Telecom Infra Project (TIP) [14] and the O-RAN alliance [15] who are working on the advancement of Open RAN realization. The TIP's OpenRAN program is an initiative that focuses on developing solutions for future RANs based on disaggregation of multivendor hardware, open interfaces, and software. O-RAN alliance is another Open RAN organization that mainly focuses on defining and enforcing new standards for Open RAN to ensure interoperability among the different vendors. At the beginning of 2020, a liaison agreement between TIP and O-RAN was made to ensure their alignment in developing interoperable Open RAN solutions. OpenRAN development of TIP has similar original goal similar to O-RAN. Thus, we use the term "Open RAN" throughout the chapter with refer to both OpenRAN and O-RAN development efforts.

However, the benefits of Open RAN come at challenges, e.g. security, deterministic latency, and real-time control [10, 16, 17]. Among these factors, the security in Open RAN is quite essential. As 5G and beyond networks are responsible for interconnecting many Internet protocol Telephony (IpT)-based critical national infrastructure, attacks on future telecommunication networks will have a ripple effect [18, 19]. Some devastating examples caused by such attacks are smart cities and factories shutting down, a complete black-out of the power grids and water supplies, a fall-out of the transportation infrastructure with crashes by autonomous vehicles [20, 21]. All these challenges demand significant effort from the research and industry communities to standardize and implement security for all the sections of 5G and beyond networks, including Open RAN networks [22]. Specially, the decentralization of control functions with Open RAN increases the number of threat vectors and the surface area for attacks.

Open RAN has distinct features that bring intelligence to future networks. While artificial intelligence (AI) helps overcome various challenges of 6G Open RANs via intelligent and data-driven solutions, it can hurt the security of RAN. Attackers can target the AI systems or even use AI-based attacks to jeopardize the operation of the Open RAN system. Thus, Open RAN will now be vulnerable to AI-related attacks such as denial-of-service (DoS) [23], spoofing [24], and malicious data injection [25] could affect the AI. For instance, AI training can be manipulated in an Open RAN spectrum access system by inserting fake signals. In addition, the integration of network softwarization will add a whole new set of security attacks related to virtualization. Similar to the 5G core and edge networks, now Open RAN needs to tackle softwarization-associated attacks such as virtual network function/cloud network function (VNF/CNF) manipulation [26], virtual machine (VM) misconfiguration [27], log leak attacks [28]. In addition, open interfaces defined in Open RAN will introduce another set of security and privacy vulnerability. Thus, it is necessary to develop correct security and privacy solutions to mitigate these new Open RAN-related security and privacy solutions at the radio network level. Existing security mechanisms, frameworks, and governance approaches will need to be upgraded to operate in open multivendor-controlled Ecosystem.

On the other hand, added features of Open RAN can bring security and privacy advantages over traditional RAN. Open RAN can also build upon the security enhancements already enabled by 5G and allow the operator to control the network's security entirely, ultimately enhancing the operational security of their network. Less hardware dependency and support for complete software control in Open RAN allow isolating security breaches quickly and intelligently, reducing the impact of security risk. In addition, these features reduce the risks associated with security mechanism upgrades. Moreover, the modularity supported by the open interface in Open RAN allows the security and privacy deployments to support continuous integration/continuous delivery (CI/CD) operating model [29]. The CI/CD model supports seamless and effective security management against the security vulnerability in Open RAN.

Moreover, Open RAN enables the possibility for zero-touch and frequent software updates [30], which is more transparent, fast, secure, and low cost than the software upgrades in a traditional network. Finally, standardization of open interfaces can also reduce the security risks to a certain extent as it can help detect incongruences and offer concrete steps to monitor the network. Thus, it is crucial to identify these new security benefits and rectify them correctly in future RAN deployments.

5.2 Open RAN Architecture

Unlike traditional RAN technology, Open RAN decouples hardware and software bonds in proprietary RAN equipment. This feature offers more flexibility for mobile operators to deploy and upgrade their RAN segment [9, 12, 31]. Figure 5.1 illustrated the key differences of traditional and Open RAN architectures.

The Open RAN architecture is proposed to enable three main goals [10, 32], i.e.

- **Cloudification**: The goal is to support cloud-native RAN functions via disaggregated hardware and software components.
- **Intelligence and automation**: The goal is to utilize advanced AI/ML capabilities to enable automated management and orchestration in RAN.
- **Open internal RAN interfaces**: The goal is to support various Open RAN interfaces, including interfaces defined by 3GPP.

As illustrated in Figure 5.1, the RAN in Open RAN architecture is disaggregated into four main building blocks, i.e. the radio unit (RU), the distributed unit

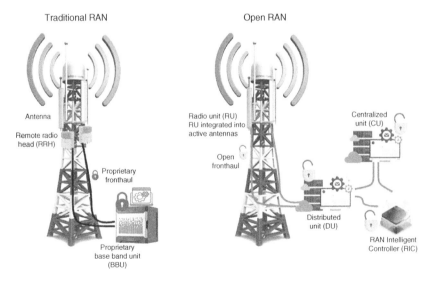

Figure 5.1 High-level comparison of Open RAN with traditional RAN.

(DU), the centralized unit (CU), and RAN Intelligence Controller (RIC). The RU is located with antennas, and it is responsible for transmitting, receiving, amplifying, and digitizing the radio frequency signals. The former BBU is now disaggregated into DU and CU. They are the computation parts of the base station. Here, DU is physically located closer to RU, while CU can be located closer to the Core. RIC is possible for taking the intelligent and automated decisions related to RAN.

O-RAN appliance has proposed a more detailed architecture for Open RAN as represented in Figure 5.2. The main elements of the Open RAN architecture include Service Management and Orchestration (SMO), RAN Intelligence Control

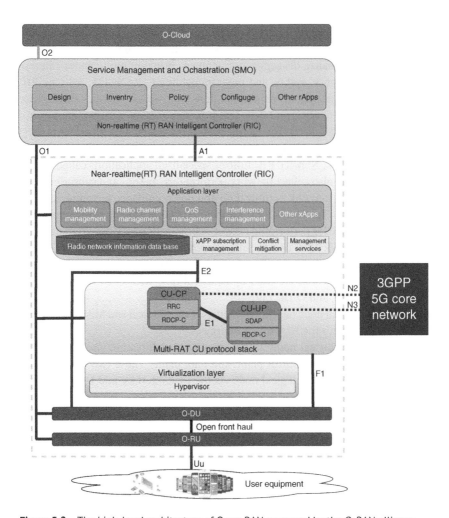

Figure 5.2 The high-level architecture of Open RAN proposed by the O-RAN alliance.

(RIC), O-Cloud, Open RAN central unit (O-CU), Open RAN distributed unit (O-DU), and Open RAN radio unit (O-RU).

- **Service Management and Orchestration (SMO)**: The SMO framework is a core component of the Open RAN architecture, whose main responsibility is to manage the RAN domain, such as the provision of interfaces with network functions, near-real-time RIC for RAN optimization, and O-Cloud computing resource and workload management [33, 34]. These SMO services can be performed through four interfaces, including A1, O1, O2, and open fronthaul M-plane.
- **RAN Intelligence Control (RIC)**: This logical function enables Open RAN to perform real-time optimization of functions and resources through data collected from the network and end-users. It is the key element in Open RAN, which helps to realize disaggregation strategy, bringing multivendor interoperability, intelligence, agility, and programmability to RANs [35, 36]. The RIC is divided into components as nonreal-time radio access network intelligence controller (Non-RT RIC) and near-real-time radio access network intelligence controller (Near-RT RIC). The NonRT RIC is integrated with Open RAN SMO Framework. It handles the control request and RAN resources within the second range. To this task, Non-RT RIC utilizes specialized applications called rApps. Non-RT RIC can also collect network performance metrics and subscriber data to offer AI-based network optimization and policy guidance recommendations for Near-RT RIC. The Near-RT RIC resides within edge servers or regional cloud as it is responsible for performing network optimization actions within milliseconds range. Near-RT RIC uses the different xApps to support these tasks [37, 38].
- **O-Cloud**: This is a physical computing platform. It creates and hosts the various VNFs and CNFs which are used by near-real-time RIC, O-CU control plane, O-CU user plane, and O-DU [39].
- **O-DU**: This logical node has functionalities of the physical and MAC layers. This element terminates the E2 with F1 interfaces.
- **O-CU**: This is a logical node in the Open RAN architecture and hosts all the functions of both the control plane and data plane. These two O-CU planes connect with the O-DU logical node via the F1-c interface and F1-u interface, respectively.
- **O-RU**: This logical node has a physical layer and radio signal-processing capabilities to connect with the SMO framework via the open fronthaul M-plane interface and connects with end-users via radio interfaces.

One of the main goals of Open RAN is "opening" the protocols and interfaces between these RAN components, such as radios, hardware, and software. The O-RAN Alliance has defined 11 different interfaces, including A1, O1, E1, F1, open fronthaul M-plane, and O2. More specifically, the open fronthaul

M-plane interface is to connect SMO Framework and O-RU, A1 is to connect nonreal-time RIC located in the SMO framework and near real-time RIC for RAN optimization, O1 is to support all Open RAN network functions when they are connected with SMO, and O2 is to connect SMO and O-Cloud for providing cloud computing resource and workflow management. According to [32], there are different deployment scenarios of the O1 interface, such as flat, hierarchical, and hybrid models, by which the SMO framework can provide numerous management services, for example, provisioning management services, trace management services, and performance management services.

5.3 Threat Vectors and Security Risks Associated with Open RAN

We start explaining the taxonomy, used to distinguish the different types of risks. Next, each of the four identified domains is further elaborated.

5.3.1 Threat Taxonomy

We categorize the risks in three main domains: process, technology, and global. First, process risks are related to rules, regulations, and oversight. Second, the technology risks correspond with the risks caused by the mechanisms for enforcing rules and procedures, as well as detecting threats. Fourth, global risks are broad risks related to the global communication instruction. Figure 5.3 provides an overview of the risk domains in Open RAN, respectively.

5.3.2 Risks Related to the Process

In the process risks, four categories are distinguished, corresponding to the preliminary assumptions or prerequisites, the general regulations, the privacy, and human-related aspects. In fact, all the process risks apply to any RAN implementation, but are in general more complex in Open RAN due to the modularity and the higher amount of stakeholders involved.

5.3.2.1 Prerequisites

The prerequisites are not under control of the RAN system, but should be carefully checked [40]. To start with, a reliable operational environment must be ensured, providing for instance reliable timestamps to be used in the audit records [41, 42]. Next, secure storage of stored logs, and credentials and secrets in external systems need to be guaranteed for instance by using hardware-based security modules like trusted platform modules (TPMs) [42, 43]. In addition, the access to

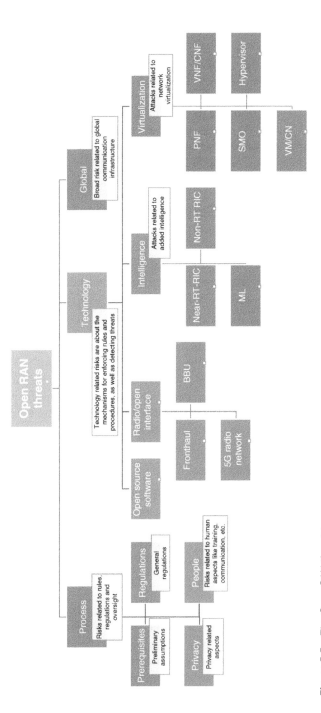

Figure 5.3 The Open RAN threat taxonomy.

this sensitive data should only be allowed by privileged users [42]. The last prerequisite is that the certificate authorities (CAs), which authenticate the network elements, are fully trusted and audited by well-established, worldwide recognized organizations [42, 44].

5.3.2.2 General Regulations

The first step in the effective Open RAN launch that needs to be done is the standardization of critical processes such as operation, administration, and management, covering the complete lifecycle of the Open RAN deployment [45]. This includes a clear description of components used for secure establishing of mutual authentication, access control, key management, trusted communication, storage, boot and self-configuration, update, recoverability and backup, security management of risks in open source components, security assurance, privacy, continuous security development, testing, logging, monitoring and vulnerability handling, robust isolation, physical security, cloud computing and virtualization, and robustness.

Next, it is also important to identify, locate, authenticate, and verify the origin of the relevant assets in the system. Furthermore, for each of the different assets, at rest and in transit and location, the type (data, component, etc.) and the security properties (confidentiality, integrity, availability – CIA) should be carefully collected. In fact, a complete and efficient supply chain process is required [46]. In particular, this is more complex for Open RAN due to the decoupling of hardware and software and the modularity. For instance, there is a risk of firms from allied states purchasing relabeled products or components from adversarial states.

Finally, when an issue arises in the network, due to the complexity of the network, it is not evident to identify and isolate the issues. Moreover, in case the issue is found, it is possible that the corresponding vendors do not take their responsibility as they can pass the blame to others because of the complexity and interdependence of the whole system.

5.3.2.3 Privacy

The privacy of end users encompasses privacy related to data, identity, and personal information [47]. Privacy sensitive data for end users are mostly leaked via communication services that are gathering all types of personal information, which are often not needed for the functioning of the services. Adversaries can even further extract more personal information about end users, such as user equipment (UE) priority, location information, trajectory, and preference. The protection of the user data is regulated by the law of the hosting country, where different jurisdictions can be applicable. There are, at least, three possible locations, the victim, the offender, or the service provider [48]. Therefore, clear guidelines should be developed in order to cope with these new interfaces, shared environments, and new players available in Open RAN.

5.3.2.4 People

First of all, it is necessary to clearly identify and authenticate the stakeholders involved in the different processes such as implementation, management, operation, and maintenance of the Open RAN system. For each of the stakeholders, their roles and responsibilities should be clearly defined and assessed. Vendors should have well established and transparent security practices built into their engineering processes [42].

Moreover, adequate training and assessments need to be organized for the different stakeholders, going from administrators, integrators, operators, and orchestrators in order to be capable to securely implement and manage the system according to the instructions provided by the Open RAN Alliance and the later to be developed standards [42].

Finally, strategies for security testing with published well-known test plans at trusted lab facilities should be defined upfront and integrated in the regular operation [42].

5.3.3 Risks Related to the Technology

The largest class of risks is related to the different components and mechanisms in the network. Here, distinction is made based on [49], considering aspects related to open software, radio/open interface, intelligence, virtualization, and general. Figure 5.4 provides an overview of the main technology-related risks in Open RAN.

5.3.3.1 Open Source Software

The open source-related risks are well-known problems available in open-source software code. Since, Open RAN is expected to be built (solely or partly) based on such open-source codes, it is in particular vulnerable for this type of attacks.

A trusted developer can intentionally insert a backdoor by injecting a few lines of malicious code into an open-source code component to be used within the Open RAN system [50]. It is then highly likely that a software project team picks it up and uses the infected open source code later, while the tools for vetting and testing of the development team do not detect the malicious code [42]. As a consequence, a vulnerability into the software code is included and can go undetected for a long period. The resulting effect on the Open RAN system can be diverse. It can either be simply annoying, but at the same time, it can significantly decrease the system performance via for instance DoS attacks, or it can even lead to serious loss of sensitive data.

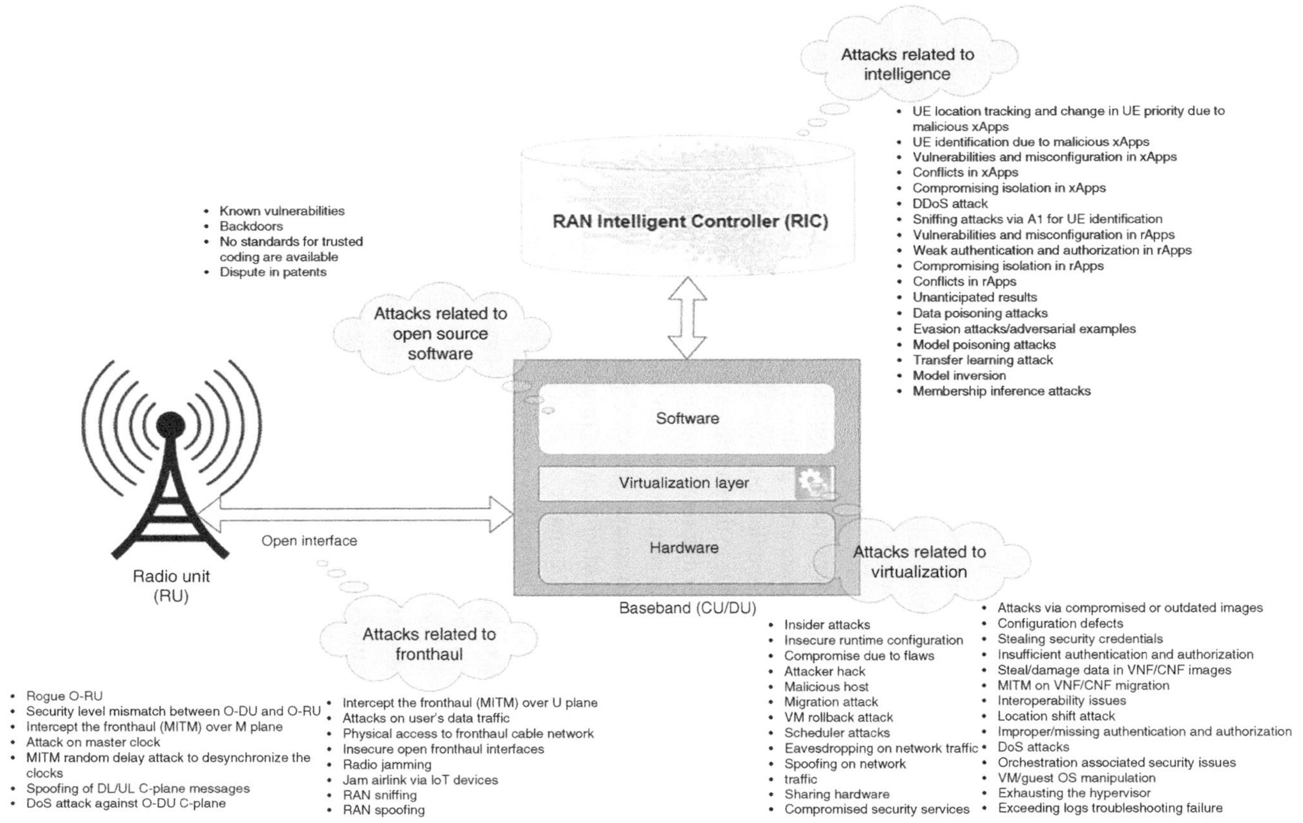

Attacks related to intelligence

- UE location tracking and change in UE priority due to malicious xApps
- UE identification due to malicious xApps
- Vulnerabilities and misconfiguration in xApps
- Conflicts in xApps
- Compromising isolation in xApps
- DDoS attack
- Sniffing attacks via A1 for UE identification
- Vulnerabilities and misconfiguration in rApps
- Weak authentication and authorization in rApps
- Compromising isolation in rApps
- Conflicts in rApps
- Unanticipated results
- Data poisoning attacks
- Evasion attacks/adversarial examples
- Model poisoning attacks
- Transfer learning attack
- Model inversion
- Membership inference attacks

Attacks related to virtualization

- Insider attacks
- Insecure runtime configuration
- Compromise due to flaws
- Attacker hack
- Malicious host
- Migration attack
- VM rollback attack
- Scheduler attacks
- Eavesdropping on network traffic
- Spoofing on network traffic
- Sharing hardware
- Compromised security services
- Attacks via compromised or outdated images
- Configuration defects
- Stealing security credentials
- Insufficient authentication and authorization
- Steal/damage data in VNF/CNF images
- MITM on VNF/CNF migration
- Interoperability issues
- Location shift attack
- Improper/missing authentication and authorization
- DoS attacks
- Orchestration associated security issues
- VM/guest OS manipulation
- Exhausting the hypervisor
- Exceeding logs troubleshooting failure

Attacks related to open source software

- Known vulnerabilities
- Backdoors
- No standards for trusted coding are available
- Dispute in patents

Attacks related to fronthaul

- Rogue O-RU
- Security level mismatch between O-DU and O-RU
- Intercept the fronthaul (MITM) over M plane
- Attack on master clock
- MITM random delay attack to desynchronize the clocks
- Spoofing of DL/UL C-plane messages
- DoS attack against O-DU C-plane
- Intercept the fronthaul (MITM) over U plane
- Attacks on user's data traffic
- Physical access to fronthaul cable network
- Insecure open fronthaul interfaces
- Radio jamming
- Jam airlink via IoT devices
- RAN sniffing
- RAN spoofing

Figure 5.4 Open RAN technology-related attacks.

Open-source vulnerabilities are normally published on the National Vulnerability Database (NVD) [51]. This database is primarily intended for developers to disclose vulnerabilities. However, this source is also used by hackers to exploit those vulnerabilities enabling backdoors to attacks on, e.g. the hypervisor, operating system (OS), virtual machine (VM), or container. Moreover, vulnerabilities frequently propagate as developers often reuse free open-source code. As a consequence, downloading open source libraries and their dependencies, as well as downloading open source code from untrusted repositories contain significant risks [42]. Open RAN vendors and operators should thus store at each moment up-to-date inventories containing the dependencies in their open-source software used in the applications. In addition, this should be complemented by a process, which receives and manages all the notifications coming from the open-source community that are related to newly discovered vulnerabilities, including newly developed patches to overcome them. This should enable a better supply chain traceability.

Existing legislation demonstrates implied security preferences but provides no explicit legislative standards, guidance, requirements, or conditions. These preferences should be explicit but transparent, reviewable, and auditable to ensure secure coding. Due to the fact that a material amount of Open RAN code is being written by firms in different countries, security audits should be mandatory making code available to security researchers [52].

Finally, the last open-source software risk is more linked to political and financial interests, instead of security interest. Both the 3GPP and Open RAN alliance operate a fair, reasonable, and nondiscriminatory (FRANS) policy when it comes to patents that are held by contributors to those respective organizations. Patents are held on aspects of the 3GPP and Open RAN Alliance specifications, but the holders of those patents agree that it is mutually beneficial for everyone if the patents are licensed with a FRANS approach. The concern in this area is politically oriented. There might be a possibility that the patents held by competing manufacturers and service providers may be withdrawn from the FRANS licensing arrangement if trade relations between different countries dramatically deteriorate [52, 53].

5.3.3.2 Radio/Open Interface

The different radio/open interface components include the Fronthaul, the central Unit/distributed unit (CU/DU) and the 5G radio network.

- **Fronthaul**: The Fronthaul of Open RAN, consisting of O1, O2, A1, and E2 are the new components, all available with open interfaces allowing the software programmability of RAN. These components and interfaces may not be secured to industry best practices, for instance containing no proper authentication and authorization processes, ciphering and integrity checks, protection against replay attacks, prevention of key reuse, validation of inputs, response to error conditions [42]. This follows often from the strict performance requirements (bandwidth, latency, fronthaul transport link length, etc.) that limit the use of some security features, enforced by the high bit rate fronthaul interface to increase the processing delay. As a consequence, different man-in-the-middle (MITM), DoS, data tampering, or even information disclosure attacks become possible.

- **CU/DU**: The shared units pool in the Open RAN cloud native deployment may suffer from insufficient isolation and impose the risk of breaking user privacy and accessing sensitive data [12].
- **5G radio network**: These attacks are classical attacks, which can be applied to any RAN system and include radio jamming, jamming via IoT devices, RAN sniffing, and spoofing [42].

Radio jamming can be on the reference signals, the synchronization signal, the Physical Broadcast Channel (PBCH), the Physical Downlink Control Channel (PDCD), the Physical Uplink Control Channel, or the Physical Random-Access Channel [54]. This would enable an attacker to disrupt the communication by deliberate jamming, blocking, or creating interference with the authorized wireless network.

In addition, due to the millions of IoT devices in the network, jamming of the air-link signals through the IoT devices can easily overload the Open RAN resources by means of Distributed denial-of-service (DDoS) attacks via a botnet army of millions to billions infected devices on which a malware instructs to reboot all devices in a specific or targeted 5G coverage area at the same time [55].

RAN sniffing and spoofing allow the attacker to decode essential network configuration details, assisting attackers to optimize and craft their attacks or to transmit fake signals pretending to be actual signals [56].

5.3.3.3 Intelligence

The different components and mechanisms that contribute to the intelligence in the Open RAN network are the Near-RT-RIC, Non-RT RIC, and machine learning (ML) algorithms. These risks are mostly specific to Open RAN as they operate on new components and new algorithms, which are currently not available.

- **Near-RT-RIC-related attacks**: xApps have the capability to manipulate behavior of a certain cell, a group of UEs, and a specific UE. The related attacks are due to either malicious xApps, xApps with vulnerabilities, misconfigured xApps, compromised xApps, or conflicting xAPPs [42].

Malicious xApps can potentially be used as a sniffer for UE identification. This follows from the fact that the A1 interface is able to point out a certain UE in the network, which creates correlations among the randomized and anonymized

UE identities between the RAN nodes. As a consequence, UE location tracking and change in UE priority become possible. In particular, identification and tracking of a certain subscriber, for instance a very important person (VIP) becomes a real threat.

Vulnerabilities can potentially exist in any xApp, since it can come from either an untrusted or unmaintained source. Such vulnerabilities can then be exploited to take over another xApp or the whole near-RT RIC and often have the purpose to degrade the performance (e.g. a DoS). It may also be possible to alter data transmitted over A1 or E2 interfaces, extract sensitive information, etc. Also, the xApp isolation can be exploited in order to break out of the xApp confinement and to deduce information from co-hosted xApps. Gaining in addition unauthorized access provides new opportunities to exploit vulnerabilities in other xApps or Open RAN components to intercept and spoof network traffic, to degrade services (DoS), etc.

Finally, since there is no clear functional split between the Near-RT RIC and the Open RAN Next Generation Node B (O-gNB), possible conflicts, including conflicts in xApps, between the decisions taken by the Near-RT RIC and the O-gNB can appear, both unintentionally or maliciously. This can have an impact on the Open RAN system functions such as mobility management, admission controls, bandwidth management, and load balancing, potentially resulting in performance degradation.

- **Non-RT RIC-related attacks**: rApps impact non-RT RIC functions such as AI/ML model training, A1 policy management, Enrichment information management, Network Configuration Optimization in the purpose of performance degradation, DoS, enrichment data sniffing (UE location, trajectory, navigation information, GPS data, etc.), etc. rApps have also the capability to manipulate behavior of a certain cell, a group of UEs, and a specific UE. The related attacks are similar as with xApps due to either malicious rApps, rApps with vulnerabilities, misconfigured rApps, compromised rApps, or conflicting rAPPs. Besides these similar ones, there are two more risks identified related to non-RT RIC [42].

Untrusted or unmaintained sources can cause vulnerabilities in any rApp. Exploitation of these vulnerabilities mostly leads to disruption of the offered network service and potentially take over another rApp or the whole non-RT RIC. As a consequence, the attacker may gain the ability to alter data transmitted over A1 interface, extract sensitive information, etc. Also, rAPP isolation can be exploited to break out of rApp confinement and to deduce information from cohosted rApps. Gaining in addition unauthorized access provides new opportunities to exploit vulnerabilities in other rApps or Open RAN components to intercept and spoof network traffic to degrade services (DoS), etc.

In addition, rApps in the non-RT RIC can cause conflicts and make conflicting decisions as they can be provided by different vendors, e.g. rApp for Carrier license scheduling, rApp for energy savings, ... These conflicts are difficult to mitigate since dependencies are impossible to observe. Conflicts can again lead to an overall network performance degradation, instabilities, etc.

There is an additional vulnerability that can appear in the case of the rApp management when it is exposed to a web front-end or REST API, whose software interfaces contain vulnerabilities or do not implement authentication and authorization in a proper way. This would allow an attacker to gain access to the rApp and pose as a tenant or to manipulate configurations, access logs, implement back doors, etc.

Finally, an attacker might also penetrate the non-RT RIC through A1/O1 interfaces or from external sources through SMO and attempt to trigger a DoS or degrade the performance of non-RT RIC.

- **ML-related attacks**: With respect to the ML algorithms, there are the data poisoning attacks, the evasion attacks, the model poisoning attacks, the transfer learning attacks, the model inversion attacks, and the membership inference attacks, which are well-known attacks typically available in a ML environment [42, 57]. Finally, due to the complexity of the models in AI/ML, the results are not yet explainable in most of the cases [58]. Therefore, its use in the RAN can potentially lead to unanticipated consequences, which might have an impact on the security or performance [59]. Examples in other domains of unanticipated consequences are for instance the racially biased facial recognition.

5.3.3.4 Virtualization

The following components, physical network function (PNF), VNF, CNF, SMO, hypervisor, virtual machine/container (VM/CN), are involved in the virtualization process. Here, we discuss the security issues associated with each of these components. Some of these attacks can be applied also in V-RAN and C-RAN [60]. However, the attack range and impact of such attacks are larger in Open RAN.

- **PNF-related attacks**: An attacker compromises a PNF to launch reverse attacks and other attacks against VNFs/CNFs. A lack of security policies to protect mixed PNF-VNF/CNF deployments, resulting in insecure interfaces, could be exploited to perform attacks against VNFs/CNFs, potentially taking advantage of legacy security used by PNFs and not provided by the virtualization/ containerization layer [42].
- **VNF/CNF-related attacks**: Because VNF/CNF images are effectively static archive modules including all components used to run a given Open RAN VNF/ CNF, modules within an image may have vulnerabilities, introducing malware, missing critical security updates, or are outdated. These images are only collections of files packaged together. Therefore, malicious files can be included

intentionally or inadvertently within them. In addition, VNF/CNF images may also have configuration defects, e.g. configuring a specific user with greater privileges than needed. This could all be used to attack other VMs/CNs or hosts within the environment. An attacker can migrate a compromised VNF/CNF to a different location which has less security or privacy policies to gain additional access to the system. Since Open RAN uses different equipment with different vendors and different configurations, they can be less-secure environments, which can lead to additional vulnerabilities if deployed in the same system [42]. Moreover, since many Open RAN VNFs/CNFs require secrets to enable authentication, access control, and secure communication between components, these secrets are embedded directly into the image file system. In addition, the images often contain also sensitive components like an organization's proprietary software and administrator credentials. Anyone with access to the image (e.g. by means of insufficient authentication and authorization) can easily parse it to extract these secrets, resulting in the compromise, stealing, or damage of the contents on the images. As a result, it can lead to intellectual property (IP) loss and expose significant technical details about an Open RAN VNF/CNF image to an attacker. Even more critically, because registries of images are typically trusted as a source of valid, approved software, compromise of a registry can potentially lead to compromise of downstream VMs/CNs and hosts [42].

Finally, there is an increased risk of MITM attacks by intercepting network traffic intended for registries in order to steal developer or administrator credentials within that traffic. This can result in fraudulent or outdated images to orchestrators [42].

- **SMO-related attacks**: Improper or insufficient authentication or authorization of Open RAN external (e.g. AI/ML, emotional intelligence (EI), Human–Machine) or internal (e.g. over O1 or O2 interfaces, with Non-RT RIC) interfaces on SMO, allow access to the SMO and in particular, the data stored on it. Besides disclosing Open RAN sensitive information, the attacker may also alter the Open RAN components [42].

DoS attacks or increased traffic can cause overload situations and thus affects availability of the SMO data and functions [42].

Finally, an attacker may exploit weak orchestrator configuration, access control, and isolation. A single orchestrator may run many different VMs/CNs, each managed by different teams, and with different sensitivity levels. If the access provided to users and groups does not conform to their specific requirements, an attacker or a careless user would be able to affect or subvert the operation of another VM/CN managed by the orchestrator. Malicious traffic from different VMs/CNs sharing the same virtual networks may be possible if VMs/CNs of different sensitivity levels are using the same virtual network with a poorly isolation of inter-VM/CN network traffic [42].

- **Hypervisor-related attacks**: An attacker can exploit the security weaknesses in the guest OS to attack the hypervisor of the hosting OS. Examples of guest OS vulnerabilities are OS command injection, SQL injection, buffer overflow, or missing authentication for critical functions [61–63].

  An attacker may also change the configurations of compromised VNFs/CNFs to consume high amounts of CPU, hard disk, and memory resources in order to exhaust the hypervisor. Another way to compromise the hypervisor is by generating an excessive amount of log entries such that it is infeasible or very difficult to analyze the log files coming from other VNFs [61, 62].

  Finally, as the hypervisor provides its own security functions and application programming interfaces (APIs) to the host system security functions, it is in full control of the security functionalities of the lower layers and thus needs to be fully trusted. When a malicious administrator has, for instance, root access to the hypervisor and by using a search operation, the user identity (ID), passwords, and Secure Shell Protocol (SSH) keys from the memory dump can be extracted, which in turn violates the user privacy and data confidentiality [61, 62].

- **VM/CN-related attacks**: VMs/CNs may be compromised due to flaws in the Open RAN VNFs/CNFs they run. For example, an Open RAN VNF/CNF may be vulnerable to cross-site scripting (SQL) injection [64] and buffer overflow vulnerabilities [65].

  Insecure VM/CN runtime configuration by the administrator can lower the security of the Open RAN system. It may expose VMs/CNs and the hypervisor/ container engine to increased risk from a compromised VM/CN. For example, it could be used to elevate privileges and attack VMs/CNs, the O-Cloud infrastructure/services [42].

  A compromised VM/CN will be able to alter that VM/CN in order to access other VMs/CNs, monitor VM/CN to VM/CN communications, attack the O-Cloud infrastructure/services, scan the network to which it is connected to in order to find other weaknesses to be exploited, etc. The container engine (in case of CN) or hypervisor (in case of VM) has access to all RAM memory, disk volumes mounted on the virtual machines and containers. This means that a malicious VM/CN or hypervisor/container engine can get access to all Open RAN network data processed in the workloads. An attacker can launch a noisy neighbor attack against the shared O-Cloud infrastructure to cause the Open RAN system performance degradation and/or its services disruption by depriving the resources required by various Open RAN running functions [42].

  An attacker hacks into VM/CN is, for instance, possible if an attacker steals VMs/CNs private key from one VM/CN and so reveals the administrator privileges. Next, all tenants' tokens and the administrator rights of the whole Open RAN system can be obtained, [42].

From the side of the application, trust is required at all levels. In case the underlying host OS is malicious, access can be obtained to all data processed in the workloads, e.g. RAM memory, disk volumes. Techniques like secure enclaves [66] have the goal to provide a trusted environment. However, the application will be hardware-instance dependent [67].

If VM/CN migration is not secured or performed over a secure channel, a MITM attacker can modify arbitrary VM/CN OS or application states during the migration. An attacker may also use an older snapshot of VM/CN without the concern of the VM/CN owner to bypass the security system and obtain access to the system. This attack is possible after an already comprised hypervisor rollback to a previous snapshot. In the scheduler attack, the vulnerabilities in the hypervisor's scheduler are exploited to acquire system resources for the malicious VM at the expense of a victim VM [61, 62].

Furthermore, due to virtualization and cloud computing, different applications might use the same hardware resources. Isolation between these applications is only at software level and not at the level of hardware. As a consequence, hardware-related vulnerabilities like the recently discovered Meltdown and Spectre attacks (https://meltdownattack.com/) can have a larger attack range [67].

Finally, besides the main functionality of the VNF/CNF itself, the administrators may also decide to deploy additional network services on their VMs/CNs in order to do extra monitoring, remote configuration, remote access to other services, such as SSH. If these additional network services are directly accessible over the Internet or from another administrator, new entry points for attackers are created, and if access is obtained to the VM/CN, more extra attacks become possible [42].

5.3.4 Global Risks

Offering the highest level of security on the network is important for a nation. We here distinguish five major types of attacks or risks that need to be taken into account [68].

- **Attack on digital economy**: Since 5G is fully integrated into the digital economy, it can result in potential life or death consequences. For instance, currently a lot of data are sent from our mobile devices, smart homes, electrical cars via a network consisting of devices, which are remotely controlled and updated and thus present a potential attack vector. The possibility of a smart city shutting down, autonomous vehicles crashing, or factories going dark due to a cyberattack are frightening situations. Also, the growing trend toward digital currency can have very bad consequences in case of a major failure.

- **Espionage**: There are currently no regulations for avoiding the collaboration between an Open RAN equipment manufacturer and an external party, like for instance a security agency of a certain country. Therefore, without any guarantee of good intentions of the equipment and software providers, possibilities for spying should be considered as viable.
- **Attacks on critical infrastructure**: Critical infrastructure typically consists for the management of power grids, water supplies, manufacturing, and transportation infrastructure. More and more, 5G is used as the backbone of communication in these infrastructures. Therefore, a dedicated cyberattack disrupting this critical infrastructure would have devastating impact on the people being dependent of this.
- **Violence against democracy**: If an actor receives the power to perform the role of big brother in all communication, there is a real threat for democracy and freedom of speech.
- **Majority attacks and supply chain concerns**: As mentioned in [69, 70], the Open RAN Alliance currently includes a wide range of high security risk companies. If the efforts in the development and standardization process for Open RAN is dominated by partners belonging to one country or even one region, it can cause an imbalance resulting in a new alliance that will still enable espionage possibilities and disrupt the intended openness.

5.4 Security Benefits of Open RAN

Besides all these risks, Open RAN brings of course a whole series of benefits. Several benefits are typical for Open RAN, others are also available in V-RAN, and some of them are common for all 5G networks. In this section, we present the security benefits of this new Open RAN architecture. These benefits can be categorized in to several categories.

5.4.1 Open RAN specific

5.4.1.1 Full Visibility

Due to virtualization and the disaggregated components connected through open interfaces, operators have direct access to all network performance data and operational telemetry data representing activities between/within the different network functions. The integrity of this data is more ensured as this data are created isolated form the functions' executing environment. Combining this data with security log data results in an earlier detection of security problems and easier detection of the root cause [71–73].

Note that full visibility can also be a risk. Due to the complexity, the root cause cannot always be easily detected, and there is a danger that different vendors will not take accountability for potential issues. Following [74], it is claimed that the time and cost to perform a complete security review would seem to be multiplied by the number of vendors the operators take on board.

5.4.1.2 Selection of Best Modules

Operators can more easily select the vendors offering the best products, meeting the required industry security standards and certifications [71–73]. Examples of industry best practices are for instance "secure by design" DevSecOps in which information security operations are integrated into DevOps workflows and automated testing in development of containerized applications [75]. The operator can also collaborate with the vendor to determine and influence CI/CD processes with continuous regression testing and software security auditing used by the supplier. Other good practices are the adoption of Supplier Relationship Management (SRM) with an inbound development process and strict security controls for Free and Open Source Software (FOSS), trust stack management with software coming from reliable supply chains and trusted, well-defined operations, intelligent vulnerability management, and multivendor system integration (SI) with continuous verification on vendors sharing the same interpretation and implementation of functions.

There are a range of industry best practices that can be adopted including Groupe Speciale Mobile Association (GSMA), National Institute of Standards and Technology (NIST), European Union Agency for Cybersecurity (ENISA), National Telecommunications and Information Administration (NTIA), Center for Internet Security (CIS), Open Web Application Security Project (OWASP), Open Standards, Open Source (OASIS), national cyber security organizations, Building Security in Maturity Model (BSIMM), Cloud Native Computing Foundation (CNCF), the Linux Foundation, and SAFECode and Cloud iNfrastructure Telco Task Force (CNTT).

5.4.1.3 Diversity

Integrating independent and individual modules decreases the risk that common coding errors or practices of one single entity impact large parts of the network and thus decrease the attack range. Consequently, diversity helps to balance the security risks. Open RAN enables an expanded pool of vendors on the market, reducing a nation's dependence on any sole vendor for wireless services [49, 76].

5.4.1.4 Modularity

Due to the modularity of the network, operators can switch to a CI/CD operating model, enabling seamless and effective patch management for fixing any detected security vulnerability. As a consequence, the vulnerabilities in the network are faster removed. In addition, updates become more transparent and have less impact on the overall network. Moreover, also operational agility is obtained making it possible to replace functional elements by new versions or capabilities [71, 73].

5.4.1.5 Enforcement of Security Controls

Due to the choice among different vendors, modularity, and open interfaces, the operator is in the position to demand strong security capabilities and control of its suppliers. For instance, in the case of a cloud architecture, the operator and the cloud infrastructure supplier have a common agreement in which this last one is responsible for the deployment of the latest security tools for detection and prevention [72, 77].

5.4.1.6 Open Interfaces

Open interfaces at the different levels give a higher exposure, resulting in more scrutiny and thus higher overall security. Thanks to the open interfaces, operators are not dependent anymore on the supplier in case of (security) issues and can do upgrades themselves, being able to react faster. It also gives the possibility to experiment with new functions and new vendors, exploring new ways to secure the network and its operation [49, 72, 78]. This is at the same time a risk as in order to explore new possibilities by the operator, sufficient qualified people are required, which is not evident due to the complexity of the overall system.

5.4.1.7 Open Source Software

Open-source software presents security challenges regarding its open nature but has the advantage of being verified by multiple independent parties, being rigorously and varied tested, and customized against threats [78].

As mentioned before, the use of open source also includes many risks. It was concluded in the Github 2020 State of the Octoverse Report that vulnerabilities remain undetected in many cases for more than four years, before being disclosed. Therefore, one cannot simply state that open-source software is faster patched than proprietary software.

5.4.1.8 Automation

The introduced intelligence in Open RAN can be used to automate the management and control via big data analysis, AI, and ML. As a consequence, closed-loop responses to changes in the network can be automatically performed. This has the advantage that no human interactions are required anymore, which inherently includes threats like humans accidentally altering the security posture of a network function or maliciously harvesting credentials, changing configurations, or implanting malware within the network [36, 49, 71].

Again, automation can bring risks as previously identified at the ML algorithms.

5.4.1.9 Open Standards

Open RAN will be developed based on open standards, defined by the Open RAN consortium. Such standards enable to align on a common approach approved by leading members in the field and coordinate all information regarding security threats, vulnerabilities, and exploits [71, 72].

A prerequisite is of course the presence of these standards, which are not fully available for the moment. In addition, these standards should be correctly implemented.

5.4.2 V-RAN Specific

5.4.2.1 Isolation

Isolation is obtained via the defined interfaces between functional elements in an Open RAN. It offers on the one hand the possibility to insert controls for monitoring and on the other hand allows software updates and patches to be installed with less risk that version dependencies will create issues [76].

5.4.2.2 Increased Scalability for Security Management

Often, there are trade-offs between application, performance, and security requirements. Due to the modularity, operators can tailor their deployments and shift more easily the resources for monitoring and control to meet better to these requirements. Also vRAN functional elements can be shifted to provide better isolation [76].

5.4.2.3 Control Trust

Since operators control the platforms on which virtualized functions run in Open RAN, they have also complete control on the trust infrastructure. The identity and provenance of each functional element are known and managed by using strong cryptographic mechanisms like signature operations. Each new version is validated by the operators and, therefore, they have control on what is running where on their networks [76].

However, it must be taken into account that the situation becomes more complex as there are more assets and stakeholders involved.

5.4.2.4 Less Dependency Between hardware [HW] and SW

In an Open RAN, there is less dependency between the network software and hardware. This makes it in the first place easier to perform the required upgrades in a faster way. Second, it also avoids risks associated to isolated security breaches [73].

5.4.2.5   Private Network

Private 5G networks will soon become the general trend as they enable companies the possibility to fully customize the network according to their specific needs with respect to speed, bandwidth, security requirements, on their locations and own timetable. It will enable companies to offer their customers a dedicated 5G experience, with applications in a large range of domains from healthcare, manufacturing, transportation, education, etc. Companies will have the option to build out and run their own private 5G network, or they can also outsource it to a mobile network operator or systems integrator [79]. One such option is via network slicing, where each slice can be seen as a complete end-to-end network and includes the security capabilities according to the needs [73].

There will be soon many players on the market to launch this innovative network as a service concept, replacing in many cases their existing Wi-Fi and fixed wireless/wired infrastructure.

5.4.2.6   More Secure Storage of Key Material

In traditional network architectures, sensitive cryptographic key material such as Access Stratum keys are more vulnerable to various threats as they are stored at the cell site. In Open vRAN, this key material can be stored deep inside the network in a secure vCU, hosted in a data center [73].

5.4.3   5G Networks Related

5.4.3.1   Edge Oriented

Due to the open interfaces, the operator is able to spread the security analysis throughout the network and include monitoring at the edge. These edge-focused analytics will facilitate the detection and prevention of attacks at the lower part in the network in order to avoid DDoS and to block malicious data from reaching the core network. This is in particular important to support mobility services like services offered by IoT [71, 80].

5.4.3.2   Simpler Security Model

In zero trust [81], nothing is trusted unless it is verified, regardless of the location. The O-RAN Alliance completely embraces this principle. Therefore, everything needs to be verified and results need to be communicated [71]. Zero trust networking can enhance security in different domains relying on robust standards. First, it enables to secure the technology and application stack including all interfaces and APIs. Second, it allows the leverage of the cloud-based nature of 5G and the deployment of cloud security functionality and telemetry. Third, it ensures the tailoring and customization of the security control via network slicing. Finally, it makes it possible to deploy multiple layers of authentication [82].

5.5 Conclusion

In order to cope with the continuous growth of mobile subscribers, mobile data and mobile services, a drastically new approach is needed in order to ensure that the network resources are used in the most optimal way. In addition, this solution should particularly take into account a thorough security protection as also the amount and impact of cybersecurity attacks are continuously increasing.

Open RAN offers all the possibilities to enable a great breakthrough in the network technology landscape and is able to address most of the current shortcomings in RANs thanks to the added openness and intelligence. However, due to this totally new approach where multiple vendors can now simultaneously integrate their technology, a more complex ecosystem exists resulting in a multitude of new risks and opportunities. We have provided a comprehensive overview of these different risks and benefits. We also discussed the best security practices to be applied. As an important conclusion, in order to fully benefit from the most essential opportunities and to avoid the most important risks, the existence of an extended standard describing in detail the different processes in Open RAN is an essential step.

References

1 A. R. Berkeley, M. Wallace, and C. Coo, "A framework for establishing critical infrastructure resilience goals," *Final Report and Recommendations by the Council, National Infrastructure Advisory Council*, pp. 18–21, 2010.

2 S. O'Dea, "Number of mobile subscriptions worldwide 1993–2021," 2021 [Online]. Available: https://www.statista.com/statistics/262950/global-mobile-subscriptions-since-1993/.

3 G. R. Faulhaber and D. J. Farber, "Spectrum management: Property rights, markets, and the commons," in *Rethinking Rights and Regulations: Institutional Responses to New Communication Technologies*. Cambridge, MA: MIT Press, 2003, pp. 193–226.

4 S. Gindraux, "From 2G to 3G: A guide to mobile security," in *3rd International Conference on 3G Mobile Communication Technologies*. IET, 2002, pp. 308–311.

5 E. Dahlman, S. Parkvall, and J. Skold, *4G: LTE/LTE-Advanced for Mobile Broadband*. Academic Press, 2013.

6 I. Parvez, A. Rahmati, I. Guvenc, A. I. Sarwat, and H. Dai, "A survey on low latency towards 5G: RAN, core network and caching solutions," *IEEE Communications Surveys & Tutorials*, vol. 20, no. 4, pp. 3098–3130, 2018.

7 K. Nguyen, P. Le Nguyen, Z. Li, and H. Sekiya, "Empowering 5G mobile devices with network softwarization," *IEEE Transactions on Network and Service Management*, vol. 18, no. 3, pp. 2492–2501, 2021.

8 M. Condoluci and T. Mahmoodi, "Softwarization and virtualization in 5G mobile networks: Benefits, trends and challenges," *Computer Networks*, vol. 146, pp. 65–84, 2018.

9 M. Yang, Y. Li, D. Jin, L. Su, S. Ma, and L. Zeng, "OpenRAN: A software-defined RAN architecture via virtualization," *ACM SIGCOMM Computer Communication Review*, vol. 43, no. 4, pp. 549–550, 2013.

10 L. Gavrilovska, V. Rakovic, and D. Denkovski, "From cloud RAN to open RAN," *Wireless Personal Communications*, vol. 113, no. 3, pp. 1523–1539, 2020.

11 A. Umesh and K. Teshima, "O-RAN Alliance trends and NTT DOCOMO's activities," *IEICE Technical Report*, vol. 120, no. 29, p. 29, 2020.

12 S. Niknam, A. Roy, H. S. Dhillon, S. Singh, R. Banerji, J. H. Reed, N. Saxena, and S. Yoon, "Intelligent O-RAN for Beyond 5G and 6G Wireless Networks," *arXiv preprint arXiv:2005.08374*, 2020.

13 L. Bonati, M. Polese, S. D'Oro, S. Basagni, and T. Melodia, "OpenRAN Gym: An Open Toolbox for Data Collection and Experimentation with AI in O-RAN," *arXiv preprint arXiv:2202.10318*, 2022.

14 Telecom Infra Project, "The Telecom Infra Project (TIP)," 2022. [Online]. Available: https://telecominfraproject.com/openran/.

15 O-RAN Alliance, "O-RAN Alliance," 2022. [Online]. Available: https://www.o-ran.org/.

16 S. K. Singh, R. Singh, and B. Kumbhani, "The evolution of radio access network towards Open-RAN: Challenges and opportunities," in *2020 IEEE Wireless Communications and Networking Conference Workshops (WCNCW)*, 2020, pp. 1–6.

17 M. Polese, L. Bonati, S. D'Oro, S. Basagni, and T. Melodia, "Understanding O-RAN: Architecture, Interfaces, Algorithms, Security, and Research Challenges," *arXiv preprint arXiv:2202.01032*, 2022.

18 G. Mantas, N. Komninos, J. Rodriuez, E. Logota, and H. Marques, "Security for 5G communications," in *Fundamentals of 5G Mobile Networks*. John Wiley & Sons, Ltd., 2015.

19 J. M. Batalla, E. Andrukiewicz, G. P. Gomez, P. Sapiecha, C. X. Mavromoustakis, G. Mastorakis, J. Zurek, and M. Imran, "Security risk assessment for 5G networks: National perspective," *IEEE Wireless Communications*, vol. 27, no. 4, pp. 16–22, 2020.

20 D. Soldani, "5G and the future of security in ICT," in *2019 29th International Telecommunication Networks and Applications Conference (ITNAC)*. IEEE, 2019, pp. 1–8.

21 I. Ahmad, T. Kumar, M. Liyanage, J. Okwuibe, M. Ylianttila, and A. Gurtov, "Overview of 5G security challenges and solutions," *IEEE Communications Standards Magazine*, vol. 2, no. 1, pp. 36–43, 2018.

22 H. Tataria, M. Shafi, A. F. Molisch, M. Dohler, H. Sjöland, and F. Tufvesson, "6G wireless systems: Vision, requirements, challenges, insights, and opportunities," *Proceedings of the IEEE*, vol. 109, no. 7, pp. 1166–1199, 2021.

23 R. M. Needham, "Denial of service," in *Proceedings of the 1st ACM Conference on Computer and Communications Security*, 1993, pp. 151–153.

24 J. R. van der Merwe, X. Zubizarreta, I. Lukčin, A. Rügamer, and W. Felber, "Classification of spoofing attack types," in *2018 European Navigation Conference (ENC)*. IEEE, 2018, pp. 91–99.

25 V. P. Illiano and E. C. Lupu, "Detecting malicious data injections in wireless sensor networks: A survey," *ACM Computing Surveys (CSUR)*, vol. 48, no. 2, pp. 1–33, 2015.

26 R. Kawashima, "A vision to software-centric cloud native network functions: Achievements and challenges," in *2021 IEEE 22nd International Conference on High Performance Switching and Routing (HPSR)*. IEEE, 2021, pp. 1–7.

27 Y. Jarraya, A. Eghtesadi, S. Sadri, M. Debbabi, and M. Pourzandi, "Verification of firewall reconfiguration for virtual machines migrations in the cloud," *Computer Networks*, vol. 93, pp. 480–491, 2015.

28 M. Wright, M. Adler, B. N. Levine, and C. Shields, "Defending anonymous communications against passive logging attacks," in *2003 Symposium on Security and Privacy*. IEEE, 2003, pp. 28–41.

29 S. Bobrovskis and A. Jurenoks, "A survey of continuous integration, continuous delivery and continuos deployment," in *BIR Workshops*, 2018, pp. 314–322.

30 B. Dutta, A. Krichel, and M.-P. Odini, "The challenge of zero touch and explainable AI," *Journal of ICT Standardization*, vol. 9, no. (2), pp. 147–158, 2021.

31 D. Johnson, D. Maas, and J. Van Der Merwe, "NexRAN: Closed-loop RAN slicing in POWDER-A top-to-bottom open-source open-RAN use case," in *Proceedings of the 15th ACM Workshop on Wireless Network Testbeds, Experimental evaluation & CHaracterization*, 2022, pp. 17–23.

32 A. Garcia-Saavedra and X. Costa-Perez, "O-RAN: Disrupting the virtualized ran ecosystem," *IEEE Communications Standards Magazine*, vol. 5, no. 4, pp. 96–103, 2021.

33 T.-H. Wang, Y.-C. Chen, S.-J. Huang, K.-S. Hsu, and C.-H. Hu, "Design of a network management system for 5G Open RAN," in *Asia-Pacific Network Operations and Management Symposium (APNOMS)*. IEEE, 2021, pp. 138–141.

34 D. Wypiór, M. Klinkowski, and I. Michalski, "Open RAN-radio access network evolution, benefits and market trends," *Applied Sciences*, vol. 12, no. 1, p. 408, 2022.

35 B. Balasubramanian, E. S. Daniels, M. Hiltunen, R. Jana, K. Joshi, R. Sivaraj, T. X. Tran, and C. Wang, "RIC: A RAN intelligent controller platform for

AI-enabled cellular networks," *IEEE Internet Computing*, vol. 25, no. 2, pp. 7–17, 2021.

36 T. Hasegawa, H. Hayashi, T. Kitai, and H. Sasajima, "Industrial wireless standardization scope and implementation of ISA SP100 standard," in *SICE Annual Conference 2011*, 2011, pp. 2059–2064.

37 O. Orhan, V. N. Swamy, T. Tetzlaff, M. Nassar, H. Nikopour, and S. Talwar, "Connection management xAPP for O-RAN RIC: A graph neural network and reinforcement learning approach," in *2021 20th IEEE International Conference on Machine Learning and Applications (ICMLA)*. IEEE, 2021, pp. 936–941.

38 M. Dryjański, Ł. Kułacz, and A. Kliks, "Toward modular and flexible open RAN implementations in 6G networks: Traffic steering use case and O-RAN xApps," *Sensors*, vol. 21, no. 24, p. 8173, 2021.

39 I. Tamim, A. Saci, M. Jammal, and A. Shami, "Downtime-Aware O-RAN VNF Deployment Strategy for Optimized Self-Healing in the O-Cloud," *arXiv preprint arXiv:2110.06060*, 2021.

40 V. Ziegler and S. Yrjölä, "How to make 6G a general purpose technology: Prerequisites and value creation paradigm shift," in *2021 Joint European Conference on Networks and Communications & 6G Summit (EuCNC/6G Summit)*. IEEE, 2021, pp. 586–591.

41 D. Palmbach and F. Breitinger, "Artifacts for detecting timestamp manipulation in NTFS on windows and their reliability," *Forensic Science International: Digital Investigation*, vol. 32, p. 300920, 2020.

42 O.R.P.C. O-Ran Alliance, "Open-RAN Security in 5G," 2021.

43 P. E. Sevinç, M. Strasser, and D. Basin, "Securing the distribution and storage of secrets with trusted platform modules," i*IFIP International Workshop on Information Security Theory and Practices*. Springer, 2007, pp. 53–66.

44 Z. Dong, K. Kane, and L. J. Camp, "Detection of rogue certificates from trusted certificate authorities using deep neural networks," *ACM Transactions on Privacy and Security (TOPS)*, vol. 19, no. 2, pp. 1–31, 2016.

45 S. A. T. Kawahara and A. U. R. Matsukawa, "O-RAN alliance standardization trends," *NTT DOCOMO Technical Journal*, vol. 1, 2019.

46 V. Hassija, V. Chamola, V. Gupta, S. Jain, and N. Guizani, "A survey on supply chain security: Application areas, security threats, and solution architectures," *IEEE Internet of Things Journal*, vol. 8, no. 8, pp. 6222–6246, 2020.

47 L. T. Sorensen, S. Khajuria, and K. E. Skouby, "5G visions of user privacy," in *2015 IEEE 81st Vehicular Technology Conference (VTC Spring)*. IEEE, 2015, pp. 1–4.

48 M. Liyanage, J. Salo, A. Braeken, T. Kumar, S. Seneviratne, and M. Ylianttila, "5G Privacy: Scenarios and solutions," in *2018 IEEE 5G World Forum (5GWF)*, 2018, pp. 197–203.

49 N. Docomo, "5G Open RAN Ecosystem Whitepaper," *White Paper*, p. 31, 2021.

50 Y. Li, B. Wu, Y. Jiang, Z. Li, and S.-T. Xia, "Backdoor Learning: A Survey," *arXiv preprint arXiv:2007.08745*, 2020.

51 H. Booth, D. Rike, and G. A. Witte, "The national vulnerability database (NVD): Overview," *ITL Bulletin, National Institute of Standards and Technology*, Gaithersburg, MD, 2013.

52 C. Balding, "Revisiting the United States Telecommunications Network Policy in a Post-Huawei World: Improving Economic Competitiveness, Addressing Security Weakness, and Building Alliances," *Addressing Security Weakness, and Building Alliances*, 2021.

53 M. Mariniello, "Fair, reasonable and non-discriminatory (FRAND) terms: A challenge for competition authorities," *Journal of Competition Law and Economics*, vol. 7, no. 3, pp. 523–541, 2011.

54 Z. Chi, Y. Li, X. Liu, W. Wang, Y. Yao, T. Zhu, and Y. Zhang, "Countering cross-technology jamming attack," in *Proceedings of the 13th ACM Conference on Security and Privacy in Wireless and Mobile Networks*, 2020, pp. 99–110.

55 A. D. Wood and J. A. Stankovic, "A taxonomy for denial-of-service attacks in wireless sensor networks," *Handbook of Sensor Networks: Compact Wireless and Wired Sensing Systems*, vol. 4, pp. 739–763, 2004.

56 A. Alina and S. Saraswat, "Understanding implementing and combating sniffing and ARP spoofing," in *2021 4th International Conference on Recent Developments in Control, Automation & Power Engineering (RDCAPE)*. IEEE, 2021, pp. 235–239.

57 Y. Siriwardhana, P. Porambage, M. Liyanage, and M. Ylianttila, "AI and 6G security: Opportunities and challenges," in *2021 Joint European Conference on Networks and Communications 6G Summit (EuCNC/6G Summit)*, 2021, pp. 616–621.

58 R. V. Yampolskiy, "Unexplainability and incomprehensibility of AI," *Journal of Artificial Intelligence and Consciousness*, vol. 7, no. 02, pp. 277–291, 2020.

59 S. Jeux, A. Zwarico, and Others, "The O-RAN ALLIANCE Security Task Group Tackles Security Challenges on All O-RAN Interfaces and Components," *Blog*, 2020.

60 M. F. Hossain, A. U. Mahin, T. Debnath, F. B. Mosharrof, and K. Z. Islam, "Recent research in cloud radio access network (C-RAN) for 5G cellular systems: A survey," *Journal of Network and Computer Applications*, vol. 139, pp. 31–48, 2019.

61 W. Yang and C. Fung, "A survey on security in network functions virtualization," in *IEEE NetSoft Conference and Workshops (NetSoft)*, 2016, pp. 15–19.

62 R. Khan, P. Kumar, D. N. K. Jayakody, and M. Liyanage, "A survey on security and privacy of 5G technologies: Potential solutions, recent advancements, and future directions," *IEEE Communications Surveys & Tutorials*, vol. 22, no. 1, pp. 196–248, 2020.

63 M. A. Ferrag, L. Maglaras, A. Argyriou, D. Kosmanos, and H. Janicke, "Security for 4G and 5G cellular networks: A survey of existing authentication and privacy-preserving schemes," *Journal of Network and Computer Applications*, vol. 101, pp. 55–82, 2018.

64 P. Tanakas, A. Ilias, and N. Polemi, "A novel system for detecting and preventing SQL injection and cross-site-script," in *2021 International Conference on Electrical, Computer and Energy Technologies (ICECET)*. IEEE, 2021, pp. 1–6.

65 G. George, J. Kotey, M. Ripley, K. Z. Sultana, and Z. Codabux, "A preliminary study on common programming mistakes that lead to buffer overflow vulnerability," in *2021 IEEE 45th Annual Computers, Software, and Applications Conference (COMPSAC)*. IEEE, 2021, pp. 1375–1380.

66 A. Brand ao, J. S. Resende, and R. Martins, "Hardening cryptographic operations through the use of secure enclaves," *Computers & Security*, vol. 108, p. 102327, 2021.

67 Ericsson, "Security Considerations of Open-RAN," *White Paper*, 2020.

68 M. Rasser and A. Riikonen, "Open Future The Way Forward on 5G," 2020. [Online]. Available: https://www.cnas.org/publications/reports/open-future.

69 D. Braeke, "A US national strategy for 5G and future wireless networks," *Information Technology and Innovation Foundation (ITIF)*, 2020.

70 E. G. Carayannis, J. Draper, and B. Bhaneja, "Towards fusion energy in the industry 5.0 and society 5.0 context: Call for a global commission for urgent action on fusion energy," *Journal of the Knowledge Economy*, vol. 12, pp. 1–14, 2020.

71 OR SFG - O-RAN Alliance, "O-RAN Security Threat Modeling and Remediation Analysis," *O-RAN.WG1.SFG.Threat-Model-v01.00, Technical specifications*, p. 57 pages, 2021.

72 M. Schwalbe, "Additive manufacturing scalability, implementation, readiness, and transition," in *Predictive Theoretical and Computational Approaches for Additive Manufacturing: Proceedings of a Workshop*. Washington, DC: The National Academic Press, 2016, Chapter 10, pp. 81–102.

73 Red Hat, "The inherent security of Open RAN," *Fierce Wireless*, 2020.

74 A. Weissberger, "Strand Consult: The 10 Parameters of Open RAN; AT&T memo to FCC," *IEEE ComSoc, Technology blog*, 2021.

75 T. H.-C. Hsu, *Hands-On Security in DevOps: Ensure continuous security, deployment, and delivery with DevSecOps*. Packt Publishing Ltd., 2018.

76 E. Hanselman, "Security Benefits of Open Virtualized RAN," p. 13 pages, 2020.

77 Altiostar, "Security in Open RAN," *White Paper*, 2021.

78 P. H. Masur and J. H. Reed, "Artificial Intelligence in Open Radio Access Network," *arXiv preprint arXiv:2104.09445*, 2021.

79 Verizon "How and why private 5G networks are taking flight, Addressing the need for enterprise network security, speed and bandwidth in the U.S." 2021.

80 M. Ge, J. B. Hong, W. Guttmann, and D. S. Kim, "A framework for automating security analysis of the Internet of Things," *Journal of Network and Computer Applications*, vol. 83, pp. 12–27, 2017.

81 S. Rose, O. Borchert, S. Mitchell, and S. Connelly, "Zero Trust Architecture," National Institute of Standards and Technology, Tech. Rep., 2020.

82 M. Liyanage, Q.-V. Pham, K. Dev, S. Bhattacharya, P. K. R. Maddikunta, T. R. Gadekallu, and G. Yenduri, "A survey on Zero touch network and service (ZSM) management for 5G and beyond networks," *Journal of Network and Computer Applications*, vol. 203, p. 103362, 2022. [Online]. Available: https://www.sciencedirect.com/science/article/pii/S1084804522000297.

6

Edge Intelligence*

* With additional contribution from Saeid Sheikhi, Center for Ubiquitous Computing, University of Oulu, Finland

This chapter has the focus on the technological evolution of edge computing, edge intelligence (EI), and their security implications. After reading this chapter, you should be able to:

- Gain an overview of EI and its relevance with 6G networks
- Identify the major security considerations ...

6.1 Overview of Edge Intelligence

The union between artificial intelligence (AI) and edge computing is instinctive since there is a close interaction [1]. In certain 6G wireless applications, it is imperative to shift the computation toward the edge of the network. Whether AI/machine learning (ML) algorithms are used to acquire, storage, or process data at the network edge, it is referred to as EI [2]. In EI, an edge server aggregates data generated by multiple devices that are associated with it. Data are shared among multiple edge servers for training models, and later used for analysis and prediction, thus devices can benefit from faster feedback, reduced latency, and lower costs while enhancing their operation. However, as data are collected from multiple sources, and the outcome of AI/ML algorithms is highly data-dependent, EI is highly prone to several security attacks. Under such circumstance, trust is also required in EI services which are critical to ensure user authentication and access control, model and data integrity, and mutual platform verification [3]. In [4], it is demonstrated how blockchain is used to secure distributed

Security and Privacy Vision in 6G: A Comprehensive Guide, First Edition.
Pawani Porambage and Madhusanka Liyanage.
© 2023 The Institute of Electrical and Electronics Engineers, Inc. Published 2023 by John Wiley & Sons, Inc.

edge services to prevent resource transactions vulnerable to malicious nodes. Blockchain ensures the consistency of decomposed tasks and the chunks of learning data required in AI implementation.

Attackers can exploit the distributed nature and the respective dependencies on edge computing to launch different attacks such as data poisoning, data evasion, or a privacy attack, thus affecting the outputs of the AI/ML applications and undermining the benefits of EI [5]. Moreover, EI may require novel secure routing schemes and trust network topologies for EI service deliveries. Security in EI is closely coupled with privacy since the edge devices may collect privacy sensitive data which contain user's location data, health, or activities records, or manufacturing information, among many others. Federated learning is one approach for privacy-friendly distributed data training in edge AI models which enable local ML models. In addition to that, secure multiparty computation and homomorphic encryption for designing privacy-preserving AI model parameter-sharing schemes in EI services are also considered by researchers.

The key architectural change in 5G which has a cloud native and microservice architecture is expected to evolve with heterogeneous aspects in the cloud transformation toward 6G [6]. The heterogeneous clouds related to numerous service delivery platforms including public, private, on-premises, and edge cloud may require proper co-ordination of communication resources and distributed computing through orchestration and network control. The security considerations may also differ based on the nature of each cloud environment and the stakeholders. Mainly, the most common security issues include the violation of access control policies, data privacy breaches, information security issues, insecure interfaces and APIs, denial of service (DoS) attacks, and loss of data [7].

5G cellular networks enable solid connectivity for an enormous number of heterogeneous devices with computational resources for intelligent and autonomous operations [8]. The future generation of wireless networks (6G) is expected to reach a fully connected world and carry forward the capacities of massive access, low latency, and enhanced broadband to provide a more robust and intelligent service than 5G networks [9]. This vision raises significant issues considering the resource limitations and challenges regarding data and network security. One of the significant technologies to fulfill the need for latency, bandwidth, and localization in future networks is edge computing in edge networks. The edge computing in 5G and beyond is used to provide cloud capabilities to the network and clients' devices. EI is utilized in 6G and 5G communications. The EI is when AI and ML algorithms are used to obtain, store, or process data at the network edge. The architecture of EI and edge computing are shown in Figure 6.1.

Usually, the edge network paradigms, including multiaccess edge computing (MEC), are prone to security issues and face attacks against their virtualization,

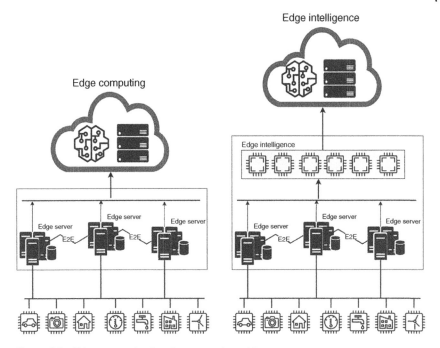

Figure 6.1 Edge computing in edge networks architectures.

network, and service infrastructures [10]. The MEC is an emerging environment that aims to converge IT and telecommunication services, and its security is crucial to keep the network secure and trusted. The edge infrastructure usually is faced with different security threats, and the most common are privacy leakage, denial of service (DoS), man-in-the-middle, and service or resource manipulation [11].

Artificial intelligence and ML algorithms are used in edge computing to enhance the security of the edge network and overcome privacy concerns [12]. ML is widely used to overcome real-world issues in various domains, including healthcare, economy, image processing, and autonomous driving. It can provide promising performance in solving issues in both classification and forecasting tasks [13]. However, the rapid evolution in network communication and upcoming technologies requires better security, greater bandwidth, and effective management of the continuous operation of edge infrastructure. Hence, AI could play a prominent role in defending large-scale edge networks from sophisticated security threats using an efficient and effective detection procedure. Therefore, this research aims to review and study the role of ML and AI in protecting the edge networks and infrastructure from various security threats in the future of wireless networks.

6.2   State-of-the-Art Related to 5G

With the progressive development of 5G communications networks in recent years, edge computing has attracted tremendous interest due to its advantages over cloud computing [14]. Edge computing deploys computing and stores resources close to endpoints, significantly decreasing data processing delay and computational load [15]. It also keeps sensitive data on edge, reduces bandwidth pressure, and maintains data privacy by processing data on edge. However, edge computing is faced with various security threats that could endanger the entire network. This section reviews some of these security challenges that threaten the edge network.

6.2.1   Denial of Service (DOS)

DoS is a large-scale cyber-attack that threatens system availability and reliability [16]. In a DoS attack, the attackers use tactics to purposefully block or degrade the availability and connection to a network or system resources [17]. The attacker takes control of large groups of compromised machines to build botnets, which he uses to damage resource servers by flooding superfluous requests and data [18]. This attack prevents genuine devices from gaining access to the server and obtaining the services. The edge servers are more vulnerable to distributed denial-of-service (DDoS) cyber threats than cloud servers due to their computational power limitation and heterogeneous firmware. Zhang et al. [19] proposed a method for detecting DDoS attacks in edge networks using reinforcement learning. In this method, authors deployed edge servers in different router locations, and then classifiers are used to classify and filter the traffic that passes through. In another work, Kozik et al. [20] presented a distributed extreme learning machine-based approach to protecting the edge networks from DDoS attacks. The suggested method analyzes and categorizes online the aggregated traffic obtained by NetFlows on selected edge nodes. Liu et al. [21] introduced the deep convolutional neural network (DCNN) Q-Learning method to protect the edge environment from large-scale low-rate distributed denial-of-service (LR-DDoS) attacks similar to Mirai botnet attack issues.

6.2.2   Man-in-the-Middle (MitM) Attack

The Man-in-the-Middle (MitM) is when attackers attempt to intercept or eavesdrop on the communications between two machines and steal data [22]. To establish a secure communication channel, the group of machines and clients inside the system is required to exchange keys. In MitM, the adversary intercepts the communications between machines and independently exchanges keys with other nodes. The attackers can take control of the communications between authorized network devices and sends out jamming signals for monitoring, eavesdropping, or modifying communication between the network's clients [23]. This attack can accomplish in various methods, such as domain name system (DNS) spoofing, address resolution protocol (ARP) cache poisoning, and dynamic host configuration protocol (DHCP) spoofing. A MitM attack is classified as an infrastructural threat in the MEC scenario. An adversary attempts to hijack a particular network segment and launches various cyberattacks, including phishing and eavesdropping on linked machines [24]. In MEC, services and applications depend heavily on virtualization, which makes launching MitM attacks on some VMs have adverse effects on other network elements [25]. Sowah et al. [26] introduced a method based on an artificial neural network (ANN) classifier to detect and prevent MitM threats on ad hoc networks (MANETs). The authors validated the classifier performance using the dataset with various traffic with different attack patterns. The method blocks and eliminates attackers after detection and reconfigures the network to respond to the attack, which helps the system perform well in protecting the network against the MitM attacks. Salem et al. [27] proposed a ML-based framework to prevent MitM attacks on the Internet of Medical Things. The goal of the methods is to detect and prevent MitM from interrupting and changing the function of the remote monitoring system. The experiment outcomes show that the approach achieves high accuracy in emergency detection with a low false rate.

6.2.3 Privacy Leakage

Privacy is one of the primary issues in different computing paradigms since the users' private information is transferred from endpoints to remote servers [28]. In edge computing, privacy-preserving is a serious concern because there are several legitimate adversaries among the services and infrastructure providers whose second objective is to obtain more sensitive data that can be exploited in egoistic ways [29]. In this case, it is inconceivable to determine trustworthy of a service provider in an open environment with several trust domains. In MEC, unauthorized access to the endpoints could compromise the confidentiality of information [30]. By separating information and access, the MEC paradigm minimizes the impact of privacy leakage in the edge networks [31]. However, some sensitive data in edge networks may leak with access to context data, including local network states, client status data, and traffic statistics according to which applications are employed to provide context-aware optimization [32]. Qiang et al. [33] proposed an ML framework to preserve test and training data privacy. The method is based on two main stages: first, the binary neural network (BNN) algorithm is trained on the cloud, and then the trained model is distributed to edge servers to protect the training data. The test results on the framework that present the framework efficiently could protect data privacy. In other work, Alguliyev et al. [34] presented a sparse denoising auto-encoder to protect data privacy in data classification and analysis. The method of the design protects data even after the training process. The validation results demonstrated the proposed method's effectiveness in protecting data privacy in data analysis.

6.3 State-of-the-Art Related to 6G

The 6G networks leverage edge intelligence to provide massive access, low latency, and enhanced broadband to connect and manage an enormous number of devices [35]. The edge intelligence is the critical enabler of the recently emerged network technologies, and it can significantly improve intelligence and autonomy in communication networks [36]. In edge intelligent computing, the central server coordinates different endpoints and uses their data to collaboratively train the ML models in edge servers [37]. The data will be shared among the various edge servers to be used for analysis and prediction. The shared data have a critical role in the development and final output of the learning models in edge servers. Hence, these data are the subject of the various security threats that corrupt the learning process. The attackers conduct poisoning and data evasion attacks on servers to manipulate the learning process and compromise clients' privacy. Therefore, in edge intelligence, the capability of controlling the security and privacy threats is critical for the platform's success.

6.3.1 Training Dataset Manipulation

The adversary groups and attackers can manipulate the performance of the learning model by controlling the compromised clients to inject the fake data to corrupt the learning process; this method is named data poisoning [38]. The data poisoning attacks involve polluting local model training data and corrupt ML models to lead them to wrong decisions. The data poisoning can reach two BlackBox and WhiteBox scenarios against the models [39]. In BlackBox, the models are updated through their learnings using user feedback, while in the WhiteBox procedure, the attackers gain access to the model and control training data using the interference in the supply chain of the data while it is collected from various endpoints.

One of the examples of this attack is poisoning data for autonomous driving, where manipulated data from a particular IoT device (e.g. camera) might mislead the model and lead the car to make wrong decisions. However, protection from such IoT endpoints against poisoning attacks is challenging since these devices own limited resources to perform sophisticated security methods. Dunn et al. [40] studied the impact of data poisoning attacks on the performance of the four ML models, including random forest, feed-forward neural network, naive Bayes, and gradient boosting machine against poisoned training data. The research examines the inherent robustness of models to the various rate of poisoned data. Chiba et al. [41] present a strategy to defend against poisoning threats and enhance the detection rate while data are obtained from different devices. The suggested approach reverses the poisoning attack and computes the removal rates of poisonous information and detection criteria of devices individually.

6.3.2   Interception of Private Information

In EI, the endpoint data need to transfer to the edge servers and remote cloud for processing. The private user information with other types of data is required to be exchanged among the clients and edge servers. In this case, the attackers can gain access to private information by intercepting the exchanged messages in the network communication [42]. The protection strategy for preventing the interception of private information is another challenging subject in intelligent edge computing.

6.3.3   Attacks on Learning Agents

In edge computing, the learning agents are distributed in the network edge. Compared to cloud computing, attacks on edge agents are straightforward. In this attack, the required interaction among learning agents is crucial. For example, reinforcement learning applications require agents to interact with their environments in order to optimize their rewards [43]. A strategically timed and luring attack can effortlessly target an agent in these conditions. When an agent is attacked, the entire edge learning process is compromised [5]. In this case, when one agent is infected with an attack vector, then, once learning agents engage with one another, all learning agents inside the edge computing system get infected with that attack vector.

6.4   Edge Computing Security in Autonomous Driving

The 6G and beyond 5G wireless networks are expected to supply a low-latency communication with high reliability ubiquitously via AI for vehicular networks [44]. Autonomous car systems have been designed to closely merge promising technologies, such as object detection and tracking, wireless offloading, and surround sensing through MEC to reduce accidents caused by human driving errors

[45]. In this context, 6G wireless networking and EI are expected to open the way for more reliable and safer autonomous cars by providing MEC with ML to autonomous driving vehicles (AVs) in close proximity [46].

The Internet of Vehicles (IoV) enables a vehicle to connect to nearby vehicles or other traffic infrastructure through vehicle-to-everything (V2X), vehicle-to-infrastructure (V2I), and vehicle-to-vehicle (V2V) wireless network connections. It is provided in intelligent transportation systems that vehicles use this system to transfer floating data to a base region that is far away [47].

Future 6G V2X services are expected to enable more secure, varied, and effective autonomous mobility (e.g. coordinated driving, intent sharing, and interactive gaming) and increase the effectiveness, reliability, and efficiency of intelligent transportation systems [48]. The IoV applications required extraordinarily high data transfer rates through V2X connections [49]. This enormous amount of data will overload infrastructures for computing and communication.

The IoV also requires considerable computational power since it must respond to real-time traffic situations with low delay. In this context, cloud computing platforms' capabilities could not be sufficient for the widespread adoption of IoV due to the threats to reliability and real-time computing in the cloud platform [50]. Therefore, vehicular edge computing (VEC) has developed to tackle the limitations of on-board computing and cloud computing latency by enhancing computing, communication, and providing resources and services to vehicular users in close proximity [51]. The VEC is the integration of MEC with roadside units (RSUs) and vehicles that provide storage capacity, high bandwidth, low delay, data, and substantial processing resources at the network edge to use for performing AI tasks [52]. The security issues in the IoV edge system are one of the most critical challenges since it has a dynamic topology, limited transmission power, vehicle mobility, and a massive volume of traffic data. The most frequent attacks which threaten the resilience of edge-centric IoV systems are illustrated in Figure 6.2.

6.5 Future and Challenges

In the future of wireless communication networks, EI and 6G with a more extensive architecture play a critical role. However, EI security research and development are still in their initial stages. The advancement in new applications and improvements in existing cryptography drives innovative ideas and implementations to secure edge computing systems in the near future. In addition, since EI is highly connected to edge endpoints, more secure solutions are expected to be designed to protect the edge endpoints with lower computational power and preserve data privacy.

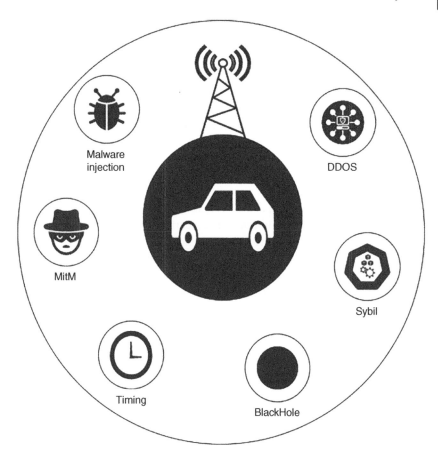

Figure 6.2 Security threats in edge-centric IoV systems.

References

1 S. Deng, H. Zhao, W. Fang, J. Yin, S. Dustdar, and A. Y. Zomaya, "Edge intelligence: The confluence of edge computing and artificial intelligence," *IEEE Internet of Things Journal*, vol. 7, no. 8, pp. 7457–7469, 2020.

2 G. Plastiras, M. Terzi, C. Kyrkou, and T. Theocharidcs, "Edge intelligence: Challenges and opportunities of near-sensor machine learning applications," in *2018 IEEE 29th International Conference on Application-specific Systems, Architectures and Processors (ASAP)*, 2018, pp. 1–7.

3 Z. Zhang, Y. Xiao, Z. Ma, M. Xiao, Z. Ding, X. Lei, G. K. Karagiannidis, and P. Fan, "6G wireless networks: Vision, requirements, architecture, and key technologies," *IEEE Vehicular Technology Magazine*, vol. 14, no. 3, pp. 28–41, 2019.

4 D. Xu, T. Li, Y. Li, X. Su, S. Tarkoma, T. Jiang, J. Crowcroft, and P. Hui, "Edge Intelligence: Architectures, Challenges, and Applications," *arXiv preprint arXiv:2003.12172*, 2020.

5 M. Mukherjee, R. Matam, C. X. Mavromoustakis, H. Jiang, G. Mastorakis, and M. Guo, "Intelligent edge computing: Security and privacy challenges," *IEEE Communications Magazine*, vol. 58, no. 9, pp. 26–31, 2020.

6 V. Ziegler, H. Viswanathan, H. Flinck, M. Hoffmann, V. Räisänen, and K. Hätönen, "6G architecture to connect the worlds," *IEEE Access*, vol. 8, pp. 173 508–173 520, 2020.

7 R. Kalaiprasath, R. Elankavi, and R. Udayakumar, "Cloud security and compliance - a semantic approach in end to end security," *International Journal on Smart Sensing & Intelligent Systems*, vol. 10, no. 5, pp. 482–494, 2017.

8 C. R. Storck and F. Duarte-Figueiredo, "A survey of 5G technology evolution, standards, and infrastructure associated with vehicle-to-everything communications by internet of vehicles," *IEEE Access*, vol. 8, pp. 117 593–117 614, 2020.

9 A. Al-Ansi, A. M. Al-Ansi, A. Muthanna, I. A. Elgendy, and A. Koucheryavy, "Survey on intelligence edge computing in 6G: Characteristics, challenges, potential use cases, and market drivers," *Future Internet*, vol. 13, no. 5, p. 118, 2021.

10 G. Gur, P. Porambage, Y. Siriwardana, R. Sedar, C. Kalalas, W. Soussi, H. N. N. MI, C. Benzaid, O. Hireche, Y. Dang et al., "D3.3: 5G security new breed of enablers," 2019.

11 Y. Siriwardhana, P. Porambage, M. Liyanage, and M. Ylianttila, "AI and 6G security: Opportunities and challenges," in *Proceedings of the IEEE Joint European Conference on Networks and Communications & 6G Summit (EuCNC/6G Summit)*, 2021, pp. 1–6.

12 Y. Zhang, Z. Feng, H. Moustafa, F. Ye, U. Javaid, and C. Cui, "Guest editorial: Edge intelligence for beyond 5G networks," *IEEE Wireless Communications*, vol. 28, no. 2, pp. 10–11, 2021.

13 S. Sheikhi, "An effective fake news detection method using WOA-xgbTree algorithm and content-based features," *Applied Soft Computing*, vol. 109, p. 107559, 2021.

14 T. Yang, M. Qin, N. Cheng, W. Xu, and L. Zhao, "Liquid software-based edge intelligence for future 6G networks," *IEEE Network*, vol. 36, no. 1, pp. 69–75, 2022.

15 H. Zeyu, X. Geming, W. Zhaohang, and Y. Sen, "Survey on edge computing security," in *2020 International Conference on Big Data, Artificial Intelligence and Internet of Things Engineering (ICBAIE)*. IEEE, 2020, pp. 96–105.

16 E. Cambiaso, M. Aiello, M. Mongelli, and I. Vaccari, "Detection and classification of slow DoS attacks targeting network servers," in *Proceedings of the 15th International Conference on Availability, Reliability and Security*, 2020, pp. 1–7.

17 R. Yaegashi, D. Hisano, and Y. Nakayama, "Light-weight DDoS mitigation at network edge with limited resources," in *2021 IEEE 18th Annual Consumer Communications & Networking Conference (CCNC)*. IEEE, 2021, pp. 1–6.

18 N. Hoque, D. K. Bhattacharyya, and J. K. Kalita, "Botnet in DDoS attacks: Trends and challenges," *IEEE Communications Surveys & Tutorials*, vol. 17, no. 4, pp. 2242–2270, 2015.

19 H. Zhang, J. Hao, and X. Li, "A method for deploying distributed denial of service attack defense strategies on edge servers using reinforcement learning," *IEEE Access*, vol. 8, pp. 78 482–78 491, 2020.

20 R. Kozik, M. Choraś, M. Ficco, and F. Palmieri, "A scalable distributed machine learning approach for attack detection in edge computing environments," *Journal of Parallel and Distributed Computing*, vol. 119, pp. 18–26, 2018.

21 Z. Liu, X. Yin, and Y. Hu, "CPSS LR-DDoS detection and defense in edge computing utilizing DCNN Q-learning," *IEEE Access*, vol. 8, pp. 42 120–42 130, 2020.

22 A. Mallik, "Man-in-the-middle-attack: Understanding in simple words," *Cyberspace: Jurnal Pendidikan Teknologi Informasi*, vol. 2, no. 2, pp. 109–134, 2019.

23 S. Singh, R. Sulthana, T. Shewale, V. Chamola, A. Benslimane, and B. Sikdar, "Machine-learning-assisted security and privacy provisioning for edge computing: A survey," *IEEE Internet of Things Journal*, vol. 9, no. 1, pp. 236–260, 2021.

24 M. A. Khan, I. M. Qureshi, I. Ullah, S. Khan, F. Khanzada, and F. Noor, "An efficient and provably secure certificateless blind signature scheme for flying ad-hoc network based on multi-access edge computing," *Electronics*, vol. 9, no. 1, p. 30, 2019.

25 A. M. Zarca, J. B. Bernabe, R. Trapero, D. Rivera, J. Villalobos, A. Skarmeta, S. Bianchi, A. Zafeiropoulos, and P. Gouvas, "Security management architecture for NFV/SDN-aware IoT systems," *IEEE Internet of Things Journal*, vol. 6, no. 5, pp. 8005–8020, 2019.

26 R. A. Sowah, K. B. Ofori-Amanfo, G. A. Mills, and K. M. Koumadi, "Detection and prevention of man-in-the-middle spoofing attacks in MANETs using predictive techniques in artificial neural networks (ANN)," *Journal of Computer Networks and Communications*, vol. 2019, p. 4683982, 2019.

27 O. Salem, K. Alsubhi, A. Shaafi, M. Gheryani, A. Mehaoua, and R. Boutaba, "Man-in-the-Middle attack mitigation in internet of medical things," *IEEE Transactions on Industrial Informatics*, vol. 18, no. 3, pp. 2053–2062, 2021.

28 V. Stephanie, M. Chamikara, I. Khalil, and M. Atiquzzaman, "Privacy-preserving location data stream clustering on mobile edge computing and cloud," *Information Systems*, vol. 107, p. 101728, 2022.

29 J. Zhang, B. Chen, Y. Zhao, X. Cheng, and F. Hu, "Data security and privacy-preserving in edge computing paradigm: Survey and open issues," *IEEE Access*, vol. 6, pp. 18 209–18 237, 2018.

30 U. Jayasinghe, G. M. Lee, Á. MacDermott, and W. S. Rhee, "TrustChain: A privacy preserving blockchain with edge computing," *Wireless Communications and Mobile Computing*, vol. 2019, p. 2014697, 2019.

31 M. B. Mollah, M. A. K. Azad, and A. Vasilakos, "Security and privacy challenges in mobile cloud computing: Survey and way ahead," *Journal of Network and Computer Applications*, vol. 84, pp. 38–54, 2017.

32 P. Ranaweera, A. D. Jurcut, and M. Liyanage, "Realizing multi-access edge computing feasibility: Security perspective," in *2019 IEEE Conference on Standards for Communications and Networking (CSCN)*. IEEE, 2019, pp. 1–7.

33 W. Qiang, R. Liu, and H. Jin, "Defending CNN against privacy leakage in edge computing via binary neural networks," *Future Generation Computer Systems*, vol. 125, pp. 460–470, 2021.

34 R. M. Alguliyev, R. M. Aliguliyev, and F. J. Abdullayeva, "Privacy-preserving deep learning algorithm for big personal data analysis," *Journal of Industrial Information Integration*, vol. 15, pp. 1–14, 2019.

35 R. Gupta, D. Reebadiya, and S. Tanwar, "6G-enabled edge intelligence for ultra-reliable low latency applications: Vision and mission," *Computer Standards & Interfaces*, vol. 77, p. 103521, 2021.

36 W. Saad, M. Bennis, and M. Chen, "A vision of 6G wireless systems: Applications, trends, technologies, and open research problems," *IEEE Network*, vol. 34, no. 3, pp. 134–142, 2019.

37 L. Welagedara, J. Harischandra, and N. Jayawardene, "A review on edge intelligence based collaborative learning approaches," in *2021 IEEE 11th Annual Computing and Communication Workshop and Conference (CCWC)*. IEEE, 2021, pp. 0572–0577.

38 V. Tolpegin, S. Truex, M. E. Gursoy, and L. Liu, "Data poisoning attacks against federated learning systems," in *European Symposium on Research in Computer Security*. Springer, 2020, pp. 480–501.

39 A. I. Newaz, N. I. Haque, A. K. Sikder, M. A. Rahman, and A. S. Uluagac, "Adversarial attacks to machine learning-based smart healthcare systems," in *GLOBECOM 2020–2020 IEEE Global Communications Conference*. IEEE, 2020, pp. 1–6.

40 C. Dunn, N. Moustafa, and B. Turnbull, "Robustness evaluations of sustainable machine learning models against data poisoning attacks in the Internet of Things," *Sustainability*, vol. 12, no. 16, p. 6434, 2020.

41 T. Chiba, Y. Sei, Y. Tahara, and A. Ohsuga, "A countermeasure method using poisonous data against poisoning attacks on IoT machine learning," *International Journal of Semantic Computing*, vol. 15, no. 02, pp. 215–240, 2021.

42 X. Xu, C. He, Z. Xu, L. Qi, S. Wan, and M. Z. A. Bhuiyan, "Joint optimization of offloading utility and privacy for edge computing enabled IoT," *IEEE Internet of Things Journal*, vol. 7, no. 4, pp. 2622–2629, 2019.

43 A. Hazra, A. Alkhayyat, and M. Adhikari, "Blockchain-aided integrated edge framework of cybersecurity for Internet of Things," *IEEE Consumer Electronics Magazine*, 2022. DOI: 10.1109/MCE.2022.3141068.

44 K. B. Letaief, W. Chen, Y. Shi, J. Zhang, and Y.-J. A. Zhang, "The roadmap to 6G: AI empowered wireless networks," *IEEE Communications Magazine*, vol. 57, no. 8, pp. 84–90, 2019.

45 J. Wang, J. Liu, and N. Kato, "Networking and communications in autonomous driving: A survey," *IEEE Communications Surveys & Tutorials*, vol. 21, no. 2, pp. 1243–1274, 2018.

46 L. Liu, C. Chen, Q. Pei, S. Maharjan, and Y. Zhang, "Vehicular edge computing and networking: A survey," *Mobile Networks and Applications*, vol. 26, no. 3, pp. 1145–1168, 2021.

47 M. Adhikari, A. Munusamy, A. Hazra, V. G. Menon, V. Anavangot, and D. Puthal, "Security and privacy in edge-centric intelligent internet of vehicles: Issues and remedies," *IEEE Consumer Electronics Magazine*, vol. 11, no. 6, pp. 24–31, 2021.

48 J. Hu, C. Chen, L. Cai, M. R. Khosravi, Q. Pei, and S. Wan, "UAV-assisted vehicular edge computing for the 6G internet of vehicles: Architecture, intelligence, and challenges," *IEEE Communications Standards Magazine*, vol. 5, no. 2, pp. 12–18, 2021.

49 S. B. Prathiba, G. Raja, S. Anbalagan, K. Dev, S. Gurumoorthy, and A. P. Sankaran, "Federated learning empowered computation offloading and resource management in 6G-V2X," *IEEE Transactions on Network Science and Engineering*, vol. 9, no. 5, 3234–3243, 2021.

50 D. P. M. Osorio, I. Ahmad, J. D. V. Sánchez, A. Gurtov, J. Scholliers, M. Kutila, and P. Porambage, "Towards 6G-enabled internet of vehicles: Security and privacy," *IEEE Open Journal of the Communications Society*, vol. 3, pp. 82–105, 2022.

51 A. Traspadini, M. Giordani, and M. Zorzi, "UAV/HAP-Assisted Vehicular Edge Computing in 6G: Where and What to Offload?" *arXiv preprint arXiv:2202.10953*, 2022.

52 X. Deng, Y. Liu, C. Zhu, and H. Zhang, "Air–ground surveillance sensor network based on edge computing for target tracking," *Computer Communications*, vol. 166, pp. 254–261, 2021.

7

Specialized 6G Networks and Network Slicing

This chapter has the focus on the evolving concept of specialized 6G networks and its close alliance with the network slicing technology as an enabler.

- Gain an overview of specialized 6G networks and the close alliance with networks slicing technology.
- Identify the security, privacy, and trust-related considerations of network slicing technology.

7.1 Overview of 6G Specialized Networks

Most of the stringent applications will be served by specialized private networks.

As introduced in [1], the trend of having vertical industries in 5G for industrial automation will continue to 6G as subnetworks (Figure 7.1). These specialized 6G networks are expected to operate as stand-alone miniaturized networks for multiple application verticals (e.g. in-body, in-car, in-robot, subnetwork of drones). When the wireless interfaces enable subnetwork owners or infrastructure to use novel applications, those external communication interfaces may impose security vulnerabilities. To avoid the unauthorized persons remotely taking control of the subnetwork functions, it will be important to use strong as well as lightweight authentication and encryption algorithms together with methods for monitoring network security by means of intrusion detection systems. Hierarchical and dynamic authorization mechanism will be more suitable to handle trust boundaries between the large networks and the miniaturized sub-networks. Use of trusted execution environments (TEE) may also guarantee the confidentiality and integrity of such closed sub-network environments.

Security and Privacy Vision in 6G: A Comprehensive Guide, First Edition.
Pawani Porambage and Madhusanka Liyanage.

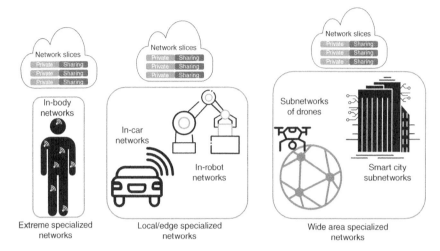

Figure 7.1 6G specialized networks.

Network slicing is a key enabling technology for the specialized networks, which are capable of catering to those specialized networking services on top of a common pool of physical infrastructure. Therefore, in the next section, we discuss the network slicing and its role in 6G together with the security, trust, and privacy aspects.

7.2 Network Slicing in 6G

Although the concept of network slicing has gained the momentum in the fifth generation (5G) networks and it has become a key enabling technology in 5G, it is expected that network slicing will continue evolving by playing an important role in coming 6G networks as well.

Network slicing allows logical separation of common network resources and infrastructure to facilitate service provisioning. With respect to the security considerations, there are multiple ways where artificial intelligence (AI) can be exploited such as secure resource allocation, slice isolation, and security management of network slice life cycle. Slice admission decision can be taken at the software defined networking (SDN) controller to maximize the resource utilization. AI-based solutions can be exploited for optimizing resource allocation and service demand prediction. In addition to that, AI-based anomaly detectors can be incorporated for the identification and detection of security attacks, predict about different security threats, and taking mitigation actions.

7.2.1 Trust in Network Slicing

Trust is a paramount property to consider in both 5G and beyond 5G systems in general. In particular network softwarization (NS) has close alliance with trust as an implicit assurance of its security-related considerations. In a zero-trust network, the devices should undergo an access control mechanism while joining the network to obtain any sort of service. In a way, the trust in NS framework can be considered as the expectation of users to avoid security threats while using NS services. This can be done with respect to life cycle (LC) security, inter- and intra-slice security, slice broker security, and zero-touch network and service management (ZSM) security. In another way, the trust in NS can be evaluated to the extent to which the network slices meet the user security requirements including the key security goals such as mutual authentication, confidentiality, integrity, authorization, and availability.

In particular, as NS supports multioperator cooperation, it is important to ensure the trust among the mobile network operators (MNOs) as well. In principle, NS enables multiple MNOs to deploy different services on top of the shared physical infrastructure allowing the inter-operator resource sharing. In the resource-sharing relationship in NS, the MNOs may play the role of resource providers (RPs) as well as the resource users (tenants). On the one hand, a RP MNO may provide slicing services to another tenant MNO with a monitoring interface, where the tenant has to trust RP in accordance with an agreed service level agreement (SLA). This can be further established with the Security SLAs with the predefined security levels. On the other hand, a RP MNO acts as a physical infrastructure provider, where the tenant has more control over virtual network function (VNF) management and orchestration. Here, the tenant has better visibility and need to trust on RP for sharing common network functions and maintaining security goals of NS. Therefore, a dynamic and efficient trust-relationship model should be carefully maintained in for the multioperator cooperation in 5G and beyond. When the network slices are created and granted to the end users, they can be certified by a trust reputation manager (TRM). The trust reputation assessment can be performed based on the historical behavior of virtual and physical networking elements. In order to demonstrate a multidomain policy enforcement ecosystem in the network slices, the trustworthiness can be evaluated to take the decisions.

In order to bring the intelligent trust enabling mechanisms with NS, there is a great potential of incorporating AI/machine learning (ML)-based solutions to evaluate trust in a quantitative manner. Particularly, since the next-generation network management is going be more intelligent and automated, trust in the NS architecture can be considered a metric to evaluate the stakeholders including tenants and RPs. The level of trust can be considered a key value indicator (KVI) for next-generation networks.

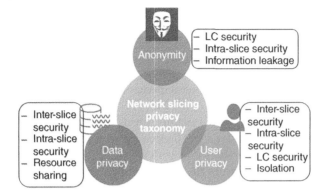

Figure 7.2 Privacy taxonomy in network slicing.

7.2.2 Privacy Aspects in Network Slicing

The key privacy goals in NS may include user privacy, data privacy, and anonymity in the 5G echo system. On the other hand, there are certain threat vectors related to the possible privacy violations in the NS architecture. For instance, the threats for inter- and intraslice security may directly affect the privacy issues with respect to user and data privacy (see Figure 7.2).

By definition of network slicing, tenants share common resources, where a slight information leak may create serious privacy violations in terms of user and data privacy by making threats on both inter- and intra-slice scenarios. One of the most common privacy issues arise with NS with regard to its close alliance of IoT use cases and the extremely large data sets they collect [2]. When there is an information leakage between the slices, this may introduce serious privacy violations by making a threat for inter-slice communication. This is a key motivation to have anonymity as a privacy goal in NS. In such situations, robust slice isolation mechanisms with the well-protected inter-slice communication via the secured channels are needed to preserve the user and data privacy [3].

When we consider different application areas, such as mission-critical vehicular-to-everything (V2X) scenarios, they may need real-time privacy-preserving mechanisms to mitigate the privacy risks that create LC security threats. In a way, NS may also provide solutions to enhance privacy by isolating networking resources to a specific set of users. However, the specification of the network slice and the LC may include to what extent the slice identification and the correlation to customer devices will assure the privacy of individual users.

PbD approaches can be used during the design stage of network slices to keep the proper slice isolation and isolate privacy of the users. Use of Homomorphic encryption and differential privacy mechanisms to manage tenant information related to NS architecture. It is also important to develop these trust-enabling technologies and privacy enhancing mechanisms related to NS in a scalable manner in such a way to support the network and application heterogeneity.

7.2.3 Solutions for Privacy and Trust in NS

For developing trust-ensuring solutions for NS, blockchain-like distributed ledger technologies are identified with a greater potential [4, 5]. Blockchain can be used to generate trust among different operators and other stakeholders and allows the collaboration among them to deploy secure end-to-end (E2E) network slices. These network slices can be certified with a certification tool in such a way that it be trusted by all the peers. Trusted blockchain-based network slices will explicitly provide distributed trust in a zero-trust network architecture.

A combination of AI and distributed ledger technologies can create the technological platform for accountability where the decisions taken by the algorithms can be recorded in a trusted distributed ledger. This may be exploited to develop consistent security policies among the network slice tenants to ensure trust. Due to the lack of transparency in the processing segments of many deep learning algorithms, it may create more issues in exploring model uncertainty and transparency. In such a circumstance, to achieve human trust in AI-empowered services related to NS in future networks, a trust broker can be incorporated [6].

Hardware-based TEEs are also used to achieve integrity and confidentiality in the networking field to bring the hardware-based root of trust. They may provide remediation to mitigate certain NS-related attacks, such as side channel attacks [7].

Privacy can be also considered with respect to the different resource allocation mechanisms in NS [8]. This may create privacy threats in slice brokering as well. In the shared-based approach, the tenant information leakages may occur while allocating total budget at individual nodes. In the reservation-based approach, a privacy violation may occur only when the information is leaked by the system while accepting or rejecting the resource requests. Introducing privacy isolation with privacy rules can be also taken as another mechanism that can be used to directly influence in the data plane by managing encryption of the slice services [9, 10]. This may allow the isolation of privacy dynamically for different privacy levels required information.

References

1 V. Ziegler, H. Viswanathan, H. Flinck, M. Hoffmann, V. Räisänen, and K. Hätönen, "6G architecture to connect the worlds," *IEEE Access*, vol. 8, pp. 173 508–173 520, 2020.

2 J. Ni, X. Lin, and X. S. Shen, "Efficient and secure service-oriented authentication supporting network slicing for 5G-enabled IoT," *IEEE Journal on Selected Areas in Communications*, vol. 36, no. 3, pp. 644–657, 2018.

3 S. Wijethilaka and M. Liyanage, "Survey on network slicing for Internet of Things realization in 5G networks," *IEEE Communications Surveys & Tutorials*, vol. 23, no. 2, pp. 957–994, 2021.

4 M. A. Togou, T. Bi, K. Dev, K. McDonnell, A. Milenovic, H. Tewari, and G.-M. Muntean, "DBNS: A distributed blockchain-enabled network slicing framework for 5G networks," *IEEE Communications Magazine*, vol. 58, no. 11, pp. 90–96, 2020.

5 M. A. Togou, T. Bi, K. Dev, K. McDonnell, A. Milenovic, H. Tewari, and G.-M. Muntean, "A distributed blockchain-based broker for efficient resource provisioning in 5G networks," in *2020 International Wireless Communications and Mobile Computing (IWCMC)*. IEEE, 2020, pp. 1485–1490.

6 C. Li, W. Guo, S. C. Sun, S. Al-Rubaye, and A. Tsourdos, "Trustworthy deep learning in 6G-enabled mass autonomy: From concept to quality-of-trust key performance indicators," *IEEE Vehicular Technology Magazine*, vol. 15, no. 4, pp. 112–121, 2020.

7 M.-W. Shih, S. Lee, T. Kim, and M. Peinado, "T-SGX: Eradicating controlled-channel attacks against enclave programs." in *NDSS*, 2017.

8 A. Banchs, G. de Veciana, V. Sciancalepore, and X. Costa-Perez, "Resource allocation for network slicing in mobile networks," *IEEE Access*, vol. 8, pp. 214 696–214 706, 2020.

9 J. P. B. Gonçalves, H. C. de Resende, R. da Silva Villaca, E. Municio, C. B. Both, and J. M. Marquez-Barja, "Distributed network slicing management using blockchains in E-health environments," *Mobile Networks and Applications*, vol. 26, pp. 2111–2122, 2021.

10 J. P. D. B. Gonçalves, H. C. De Resende, E. Municio, R. Villaça, and J. M. Marquez-Barja, "Securing E-health networks by applying network slicing and blockchain techniques," in *2021 IEEE 18th Annual Consumer Communications & Networking Conference (CCNC)*. IEEE, 2021, pp. 1–2.

8

Industry 5.0*

* With additional contribution from Tharaka Hewa, University of Oulu, Finland

This chapter has the focus on the evolving concept of Industry 5.0 and its security and privacy issues. After reading this chapter, you should be able to:

- Understand the network level requirements to enable Industry 5.0, Collaborative Robots, and Digital Twin.
- Understand the relevance of 6G networks for realization of Industrial Automation.
- Understand the importance of 6G technologies for Industry 5.0.
- Understand the security and privacy issues related to Industry 5.0.

8.1 Introduction

Vehicles, clothing, houses, and weapons have been designed and manufactured by humans and/or with the help of animals in the past centuries. With the emergence of Industry 1.0 in 1974, industrial production began to change significantly. Figure 8.1 shows an overview of the evolution of Industrial $X.0$ [1]. The development time for the first three revolutions was around 100 years, and it took only 40 years to reach the fourth from the third. In 1800s, Industry 1.0 evolved through the development of mechanical production infrastructures for water and steam-powered machines. There is a massive gain in the economy as production capacity has increased. Industry 2.0 evolved in the year of 1870 with the concept of electric power and assembly line production. Industry 2.0 focused primarily on

Security and Privacy Vision in 6G: A Comprehensive Guide, First Edition.
Pawani Porambage and Madhusanka Liyanage.

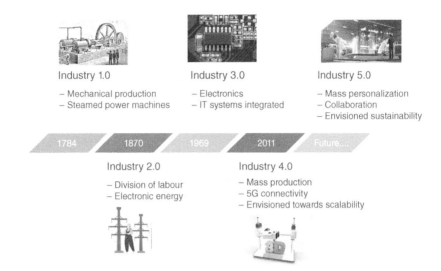

Figure 8.1 Illustration of industrial evolution. Source: GJ/Pixabay.com; manfredxy/Adobe Stock; jamesteohart/Adobe Stock.

mass production and distribution of workloads, which increased the productivity of manufacturing companies. Industry 3.0 evolved in 1969 with the concept of electronics, partial automation, and information technologies. Industry 4.0 evolved in 2011 with the concept of smart manufacturing for the future. The main objective is to maximize productivity and achieve mass production using emerging technologies [2, 3]. Industry 5.0 is a future evolution designed to use the creativity of human experts working together with efficient, intelligent, and accurate machines [4].

The evolution of Industry 5.0 is aligned toward prominent environmental considerations including "sustainable manufacturing," "global warming," "zero carbon emission," and so on. Especially, the global logistics chain will be reshaped with the blessings of computer science and communication technologies. The remote manufacturing technologies are one of the prominent examples that reduce the gap between consumer and manufacturer using Internet integrated production. However, the future industrial paradigm is more sophisticated when compared with the previous generations as the industrial ecosystems comprise multitudinous endpoint Internet of Things (IoT) nodes, cloud systems, consumer applications, mobile applications, and so on. Ensuring the primary security standards including authentication, integrity, privacy, access control, and audit is required while preserving the performance requirements, computational limitations, and data protection regulations.

8.2 Motivations Behind the Evolution of Industry 5.0

Industry 4.0 standard has revolutionized the manufacturing sector by integrating several technologies, such as artificial intelligence (AI), the Internet of Things (IoT), cloud computing, cyber physical systems (CPSs), and cognitive computing. The main principle behind Industry 4.0 is to make the manufacturing industry "smart" by interconnecting machines, devices that can control each other throughout the life cycle [5–8]. In Industry 4.0, the main priority is process automation, thereby reducing the intervention of humans in the manufacturing process [9, 10]. Industry 4.0 focuses on improving mass productivity and performance through the provision of intelligence between devices and applications using machine learning (ML) [11–13]. Industry 5.0 is currently conceptualized to leverage the unique creativity of human experts to collaborate with powerful, smart, and accurate machinery. Many technical visionaries believe that Industry 5.0 will bring back the human touch to the manufacturing industry [14]. It is expected that Industry 5.0 merges the high speed and accurate machines and critical, cognitive thinking of humans (Figure Industry 50). Mass personalization is another important contribution of Industry 5.0, wherein the customers can prefer personalized and customized products according to their taste and needs. Industry 5.0 will significantly increase manufacturing efficiency and create versatility between humans and machines, enabling responsibility for interaction and constant monitoring. The collaboration between humans and machines aims to increase production at a rapid pace. Industry 5.0 can enhance the quality of the production by assigning repetitive and monotonous tasks to the robots/machines and the tasks which need critical thinking to the humans.

Industry 5.0 promotes more skilled jobs compared to Industry 4.0 since intellectual professionals work with machines. Industry 5.0 focuses mainly on mass customization, where humans will be guiding robots. In Industry 4.0, robots are already actively engaged in large-scale production, whereas Industry 5.0 is primarily designed to enhance customer satisfaction. Industry 4.0 focuses on CPS connectivity, while Industry 5.0 links to Industry 4.0 applications and establishes a relationship between collaborative robots (cobots). Another interesting benefit of Industry 5.0 is the provision of greener solutions compared to the existing industrial transformations, neither of which focuses on protecting the natural environment [15]. Industry 5.0 uses predictive analytics and operating intelligence to create models that aim at making more accurate and less unstable decisions. In Industry 5.0, the majority of the production process will be automated, as real-time data will be obtained from machines in combination with highly equipped specialists. Figure 8.2 reflects the composition of distinguishing features of the Industry 5.0.

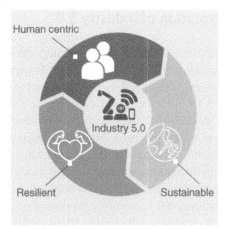

Figure 8.2 The core elements of Industry 5.0.

8.3 Key Features of Industry 5.0

Industry 5.0 is the enhanced version of the fourth industrial revolution. The added features of Industry 5.0 are discussed in this section.

8.3.1 Smart Additive Manufacturing

The most popular cost-effective approach for current manufacturing industries, which support producers to execute development plans, reduces pollution and resource utilization throughout the development lifecycle, is sustainable manufacturing [16]. Additive manufacturing is the sustainable approach adopted for industrial production, which builds the product part layer by layer instead of a solid block, thereby developing lighter but more robust parts one layer by layer. It adds up material layer by layer on the 3D objects. Smart additive manufacturing (SAM) applies AI algorithms and computer vision to add more accuracy and better graphical representation of product design in 3D printing. Now, 5D printing, a new subset of additive manufacturing, is employed for better compositions. The recent enterprises and researchers are focusing on deploying smart manufacturing products in their research and industrial domains. With the recent advancement of technologies such as AI, IoT, Cloud computing, Big Data, CPS, 5G, DT, EC and manufacturing, smart-empowering technologies are becoming popular and remarkably strengthened the development of smart manufacturing. Sustainability, profitability, and productivity are the main advantages of smart manufacturing industry. From the last decade, SAM has become the emerging technology in smart manufacturing domain [17]. One of the prominent features of Industry 5.0

is additive manufacturing referred to as 3D printing which is applied to make manufacturing products more sustainable. Additive manufacturing in Industry 4.0 focused on customer satisfaction by including benefits in products and other services. It also facilitates transparency, interoperability, automation, and practicable insights [18]. SAM defines the various processes in which the component to be manufactured is developed by adding materials, and the development is executed in various layers. SAM has the capability to save energy resources, helps to reduce material and resource consumption which leads to pollution-free environmental production. To obtain the complete benefits of Industry 5.0, SAM is merged with integrated automation capability to streamline the processes involved in supply chain management and reduces the delivery time of the products.

8.3.2 Predictive Maintenance

As the economy of the world is moving toward globalization, the industries are facing many challenges. This is forcing the manufacturing units move to upcoming transformation such as predictive maintenance (PdM). To enhance the productivity and efficiency, the manufacturers started utilizing evolving technologies, such as CPS approaches and advanced analytical methods [19]. Transparency is the capability of industry to uncover and assess the uncertainties in order to estimate the manufacturing ability and availability. Basically, most of the manufacturing schemes assume the availability of equipment continuously. However, it never practically happens in the real industries. Thus, the manufacturing units should transform themselves to the PdM to acquire transparency. This transformation needs application of state-of-the-art prediction tools in which the data are processed to information systematically and define the uncertainties to allow the manpower in taking smart decisions. The implementation of IoT provides the basic framework for PdM with the utilization of smart machines and smart sensor networks. Enabling self-conscious capability for systems and machines is the main goal of PdM. Smart computational agent is the key technology for PdM which includes smart software to provide functionalities for predictive modeling. In Industry 5.0, PdM helps to perform maintenance activity for avoiding problems instead of performing planned and scheduled maintenance and when a problem arises [20].

8.3.3 Hyper Customization

Industry 4.0 targeted linking machines, created intelligent supply chains, promoted the production of smart products, and isolated the manpower from automated industries. But Industry 4.0 has failed to manage the growing demand for customization, whereas Industry 5.0 does it using hyper customization. Hyper customization is a personalized marketing strategy which applies cutting-edge technologies such as AI, ML, cognitive systems, and computer vision to real-time data in order to provide more specific product, service, and content to every customer. The integration of human intelligence with robots helps manufacturers to customize the products in bulk. In order to achieve this, many variants of the functional material is shared with other personnel with the motive of customizing the product with different variants for customers' choice. Industry 4.0 aimed at huge production with low wastage and maximum efficiency, whereas Industry 5.0 aims at mass customization with minimum cost and maximum accuracy. The collaboration between manpower and robots along with cognitive systems enable the industries to coordinate the processes in the manufacturing to implement the customer needs and market changes. The first step in hyperpersonalization is transition to agile manufacturing process and supply chain. This also needs human intervention, production team, and customer preferences. Also, the applicability of hypercustomization depends strongly on the cost-effectiveness of the developed products [21].

8.3.4 Cyber-Physical Cognitive Systems

Due to the advancement of technologies such as smart wearable devices, IoT, cloud computing, and big data analytics, CPS has become popular now a days. The fourth industrial revolution has transformed the manufacturing process from complete manual systems to CPS [22, 23]. The framework for Industry 4.0 is established on the communication between CPS with the help of IoT. Cloud technology is used for huge amount of efficient, secure data storage and exchange [24]. Also, cognitive methods are used in several applications such as surveillance, industrial automation, smart grid, vehicular networks, and environment monitoring to increase the performance of the system and thus called as cyber-physical cognitive system (CPCS) [25, 26]. Cognitive capabilities such as observe/study the environment and take actions accordingly are contained in the nodes of CPCS. Learning and knowledge are the primary components of decision-making in CPCS. The CPCS has been introduced for human robot collaborative (HRC) manufacturing. The HRC executes the assembly of components in manufacturing division in collaboration with a robot and a human. The integration of machine–human cognition is modeled and applied for this collaboration work in real time. The fifth industrial revolution confined the merits of fourth industrial revolution and brings back the human labor for production. The fifth revolution facilitates the robots and skilled labor to work together in order to produce customized products and services in Industry 5.0 [27].

8.4   Security of Industry 5.0

Industry 5.0 will face critical security issues during the deployments. Similar to the tradition CPSs, Industry 5.0 will also need to provide security needs such as integrity, availability, authentication, and audit aspects [28].

- **Authentication**: The authentication of massive number of different stakeholders such as IoT nodes, machines, fog nodes, communication nodes, and collaborative partner nodes is critical requirement to establish the mutual trust in the ecosystem. The authentication mechanisms used in Industry 5.0 should be scalable to connect billons of devices, quantum resistance to stand against future quantum computing applications, and lightweight to deploy with IoT nodes [29]. The attacks to break the authentication of integrated cloud infrastructure and endpoint nodes are anticipated to be more sophisticated in the Industry 5.0. Hence, automated and dynamic authentication frameworks that preserve performance are the ideal candidates to facilitate Industry 5.0 authentication.
- **Integrity**: The integrity is a primary concern in the perspective data security [30] in Industry 5.0 as controlling commands and monitoring data will be transferred over third-party networks. However, the integrity validation must not affect the performance features in the system. The integrity related state-of-art cryptographic operations include digital signature generation, hashing, and validation. As the smart infrastructure of Industry 5.0 varies from lightweight sensor nodes to high-end cloud servers, the integrity verification techniques anticipate cross-infrastructure functional capability. Furthermore, the evolution of quantum computing requires redefining of the state-of-art mechanisms to establish integrity of data.
- **Access control**: The establishment of access control mechanisms is an essential security measure in the future Industry 5.0 ecosystems to ensure that access to the sensitive resources such as intellectual properties restricted only to authorized stakeholders. The establishment of access control mechanisms with the demand expansion is challenging in most of the computing implementations. Therefore, the classical access control techniques must be redefined by integrating intelligent and automated access control mechanisms. The access control can be deployed as a distributed microservice that is integrated with intrusion detection systems (IDS) to revoke access in real time upon identification of potential malicious actions executed by the components in the Industry 5.0 ecosystems.
- **Audit**: Auditability is a primary consideration to evaluate the alignment of service operation along with the regulatory compliance definitions. Furthermore, the audit logs require investigation of the dispute resolution cases. The log

management in Industry 5.0 must support the scalability requirements in the massive connectivity anticipated in the future Industry 5.0 systems. The auditing frameworks require supporting of scalability in terms of storage as the massive number of smart computational endpoints and communication channels are integrated in the future Industry 5.0 ecosystems. Automated and intelligent audit services that reflect important insights for the compliance will extend the value of Industrial 5.0 audit services.

- **Process and data consistency**: Industry 5.0 seamlessly integrates humans, machines, and AI into the service value chain. The availability is applicable in general to the service and the information exchanged in the Industry 5.0 systems. The security services must be hardened to defend the network, services, and data against more sophisticated and intelligent attacks launched with the integration of novel technologies such as AI. The inconsistency of process workflow and data are time-critical, and unavailability of data due to attacks such as ransomware attacks or jeopardized process workflow will end up with catastrophic losses of human life.

- **Backup and data recovery**: Industry 5.0 is integrated with data-driven intelligent technologies to leverage the cross-industrial business processes. The training materials for the AI-based algorithms require persistent availability within the execution of business workflows. The data loss due to malicious parties as well as the hardware-related failures compromises the associated business operations. The automated and intelligent data backup and recovery services are mandatory to establish the data resilience in the Industry 5.0.

8.4.1 Security Issues of Industry 5.0

The use of AI and supporting automation in Industry 5.0 will open up a new threat vector. For AI/ML functions, it is important to have trusted execution for security. Specially, the integrity of data set used for ML model training and also AI algorithm should be protected for proper operation in Industry 5.0 applications [31]. For instance, different tenants in Industry 5.0 should securely share empirical data for AI model training or incremental model updates as in federated learning. Also, the significant Industry 5.0 applications are highly dependent on information and communication technologies (ICT) systems that will lead to new security requirements such as deployment of proactive security mechanisms [32, 33] and mitigation of Zero day attacks [34, 35]. Moreover, the development of quantum computing, may lead Industry 5.0 to operate in Quantum computing era. The protection of legacy security mechanisms will be dramatically simplified by a quantum computer [36]. In that case, Industry 5.0 systems should utilize quantum-resistant cryptography or postquantum cryptography mechanism to provide required level of protection [28].

- **Intelligent attacks**: The recent evolution of AI advanced the capabilities of Industry 5.0 ecosystems through automation and intelligence. In contrast, the attack landscape has been revamped with the incorporation of AI to advance the attack capabilities and bypass incident response mechanisms. Through the integration of ML, the network tenants can be impersonated using the previously generated data. Furthermore, AI can be utilized to scope the vulnerabilities of Industry 5.0 ecosystems and formulate more precise attack models. In addition to that, intelligent malware may disrupt the anticipated business workflow without the system stakeholders' notice.

- **Process workflow compromise**: Industry 5.0 business processes rely on the computational services that are encoded with the business logic. The end-to-end business process of Industry 5.0 is a composition of individualized micro-operations that execute on interconnected computing infrastructure with different roles. The compromise of an intermediary computing node to a malicious party jeopardizes the consistent business workflow. The inconsistency of business workflow will eventually end up in a massive financial losses to the business stakeholders.

- **Physical layer attacks**: The data exchanged across the network and the computational infrastructure integrated in the Industry 5.0 are potential for physical-layer attacks. The physical tampering of hardware such as cryptographic key storage will expose the Industry 5.0 systems into different threats such as impersonation and privacy compromise of the communication channels. Furthermore, as a safety measure of specialized hardware to store the cryptographic keys, the systems erase the cryptographic keys upon detection of physical tamper attempts. The business processes integrated with the cryptographic keys will be hindered until the system is being restored. Furthermore, the communication channels are vulnerable for physical layer attacks as the Industry 5.0 heavily relies on the 6G communication techniques. The key technologies of 6G include TeraHertz (THz) communication and visible light communication (VLC) that are potential for eavesdropping and interceptions.

- **Supply chain-related attacks**: Industry 5.0 anticipates advanced supply chain management through the incorporation of AI and automation technologies. The supply chain workflow includes sequential steps and to collaborate to the execution of end-to-end business process. Attacks for the computational services that execute the intermediary steps, the supply chain will affect the process consistency of the supply chain. Industry 5.0 anticipates more transparent supply chain that delivers the strategic objectives of Industry 5.0 such as sustainability. To ensure the supply chain consistency, different active services such as automated processes as well as passive services such as monitoring will be integrated in future supply chain management in the Industry 5.0. The attacks that affect the intermediary computational services, monitoring services, and data will disrupt the business processes of Industry 5.0.

8.5 Privacy of Industry 5.0

As the entire Industry 5.0 ecosystem functions with expensive intellectual assets, expensive manufacturing materials, and subscription management, privacy is a vital requirement of Industry 5.0 applications [37]. In Industry 5.0, the data are exchanged over the Internet to connect machines with humans, designers with other collaborators, and also to exchange monitoring and control information. Such data must not be visible to the malicious users in the Internet to ensure the trust of cloud manufacturing ecosystem [38].

Upon implementing AI, specific societal and ethical implications should be adhered to avoid the negative societal impact. It is common for a human labor to think that AI may lead to the displacement of their jobs, but Industry 5.0 will increase the job opportunities. The ethical issues with AI [39] and its impact on humans must be alleviated for the seamless collaboration of humans with cobots for co-production. The social choices, ethical values, relations, and cultural patterns must be integrated into cobots [40]. The policymakers for the industrial revolution must account for the ethical issues concerned with the human on human–machine coworking. The privacy issues include human data protection rights, i.e. humans have control over their data. They have the right to claim for any data theft incurred on their private and sensitive information. Therefore, data privacy must be ensured by safeguarding the user data [41] while using it for cognitive analysis for PdM.

Blackbox AI is an automated decision-making system using ML over a large volume of data that maps the features for predicting the individual's behavioral traits without exposing the reason behind the analysis. The issues concerned with Blackbox AI, i.e. lack of transparency observed with the increase in AI adoption from the dawn of the fourth industrial revolution, must be considered upon its integration with human intelligence [42]. Sometimes, a faster production process may lead to overproduction and wastage of goods [43]. So implementation transparency should be accounted. Though the explainable AI (that expands human intelligence) does not replace humans, gaining human trust in AI applications (making crucial decisions) is wearisome [44]. Explainable AI should be embarked to pursue enhanced trust in AI systems through transparent AI systems. Therefore, privacy should be adapted as a design concept to ensure AI-based systems' steady progress with powerful predictions. Although many enterprises are still struggling to implement Industry 4.0, the dawn of Industry

5.0 will be even more challenging. Blockchain is a distributed ledger technology that is transparent, immutable, decentralized, and records the information in a more secure way [45]. Also, it is more resource-consuming as it requires more energy for mining the data at its nodes. But it ensures the security of the data through digital hashes of the previous records. Blockchain can add a significant contribution to security and privacy issues in Industry 5.0. Since the blockchain is resource-consuming, when the number of nodes in the blockchain-based Industry 5.0 applications increase, it may slow down. So to avoid this, a lightweight blockchain framework can be deployed by segregating the rarely used information from the blockchain to its sidechain. Furthermore, quantum computing can be used for securing the CPS [46] or cyber-physical process systems (CPPS) without any downtime.

The scope of privacy in the Industry 5.0 is massive and specific to the industry verticals. Especially, the privacy of data associated with the business workflows of the industries is classified with application-specific privacy standards. At high level, the data can be categorized into three major categories as real-time data generated from the smart equipment, data exchanged across the network, and data stored in the storage services. The data exchanged across the network include application-specific sensitive data, such as trade secrets, personal healthcare-related information, and so on. The data exchanged over the network are vulnerable to the malicious parties that listen to the data and derive insights. The end-to-end encryption is one of the most prominent solutions to establish privacy over the communication channels. The encryption is computational resource-intensive operation, and it may be challenging to execute encryption on lightweight smart devices incorporated in the Industry 5.0 ecosystems. With the evolution of Industry 5.0 ecosystems into massive networks composed with enormous volume of smart devices, the real-time data generated from those devices form a massive data volume which will be required to be stored. The important consideration on the data storage is that the data are usable as inputs to train the AI-based algorithms that leverage Industry 5.0 business processes. The data privacy as well as learning capabilities must be well balanced in designing the architecture of Industry 5.0 systems. The distributed learning mechanisms such as federated learning are prominent candidates to ensure privacy-oriented Industry 5.0 systems development. The data storage must be ensured to be privacy protected against corporate espionage that will compromise the business consistency. Finally, the privacy specifications will be exercised once they have been declared as legal definitions. The most prominent examples are General Data Protection Regulations (GDPR) [47] and Health Information Portability Act (HIPAA). These regulations legally interpret the privacy specifications and eliminate the gap between technical specification and governing authorities. Figure 8.3 summarizes the privacy focus on the Industry 5.0.

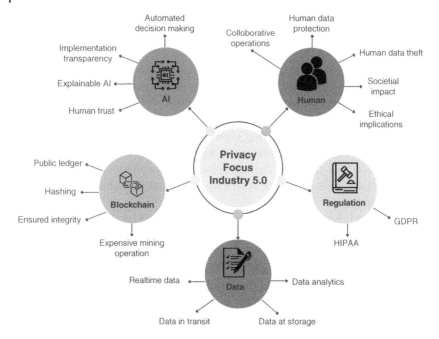

Figure 8.3 Privacy focus on Industry 5.0.

References

1 F. Aslam, W. Aimin, M. Li, and K. Ur Rehman, "Innovation in the era of IoT and Industry 5.0: Absolute innovation management (AIM) framework," *Information*, vol. 11, no. 2, p. 124, 2020.

2 Y. Lu, "Industry 4.0: A survey on technologies, applications and open research issues," *Journal of Industrial Information Integration*, vol. 6, pp. 1–10, 2017.

3 S. Echchakoui and N. Barka, "Industry 4.0 and its impact in plastics industry: A literature review," *Journal of Industrial Information Integration*, vol. 20, p. 100172, 2020.

4 O. A. ElFar, C.-K. Chang, H. Y. Leong, A. P. Peter, K. W. Chew, and P. L. Show, "Prospects of Industry 5.0 in algae: Customization of production and new advance technology for clean bioenergy generation," *Energy Conversion and Management: X*, vol. 10, p. 100048, 2020.

5 M. Parimala, R. M. Swarna Priya, Q.-V. Pham, K. Dev, P. K. R. Maddikunta, T. R. Gadekallu, and T. Huynh-The, "Fusion of Federated Learning and Industrial Internet of Things: A Survey," *arXiv preprint arXiv:2101.00798*, 2021.

6 L. D. Xu, "Industry 4.0 – frontiers of fourth industrial revolution," *Systems Research and Behavioral Science*, vol. 37, no. 4, pp. 531–534, 2020.

7 L. D. Xu, "The contribution of systems science to Industry 4.0," *Systems Research and Behavioral Science*, vol. 37, no. 4, pp. 618–631, 2020.

8 L. Li, "China's manufacturing locus in 2025: With a comparison of "Made-in-China 2025" and "Industry 4.0"," *Technological Forecasting and Social Change*, vol. 135, pp. 66–74, 2018.

9 H. Lasi, P. Fettke, H.-G. Kemper, T. Feld, and M. Hoffmann, "Industry 4.0," *Business & Information Systems Engineering*, vol. 6, no. 4, pp. 239–242, 2014.

10 V. Priya, I. S. Thaseen, T. R. Gadekallu, M. K. Aboudaif, and E. A. Nasr, "Robust attack detection approach for IIoT using ensemble classifier," *Computers, Materials & Continua*, vol. 66, no. 3, pp. 2457–2470, 2021.

11 I. de la Pe na Zarzuelo, M. J. F. Soeane, and B. L. Bermúdez, "Industry 4.0 in the port and maritime industry: A literature review," *Journal of Industrial Information Integration*, vol. 20, p. 100173, 2020.

12 M. Azeem, A. Haleem, and M. Javaid, "Symbiotic relationship between machine learning and Industry 4.0: A review," *Journal of Industrial Integration and Management*, vol. 7, no. 03, p. 2130002, 2021.

13 C. Zhang and Y. Chen, "A review of research relevant to the emerging industry trends: Industry 4.0, IoT, blockchain, and business analytics," *Journal of Industrial Integration and Management*, vol. 5, no. 01, pp. 165–180, 2020.

14 S. Nahavandi, "Industry 5.0 – a human-centric solution," *Sustainability*, vol. 11, no. 16, p. 4371, 2019.

15 K. A. Demir, G. Döven, and B. Sezen, "Industry 5.0 and human–robot co-working," *Procedia Computer Science*, vol. 158, pp. 688–695, 2019.

16 M. Sanchez, E. Exposito, and J. Aguilar, "Autonomic computing in manufacturing process coordination in Industry 4.0 context," *Journal of Industrial Information Integration*, vol. 19, p. 100159, 2020.

17 A. Majeed, Y. Zhang, S. Ren, J. Lv, T. Peng, S. Waqar, and E. Yin, "A big data-driven framework for sustainable and smart additive manufacturing," *Robotics and Computer-Integrated Manufacturing*, vol. 67, p. 102026, 2020.

18 A. Haleem and M. Javaid, "Additive manufacturing applications in Industry 4.0: A review," *Journal of Industrial Integration and Management*, vol. 4, no. 04, p. 1930001, 2019.

19 T. Zonta, C. A. da Costa, R. da Rosa Righi, M. J. de Lima, E. S. da Trindade, and G. P. Li, "Predictive maintenance in the Industry 4.0: A systematic literature review," *Computers & Industrial Engineering*, vol. 150, p. 106889, 2020.

20 M. Compare, P. Baraldi, and E. Zio, "Challenges to IoT-enabled predictive maintenance for Industry 4.0," *IEEE Internet of Things Journal*, vol. 7, no. 5, pp. 4585–4597, 2019.

21 H. Yetış and M. Karaköse, "Optimization of mass customization process using quantum-inspired evolutionary algorithm in Industry 4.0," in *2020 IEEE International Symposium on Systems Engineering (ISSE)*. IEEE, 2020, pp. 1–5.

22 Y. Lu, "Cyber physical system (CPS)-based Industry 4.0: A survey," *Journal of Industrial Integration and Management*, vol. 2, no. 03, p. 1750014, 2017.

23 L. D. Xu and L. Duan, "Big data for cyber physical systems in Industry 4.0: A survey," *Enterprise Information Systems*, vol. 13, no. 2, pp. 148–169, 2019.

24 C. S. de Oliveira, C. Sanin, and E. Szczerbicki, "Visual content representation and retrieval for cognitive cyber physical systems," *Procedia Computer Science*, vol. 159, pp. 2249–2257, 2019.

25 O. A. Topal, M. O. Demir, Z. Liang, A. E. Pusane, G. Dartmann, G. Ascheid, and G. K. Kur, "A physical layer security framework for cognitive cyber-physical systems," *IEEE Wireless Communications*, vol. 27, no. 4, pp. 32–39, 2020.

26 S. Wang, H. Wang, J. Li, H. Wang, J. Chaudhry, M. Alazab, and H. Song, "A fast CP-ABE system for cyber-physical security and privacy in mobile healthcare network," *IEEE Transactions on Industry Applications*, vol. 56, no. 4, pp. 4467–4477, 2020.

27 X. Chen, M. A. Eder, and A. Shihavuddin, "A concept for human-cyber-physical systems of future wind turbines towards Industry 5.0," 2020. [Online]. Available: 10.36227/techrxiv.13106108.v1.

28 P. Porambage, G. Gür, D. P. M. Osorio, M. Liyanage, A. Gurtov, and M. Ylianttila, "The roadmap to 6G security and privacy," *IEEE Open Journal of the Communications Society*, vol. 2, pp. 1094–1122, 2021.

29 M. Liyanage, A. Braeken, P. Kumar, and M. Ylianttila, *IoT Security: Advances in Authentication*. John Wiley & Sons, 2020.

30 X. Xu, "From cloud computing to cloud manufacturing," *Robotics and Computer-Integrated Manufacturing*, vol. 28, no. 1, pp. 75–86, 2012.

31 N. V. Korneev, "Intelligent complex security management system FEC for the Industry 5.0," *IOP Conference Series: Materials Science and Engineering*, vol. 950, p. 012016, 2020.

32 Z. Kotianová, "Aspects of safety and security in Industry 4.0," *Industry 4.0*, vol. 4, no. 6, pp. 319–321, 2019.

33 P. Porambage, G. Gür, D. P. M. Osorio, M. Liyanage, and M. Ylianttila, "6G security challenges and potential solutions," in *2021 Joint European Conference on Networks and Communications (EuCNC) and 6G Summit*. IEEE, 2021, pp. 1–6.

34 L. Bilge and T. Dumitraş, "Before we knew it: An empirical study of zero-day attacks in the real world," in *Proceedings of the 2012 ACM Conference on Computer and Communications Security*, 2012, pp. 833–844.

35 Y. Siriwardhana, P. Porambage, M. Liyanage, and M. Ylianttila, "*AI and 6G security: Opportunities and challenges,*" in *Proceedings of the IEEE Joint European Conference on Networks and Communications & 6G Summit (EuCNC/6G Summit)*, 2021, pp. 1–6.

36 C. Cheng, R. Lu, A. Petzoldt, and T. Takagi, "Securing the Internet of Things in a quantum world," *IEEE Communications Magazine*, vol. 55, no. 2, pp. 116–120, 2017.

37 C. Esposito, A. Castiglione, B. Martini, and K.-K. R. Choo, "Cloud manufacturing: Security, privacy, and forensic concerns," *IEEE Cloud Computing*, vol. 3, no. 4, pp. 16–22, 2016.

38 L. J. Wells, J. A. Camelio, C. B. Williams, and J. White, "Cyber-physical security challenges in manufacturing systems," *Manufacturing Letters*, vol. 2, no. 2, pp. 74–77, 2014.

39 B. C. Stahl, "Ethical issues of AI," in *Artificial Intelligence for a Better Future*, SpringerBriefs in Research and Innovation Governance. Cham: Springer, 2021, pp. 35–53.

40 L. Vesnic-Alujevic, S. Nascimento, and A. Polvora, "Societal and ethical impacts of artificial intelligence: Critical notes on european policy frameworks," *Telecommunications Policy*, vol. 44, no. 6, p. 101961, 2020.

41 E. Pauwels, "The new geopolitics of converging risks: The UN and prevention in the era of AI," *United Nations University-Centre for Policy Research*. https://i.unu.edu/media/cpr.unu.edu/attachment/3472/PauwelsAIGeopolitics.pdf, 2019.

42 R. Guidotti, A. Monreale, S. Ruggieri, F. Turini, F. Giannotti, and D. Pedreschi, "A survey of methods for explaining black box models," *ACM Computing Surveys (CSUR)*, vol. 51, no. 5, pp. 1–42, 2018.

43 S. Nahavandi, "Industry 5.0-A human-centric solution," *Sustainability*, vol. 11, no. 16, p. 4371, 2019.

44 A. Adadi and M. Berrada, "Peeking inside the black-box: A survey on explainable artificial intelligence (XAI)," *IEEE Access*, vol. 6, pp. 52 138–52 160, 2018.

45 M. E. Peck and S. K. Moore, "The blossoming of the blockchain," *IEEE Spectrum*, vol. 54, no. 10, pp. 24–25, 2017.

46 D. Tosh, O. Galindo, V. Kreinovich, and O. Kosheleva, "Towards security of cyber-physical systems using quantum computing algorithms," in *2020 IEEE 15th International Conference of System of Systems Engineering (SoSE)*. IEEE, 2020, pp. 313–320.

47 P. Voigt and A. Von dem Bussche, "The EU general data protection regulation (GDPF)," in *A Practical Guide*, 1st Ed., Cham: Springer International Publishing, 2017.

Part III

Security in 6G Use Cases

9

Metaverse Security in 6G

This chapter has the focus on the Metaverse, how it will drive toward a reality with the help of 6G, and the security and privacy concerns over that. After reading this chapter, you should be able to:

- Gain an overview of Metaverse and its relevance with 6G networks.
- Identify the major security considerations on Metaverse.

9.1 Overview of Metaverse

Few years ago, we watched in movies and science fictions how Metaverse is incorporated with the human life activities. However, the incredible advancements of virtual and augmented realities and the other renowned technologies such as artificial intelligence and blockchain, Metaverse is actually stepping ahead from science fiction to reality in the coming years. With the very optimistic forecasts of 6G, many scholars and scientists expect that 6G will be an important support for Metaverse. After the revolution of World Wide Web and the mobile Internet during the last few decades, Metaverse will be the next evolutionary paradigm in the coming years in which the users can experience a fully digitalized life virtually. Throughout this journey, information is the key resource in the Metaverse which maintains the coordination and interaction between physical, human, and digital worlds (Figure 9.1). In a nut shell, Metaverse embraces a social element with a strong virtual narrative with the acceleration through novel technologies such as Web 3.0, AI, XR, and 6G.

Security and Privacy Vision in 6G: A Comprehensive Guide, First Edition.
Pawani Porambage and Madhusanka Liyanage.

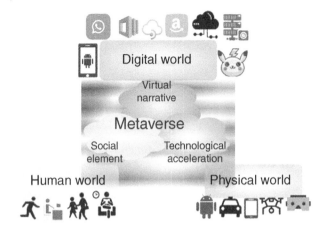

Figure 9.1 The convergence of digital, physical, and human worlds toward a Metaverse. Source: Meta; Amazon Web Services, Inc.; Microsoft Corporation; Amazon.com, Inc.; Google LLC.

9.2 What Is Metaverse?

The term "Metaverse" is coined with transcendence (meta) and universe (verse) terms which represent the next-generation immersive interaction between digital and physical worlds through digital avatars of users by augmented reality (AR)/virtual reality (VR) equipment and the integration of variety of technologies. This may also turn into the next-generation Internet that embraces a social element. It will be accelerated through many novel technologies such as digital twins, networking, artificial intelligence (AI), machine learning (ML), Web 3.0, AR/VR/XR, blockchain, and nonfungible token (NFT). A Metaverse platform may include multiple segments such as networking infrastructure, human interfaces, decentralized technologies, spatial computing techniques, creator economy, discovery tools, and the types of experience(Figure 9.2).

With the advancement of wearable devices and their adjacent sensors and actuators and the advanced learning algorithms, they provide the projections of human interactions and movements from the physical world to the digital world. This allows the human users to fully control their digital avatars which are interacting with the other avatars and objects in the Metaverse easily. In order to make this Metaverse platform a reality, the 6G networking infrastructure together with many ML algorithms are expected to support (and not limited to) traffic off-loading, security attack prevention, fault detection, channel estimation, automatic resource allocation, and efficient spectrum monitoring.

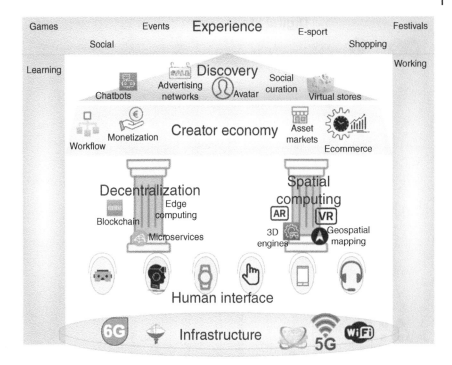

Figure 9.2 The seven segments of a Metaverse platform.

9.2.1 Metaverse Architecture

There are two main standards related to Metaverse: ISO/IEC 23005 and IEEE 2888. These explain how the sensory data from the real world are converted into the virtual world for its object characteristics. The general architecture for Metaverse is proposed as a convergence of digital, physical, and human worlds, with the characteristics such as 3D immersive virtual shared space, hyper spatiotemporality, and self-sustainability (Figure 9.3). Metaverse is a human-centric paradigm which allows humans to generate their digital avatars in a synthesized digital world and to use their digital identities through smart devices to play, work, socialize, or communicate. Physical infrastructure provides the Metaverse to support many services such as sensor data perception, transmission, data storage, caching, and processing. They facilitate the human users to interact with Metaverse engines through interactivity, AI, digital twin, and blockchain technologies. Then the digital worlds can be composed as interconnected virtual worlds which include users as digital avatars and different types of virtual goods/services.

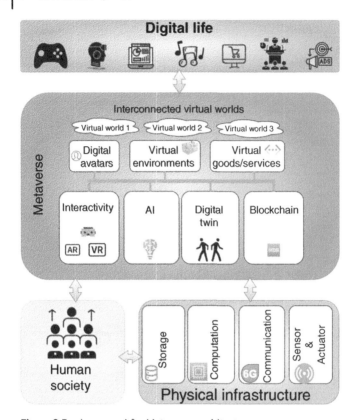

Figure 9.3 A proposal for Metaverse architecture.

9.2.2 Key Characteristics of Metaverse

When Metaverse is bringing together virtual and digital worlds to the economic interactions of human world through the physical infrastructure, it exhibits several key features with the perspectives of immersiveness, hyperspatiotemporality, sustainability, interoperability, scalability, and heterogeneity (Figure 9.4).

Immersiveness: This is the virtual space generated by the computers and digital appliances with a nearly realistic atmosphere where users can be mentally and emotionally immersed. It allows the humans to interact with the environment through their bodies, senses, and expressions.

Hyper spatiotemporality: The users may no longer feel the finiteness of the real world and the irreversibility limitations of time. In a Metaverse, the users can freely travel across space and experience the difference of spatiotemporal dimensions.

Figure 9.4 Key Characteristics of Metaverse.

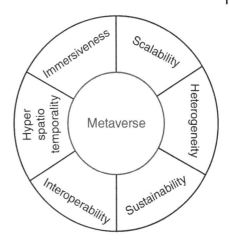

Sustainability: The Metaverse is expected to maintain the persistence in a decentralized networking paradigm to allow the coexistence of multiple underlying technologies to collaborate with different types of stakeholders.

Interoperability: The users can freely move across virtual worlds or the submetaverses without being interrupted in the immersive experience. Digital assets are expected to reconstruct the virtual worlds in such a way to facilitate the interchangeability across different platforms.

Scalability: Metaverse should have the capacity to withstand with the increment and complexity of concurrent users or avatars and maintain their interactions in terms of scope, type, and range.

Heterogeneity: This stands for the heterogeneous virtual spaces, physical devices, communication modes, and data types which will contribute to Metaverse to accommodate a diversified human psychology.

9.2.3 Role of 6G in Metaverse

It is hard to forecast that a fully functional Metaverse may exist by the time of 6G networks. However, 6G will be an integral support for Metaverse to move forward with the seamless connectivity between human, physical, and digital worlds which lead to a holographic society. Metaverse may facilitate a broader variety of human activities including games, social spaces, conferences, and social engagement with a strong virtual narrative. 5G communication technologies are already providing networking infrastructure for the operations in Metaverse. However, Metaverse still requires high throughput, high reliability and extremely low latency to provide a real-time experience. The challenges are yet to be addressed from 5G to beyond-5G with respect to spectrum allocation, network densification, increased capacity, and the co-existence of heterogeneous networks. In addition to the performance advancements, 6G is also expected to encourage Metaverse toward more sustainable, reliable, and human-centric paradigm. Certainly, the sustainability is expected from the lowest energy consumption and the highest resource efficiency.

Among the multiple core technologies which accelerate Metaverse, AI plays an integral role by bringing intelligence for speech recognition and natural language processing to understand the user intent and enabling a fully automated end-to-end process. Web 3.0 with the contribution of blockchain and dApps construct completely distributed applications for numerous purposes including browsing (Brave), storage (IPFS), video and audio calls (Experty), operating systems (EOS), social networks (Steemit), messaging (Status), and remote jobs (Ethlance). This may even bring a new OSI-stack with three layers: a blockchain-based infrastructure layer for the web3.0 ecosystem; token layer to build coins out of basic blockchain with the ways of paying with wallets, assets, NFTs, and market places which are interoperable with other layers; dApp layer is for the decentralized Metaverse applications which can use assets in any platform. However, the biggest challenge in such a distribute eco system is the energy. In addition to that Metaverse is getting the touch of digital twin which is accomplished with 3D simulation and 3D modeling as well as the interactivity for immersive experience with AR lenses, VR, mixed reality (MR), extended reality (XR), Brain–Computer Interaction, and holographic. Especially as the gateways to Metaverse, these technologies need to provide an extremely high fidelity in a controlled manner. Since traffic offloading is significant in such technologies, with 6G integrated AR, it is expected to experience the extreme offload with the down-links and up-links in Mbps range (latency in ms). For instance, a more immersive engagement is expected with 12K 3D (11520 × 64800) resolution, 120 FPS frame rate, and H.266.

9.3 Security Threats in Metaverse

By nature, Metaverse needs to always interact with the users and avatars where authentication is needed to prevent unauthorized access. The identities of the users or avatars can be stolen illegally, impersonated, or encountered with cross-domain authentication issues. Adversaries may create forge avatar identities with AI bots to behave identical to human users. Unauthorized data access can be performed by such malicious entities to misuse the data coming from users and avatars. The data generated by the wearable devices and users/avatars and collected for later usage can be subjected to different types of attacks such as data-tampering attacks, false data injection attacks, provenance tracing, or intellectual property violations. Moreover, managing (in the process of data collection and data transmission) new types of sensitive data in Metaverse and threats to data quality will create poor user experiences as well as the privacy threats. In particular, privacy leakages may occur in data collection, data transmission, data processing, or data storage. The compromised end devices may also reveal privacy sensitive data of the users and reveal their seamless connectivity to physical, human, and virtual worlds. In addition to this, the conventional threats to the wireless communication networks may be also effective to Metaverse. For instance, the most common attacks such as DDoS and Sybil may occur in Metaverse. When one or a group of compromised avatars are sending repeated connectivity requests, the resources may not be available for the legal avatars of the legal users. In a sybil attack, an attacker may generate sybil avatars to compromise Metaverse services by taking overs the control of digital voting services. Some further threat types in Metaverse may also include trust violations, threats to economic fairness, threats to personal, infrastructure and social safety violations, and governance-related threats.

9.4 Countermeasures for Metaverse Security Threats

As security countermeasures to Metaverse authentication, efficient identity management plays a vital role as a basis for the user–avatar interaction and service provisioning. The digital identities can be managed by a centralized entity, multiple institutes or federations, or as self-controlled. In Metaverse, the identity management needs be resilient to node damage, scalable to cope up with massive number of users/avatars, and interoperable among multiple sub-metaverse. Key management is necessary to establish secure communication links between wearable devices which are integrated with physical world devices. Identity authentication for devices and cross-domains are also important to assure device and user interaction in Metaverse. In order to eliminate unauthorized exposures, fine-grained high precise access control mechanisms and usage audit schemes can be introduced. In Metaverse data management, it is important to consider the security countermeasures for improving data reliability, data quality, and provenance (data sharing). The enhanced privacy preservation techniques, confidentiality protection mechanisms, digital foot print protection, and personalized privacy preserving are also getting higher priority in privacy countermeasures in Metaverse. Similar to the need of having personal borders in the real human world, the avatars in the virtual world require their own personal space. Both global and local situational awareness in Metaverse will be an effective tool for monitoring security threats and their early predictions. In addition to the technical security countermeasures in Metaverse, the economic fairness should be introduced to Metaverse for manipulation prevention with game theory, blockchain, auctioning, and ML like tools.

9.5 New Trends in Metaverse Security

Metaverse itself creates many novel research directions in many different aspects toward its concept to the actual realization. Metaverse security and privacy are equally getting higher attention from the research community to define new threat models, security vulnerabilities, and developing security countermeasures. Unlike the later-added security features in the current systems, Metaverse need an endogenous security which offers built-in security or security-by-design for the complex systems. Advanced security mechanisms like quantum key distribution and quantum resistant cryptography will be promising research approaches to achieve quantum safe Metaverse. When Metaverse collects extremely high volumes of data to bring users a 3D immersive experience, they require enormous interactions with cloud, edge, and end for caching and data processing. More work needs to be done in this direction to provide uninterrupted services with cloud-edge-end orchestrated security in Metaverse. In addition to these, there are more research directions with Metaverse security in terms of improving energy efficiency toward green Metaverse, content/human-centric nature of Metaverse and achieving cross-chain inter-operable and regulatory Metaverse. All in all, Metaverse will be an evolutionary paradigm in the next generation of human-centric cyber-physical continuum, and it will create many broader research topics which are still to be investigated by security researchers to make it more safe and trustworthy.

10

Society 5.0 and Security*

* By Onel L. Alcaraz López, Center for Wireless Communications, University of Oulu, Finland, and Pawani Porambage (Editor)

This chapter has the focus on the concept of Society 5.0, its technological evolution, and security considerations.

- Gain an overview of industry and society evolution toward Society 5.0 and the role of 6G technologies.
- Identify the major security considerations of Society 5.0.

10.1 Industry and Society Evolution

Our society is becoming increasingly digitized, hyper-connected, and globally data-driven [1]. Academy, industry, and governmental and regulatory bodies are continuously coordinating efforts to meet the most ambitious and challenging societal needs, including economical, well-being, environmental, and sustainability aspects. Indeed, many radical and transformative changes are taking place in a short period of time. This can be easily understood by analyzing the evolution of society and industry, and its acceleration in the last centuries/decades as depicted in Figure 10.1.

Societies are determined by context, including culture, economy, politics, communication, and many other social levels. During the transition from one society to another, changes occur at all these levels. Nomads, for instance, led a "first society," where women gathered food and men hunted in harmonious coexistence with nature. That was the panorama for a very long time. The "second society" was only triggered many thousand years later by the development of the agriculture, when humans became nonnomad and self-sufficient, and organization and nation-building efforts gained strength. The "third society" did not come until

Security and Privacy Vision in 6G: A Comprehensive Guide, First Edition.
Pawani Porambage and Madhusanka Liyanage.

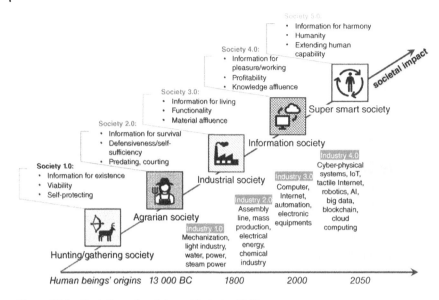

Figure 10.1 Evolution of society and industry [2, 3].

several thousand years later after the Industrial revolution in the eighteenth century, when the advent of factories and machines replacing manual labor led to production massification and currency exchange. During such an epoch, social class structures gained more distinction, people started fighting for and gaining social/economic/human rights, and transportation evolution propitiated rapid cultural and economical interactions. Just few hundred years later, the "fourth society" was triggered by the rise of the modern information and communication technologies (ICTs) and the era of the Internet. Our society, Society 4.0, is an information society that realizes increased added-value by connecting intangible assets as information networks [2]. The massive access to information and virtual interaction among individuals influence social, cultural, economic activities at numerous levels, which reflect in every aspect of our daily lives. Note that technological and industrial changes are the ones potentially triggering society evolution, as they have the potential to influence many social levels in the short and long term [4].

We are currently immersed in a new industrial revolution, known as Industry 4.0, with the potential to lead to a new society: the "fifth society" or simply Society 5.0. We deepen into this in the following:

10.1.1 Industry 4.0

Industry 4.0 paradigm constitutes the new evolutionary organization and management level of the entire value chain process involved in the manufacturing

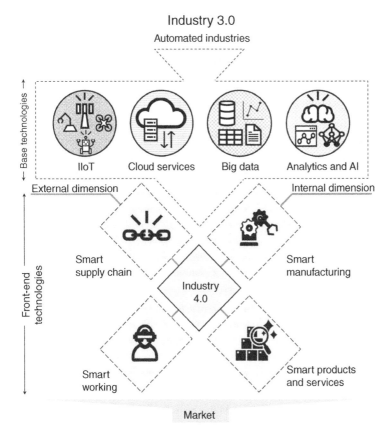

Figure 10.2 Theoretical framework of Industry 4.0 driving digital transformation. Source: Adapted from [5, 6].

industry. It revolves around cyber–physical industrial systems, way beyond advanced automation/production. A conceptual Industry 4.0 framework is illustrated in Figure 10.2. Noteworthy, although Industry 4.0 is commonly only associated with smart manufacturing in the literature, there are other three key components: smart supply chain (new ways raw materials and products are delivered), smart product (new ways products are offered), and smart working (new ways workers perform their activities), with inherent cross-synergies. All these are referred to as "front-end technologies" as they are concerned with operational and market needs, thus, they have an end-application purpose for the corresponding value chain [5].

Meanwhile, a set of "base technologies" supports the previous components, and their enablers (see Table 10.1), by providing connectivity and intelligence, while

Table 10.1 Enablers of front-end technologies in Industry 4.0 [5].

Front-end technical	Category	Enablers
Smart manufacturing	Vertical integration	• Sensors, actuators, and programmable logic controllers (PLC)
		• Supervisory control and data acquisition (SCADA)
		• Enterprise resource planning (ERP)
		• Machine-type communication (MTC)
	Virtualization	• Virtual commissioning
		• Simulation of processes (e.g. digital manufacturing)
		• Artificial intelligence (AI) for predictive maintenance
		• AI for planning of production
	Automation	• MTC
		• Robots (e.g. industrial robots, autonomous guided vehicles (AGV), or similar)
		• Automatic nonconformities identification in production
	Traceability	• Identification and traceability of raw materials
		• Identification and traceability of final products
	Flexibility	• Additive manufacturing
		• Flexible and autonomous lines
	Energy management	• Energy efficiency monitoring and improving system
Smart product	—	• Product's connectivity, monitoring, control, optimization, and autonomy
Smart supply chain	—	• Digital platforms with suppliers, customers, and other units
Smart working	—	• Remote monitoring/operation of production
		• Augmented reality for maintenance
		• Virtual reality for workers training
		• Augmented and virtual reality for product development
		• Collaborative robots

enabling their integration into a complete integrated manufacturing system. They are

- Industrial Internet of Things (IIoT) is an IoT that focuses on providing autonomous, robust, secure, dependable, and potentially wireless, connectivity in industrial environments. IIoT connects objects, machines, and products by exploiting a variety of networking technologies, human–machine interactions, sensors and actuators, and applications [1, 7].
- Cloud services include storing devices' data in Internet servers, computing, and data retrieving, potentially through remote access [8]. They enable network access to a shared pool of computing resources, thus, the possibility to run heavier applications at the network edge and facilitate the operation and coordination among the devices as they do not need to be physically near to share information and coordinate activities [5].
- Big data originate from the massive data gathering from systems and objects, e.g. sensor readings [5, 9]. Indeed, the exploitation of cloud services by massive IIoT deployments leads to the collection and processing of huge amounts of data that can generate relevant information, e.g. for the generation of digital twins of a factory/system.
- Analytics refers to the data processing and the corresponding information/ knowledge extraction, e.g. via data mining and machine learning (ML) tools and outputs. Remarkably, analytics can enable advanced predictive capacity and intelligent decision-making which are key for real-time responses/reactions/ optimization, e.g. identifying and mitigating events that can affect a given system in the value chain.

In a nutshell, IIoT connects objects and systems in industrial environments, while cloud services provide them access to nonlocal information and services. Meanwhile, big data and analytics refer to the corresponding massive data gathering and processing, which enables intelligent monitoring, scheduling, planning, and control of industrial processes. Noteworthy, Industry 4.0 envisions [10, 11]:

1. Vertical integration of ICT systems in production and automation engineering, and horizontal integration across the value chain.
2. Consistency of engineering over the complete life cycle.
3. Customization of products through small lots or even "lot size one".[1]
4. New social infrastructures for work.

The above inevitably includes and/or leads to a thorough end-to-end digitization of all physical assets together with the integration of the value chain partners into digital ecosystems.

---

1 "Lot size one" refers to the customization of a single product for a single customer. The product doesn't exist until the consumer defines what it should be.

Table 10.2 Benefits of Industry 4.0 in terms of sustainable development.

Economical	Social	Environmental
• Higher quality and flexible production	• New work places	• Energy consumption reduction
• Ad-hoc and rapid reaction to changes in demand, stock level, errors	• Safer work conditions	• Reduction of transportation and travel effort
• Innovative image of the company	• New level of customer satisfaction	• Saving of natural resources
• Driver for further innovation (including new business models)	• Work-life balance	• Reduction of overproduction and wastes
• Increased productivity	• Personalized products	
• Increased net revenue	• More nonroutine jobs	

All this leads to many (direct and indirect) benefits in terms of sustainable development,[2] which are summarized in Table 10.2. Specifically, from the business perspective, Industry 4.0 enhances industrial dependability, operational efficiency, and productivity, which may potentially decrease the products' time-to-market (i.e. by reducing unplanned downtime) and promote unprecedented levels of economic growth and productivity efficiency [13].

10.1.2 Society 5.0

The main focus and concern of Industry 4.0 lies in the economical growth aspect, while (mostly) indirectly empowering humans and society (see Table 10.2). In addition, and also as a consequence. Industry 4.0 opens the way for an evolutionary human-centered philosophy, the so-called "Society 5.0." Notice that the term "Society 5.0" also refers to an initiative of the Fifth Science and Technology Basic Plan taken by the government of Japan [2], but here we consider it more broadly and global, as the society triggered by Industry 4.0 (see Figure 10.1).

Society 5.0 focuses on the human well-being in all the dimensions. Specifically, Society 5.0 aims to not simply provide the minimum services needed for individuals' survival but to make life more meaningful and enjoyable through the holistic integration of the physical world and the cyberworld. The relationship

---

2 Notice that a well-accepted notion of sustainability is that of an assemblage of mutual interests between social equity, viable economics, and a healthy environment [12]. Refer also to our discussions around Figure 10.3 in Section 10.1.2.

Figure 10.3 UN's SDGs for a sustainable (green, profitable, and fair) society, the support for Society 5.0.

with the sustainable society vision of the United Nations (UN), and specially with the corresponding sustainable development goals (SDGs) set for 2030 [14] as illustrated in Figure 10.3, is somewhat intuitive: Society 5.0 constitutes a highly sustainable and harmonious society. Notice that the SDGs constitute a universal call for the ending of poverty, the protection of the planet, and to guarantee the enjoyment of peace and prosperity by all people [15, 16].

Both Industry 4.0 and Society 5.0 rely on and promote increasingly sophisticated cyber–physical systems (CPSs). Basically, in a CPS, a mechanism or an object in the physical space is represented in the cyberspace and monitored/controlled according to the corresponding application via computer-based algorithms as depicted in Figure 10.4. Therefore, CSPs rely on embedded, decentralized, real-time computations and interactions, where physical and software components are deeply

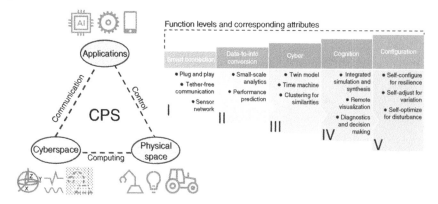

Figure 10.4 CPS: structural representation and function levels and corresponding attributes.

intertwined [17, 18]. In addition, all the components that are able to operate on different spatial and temporal scales can exhibit multiple and distinct behavioral modalities and interact with each other in ways that change with context. In the case of Society 5.0, the notion of cyber–physical–social system (CPSS) is more accurate, as it refers to a system where human beings, social robots, AIs, and/or other entities, are functionally integrated into a CPS at the social, cognitive, and physical levels [19]. Notice that the members of a CPSS may engage in *cyber–physical–social behaviors* within cyber–physical spaces and may give rise to *cyber–physical–social networks*, whose topologies follow the members' social connections [20, 21].

CPSSs will invade every aspect of our lives in Society 5.0. Notice that there are numerous sectors that, or can, benefit from CPS designs in our future societies. Basically, every society sector can incorporate CPSs for further enhancements. We discuss briefly a few examples as follows:

Transportation: Modern transportation systems are already CPSs. Indeed, modern vehicles are equipped with enhanced displays, provide numerous forms of data and entertainment, and their motion and energy consumption/charging are manageable and/or autonomously optimized. Also, road monitoring and control are becoming increasingly intelligent and autonomous. Soon, we'll travel in driverless cars that communicate securely with each other on smart roads and in planes that coordinate to reduce delays. The work in the coming years will mostly focus on improving the level of driving automation (from level 3 to level 4 or level 5), which requires deploying a huge amount of sensors and edge processing systems [1]. These systems require real-time wireless connectivity to its sensor/actuator nodes and AI-enabled external processing units. Additionally, the three-dimensional (3D) connectivity landscape will be strengthened with the increasing number of applications relying on unmanned and flying vehicles. An interesting example is that of drones checking infrastructures for damage and providing connectivity to disaster zones.

Health care and medicine: There are many CPS application fronts in the health-care and medicine systems, e.g. smart operation room, smart and autonomous medical devices, and remote medical assistance. One interesting trend lies in enabling real-time visualization of patient data along with seamless capability of plugging in new sensors with interoperable medical devices [22]. More futuristically, sensors in the home will detect changing health conditions; new operating systems will make personalized medical devices interoperable; and robotic surgery and bionic limbs will help heal and restore movement to the injured and disabled and even augment human abilities. All in all, CPSs in the health care and medicine systems enable health awareness and consequently help people detect and address diseases that may arise in their daily life.

Smart homes and buildings: Here, common CPSs are related to electronic equipment and heating ventilation and air conditioning (HVAC) control, healthcare, and energy management [23]. In smart homes/buildings, the sensors and actuators are configured such that they can be remotely controlled through the Internet. A related concept proposed in [24] is that of a smart community, where the individual homes/buildings are modeled as multifunctional sensors, and whenever necessary, automatic or human-controlled physical feedback is given to improve community safety, health care quality, and home security. Another, more futuristic, CPS example is that of the "consumer robots," which constitute home robots performing domestic chores, e.g. tasks that elderly people cannot perform any more [25], for which holistic interactions with the smart homes/buildings and the humans are needed. Finally, CPSs can be also designed to provide rapid access to information and autonomous interaction with the grid and to enable the homes/buildings to participate in the utility markets.

Immersive telepresence: This service aims to blur the boundary between the digital and physical worlds [25]. The concept is rapidly gaining attention due to the need of performing extreme and immersive remote collaborative work, shopping, gaming, and other activities, and recent technological advances on brain–computer interactions and multisensory augmented/virtual/mixed reality (XR) together with a massive proliferation of cost-effective miniaturized smart wearable and implant sensors with Quality of Service (QoS) guarantees [1]. Specifically, the user will experience being in a different place of his/her actual location, with enhanced rendering of the interactions (e.g. gesture, touching the objects). A fully immersive experience may be developed by the involvement of the senses of taste and smell, which may bring some new services, e.g. in food and texture industries [26].

Noteworthy, since humans are a key component in all these sectors, as they are mostly influenced by human behavior and dynamics, these CPSs will naturally evolve to become CPSSs. The main challenge lies in how to accurately describe the relationships between different forms of data/information and the human behavior, together with the corresponding risk quantification.

10.2 Technical Enablers and Challenges

Notice that Society 5.0 rests on ICT infrastructures, i.e. devices, networks, cloud computing, and data centers. Therefore, realizing Society 5.0 necessarily requires pushing the performance limits of such ICT infrastructures far beyond that we can imagine nowadays. For this, there is a vast set of technologies and technological trends rapidly gaining the attention of academy and industry communities in the last years. In this section, we discuss some of such technical enablers and associated challenges.

10.2.1 Dependable Wireless Connectivity

Dependability is the ability to provide services that can defensibly be trusted within a given time-period, being availability, reliability, safety, integrity, and maintainability, some key dependability attributes [27, 28]. Therefore, enabling dependable wireless connectivity is essential for a thorough and holistic digitization of the society at large, and as such, it is being targeted by the 6G research worldwide [1, 29, 30].

Next, we briefly discuss some of the relevant technological enablers for realizing high-performance wireless connectivity as illustrated in Figure 10.5. Notice that the latter can be directly translated into high-performance availability/reliability. Safety and integrity issues receive dedicated attention in Chapter 20 on physical layer security (PLS).

10.2.1.1 New Spectrum and Extreme Massive MIMO
The exploitation of new spectrum bands opens the path to support new and challenging services.[3] Indeed, 5G cellular systems already leverage centimeter and millimeter frequency bands for ultra-wideband communication in addition to the traditional sub-6 GHz frequency bands, which are already crowded. Since global leaders are becoming fully aware of the relevance of advanced mobile networks

---

3 Low-frequency bands are often preferable in terms of spectral efficiency, reliability, mobility support, and connectivity density, while high-frequency bands are often preferable in terms of peak data rates and latency [31].

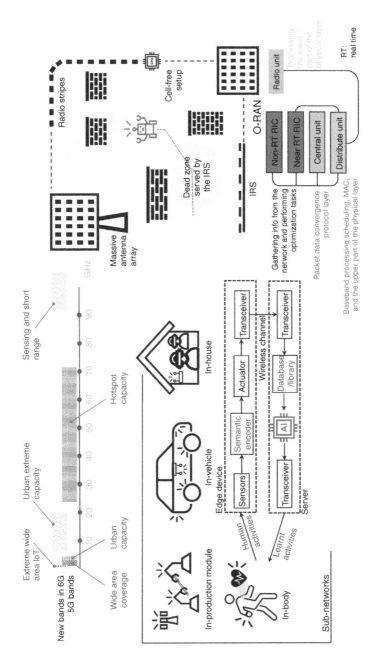

Figure 10.5 Selected technological enablers for realizing high-performance wireless connectivity.

for powering national economic growth, there is plenty of incentive to clear new spectrum for mobile use, especially between 7 and 20 GHz [32]. In the coming years, the exploitation of the spectrum will also expand to sub-THz and visible light regions. By moving into higher-frequency bands, denser antenna arrays can be supported and significant capacity increases are achievable by sending many more streams of data simultaneously. However, further technology evolution, especially in terms of new, scalable, and low-power radio-frequency (RF) and digital front-ends, more sophisticated beamforming algorithms, and high-capacity fronthaul and baseband processing, is required to make these high-performance extreme massive multiple-input multiple-output (MIMO) arrays possible.

10.2.1.2 In-X Subnetworks

Wireless networks cannot currently compete with wired solutions in terms of reliability and QoS support. Hence, wired-connectivity solutions, which are in general inflexible and invasive, are nowadays still necessary/used for supporting life-critical services. This panorama may change soon with the so-called "in-X subnetworks" in 6G research. In-X subnetworks are highly specialized radio cells installed within the entity where the application runs, e.g. in robots, production modules, vehicles, and even human bodies [32]. They can operate autonomously even when out of coverage of an overlay wide area network, though they can benefit when those connections are available. The challenges here are mostly related to enable huge diversity gains that efficiently combat detrimental effects such as fading, noise, and interference, and turn the wireless channels into virtual wires.

10.2.1.3 Semantic Communication

Going beyond the Shannon paradigm, semantic communication aims at the successful transmission of semantic information conveyed by the source rather than the accurate reception of each single symbol regardless of its meaning [33]. In semantic communication, only information relevant to the specific task at the receiver is transmitted, which leads to a truly intelligent system with significant reduction in data traffic. A very related concept is that of the human-to-machine semantic communication, which the transmission of messages that can be understood not only by humans but also by machines, such that they can have dialogue or the latter can assist or care for the former [34]. Recent advancements in deep learning (DL) and its applications, such as natural language processing (NLP), speech recognition, and computer vision provide significant insights on developing semantic communications. In the coming years, research may focus on quantifying performance gains and defining relevant metrics (e.g. semantic entropy and semantic channel capacity?), materializing semantic level joint source channel coding for many different types of sources, developing intelligent semantic communication systems with reasoning [35], investigating efficient resource allocation mechanisms in semantic-aware networks, and others.

10.2.2 Integrated Communication, Control, Computation, and Sensing

Communication, control, computing, and sensing are key processes in a CPS as discussed in Section 10.1.2. Hence, a thorough joint optimization of such processes enhances the performance of CPSs, which is instrumental for enabling Society 5.0. In the following, we overview the main related research trends in the coming years, namely communication and control co-design (CoCoCo), quantum computing and communication, and joint communication and sensing (JCAS), which are exemplified in Figure 10.6.

10.2.2.1 CoCoCo

Currently, real-time control is already critical for many industrial use cases. In the coming years, with the proliferation of CPSs,[4] the set of use cases will expand to other sectors and applications, e.g. remote surgery in health care (refer to Section 10.1.2 for few other examples). For this, traditional real-time control systems, which use wired connections between the sensors, actuators, and controllers, must become wireless such that systems can properly scale and become flexible and sustainable.

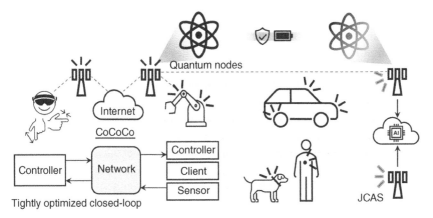

Figure 10.6 Illustration of integrated communication, control, computation, and sensing technologies.

---

4 Notice that communication and control constitute two out of the three edges in a CPS structural representation as shown in Figure 10.4.

Noteworthy, supporting real-time control via a wireless communication network where communication and control systems are designed and operate independently leads to substantial wireless resource consumption [36]. This is basically because control requirements for the wireless communication system are set based on the worst-case control performance. This motivates CoCoCo approaches, which treat the system as a whole to efficiently capture the tight interaction between communication and control processes. For this, wireless channel dynamics/impairments, which may lead to that received state observations and actuating commands may be outdated and distorted, must be thoroughly considered.

10.2.2.2 JCAS

Radio signals are currently transmitted with the main purpose of conveying information data (from transmitter(s) to receiver(s)). Another, so far secondary, application of radio signals is that of sensing. The positioning service is the most popular sensing enabled by radio signals nowadays, but in the future, this will be extended to much more sensing-related services [37]. The premise is that our awareness is no longer limited to our surroundings if networks are given the ability to sense, i.e. the network becomes our sixth sense [32].

Notice that wireless propagation channels constitute a source of situational information. Received signals contain coded information about the presence of an object, its type and shape, its relative location and velocity, and even its material properties. By exploiting this and given network's ubiquity, our senses become ubiquitous. In the future, this will allow us to directly interact with machines and robots remotely and sense what they sense, while directing their actions through simple hand gestures captured by the network.

The most immediate step moving forward lies in extending the current radio sensing capabilities, which are only enabled between specialized machines and applications, to every cell and node of the wireless communication infrastructure that surrounds us today. Indeed, the functions required for a tight integration of communication and sensing technologies, which is often referred to as JCAS in the literature, will be natively designed into 6G networks [32]. Noteworthy, sensor fusion, which combines network-sensing data with that of other sensors such as location tags or sensors ubiquitously employed in devices such as accelerometers, gyroscopes, and cameras, will be used to provide the complete sensing solution needs of the 6G era [38]. Since JCAS systems are in their infancy, lots of research in this area is expected in the next few years, especially in terms of (i) handling synchronization and processing issues in distributed sensing setups; (ii) optimizing multi-band JCAS systems; (iii) developing more accurate channel modeling for JCAS; (iv) thoroughly exploiting sensing for improving communication capabilities; and (v) investigating more suitable waveforms.

10.2.3 Intelligence Everywhere

Connectivity, processing, actuating, sensing, data, and "intelligence" everywhere are key vision elements in Society 5.0. Indeed, AI reasoning and inferring capabilities will embedded be everywhere in Society 5.0, thus, motivating the notion of "Internet of Intelligence," with connected people (and their senses), things, and intelligence, within the 6G research [39]. The "Internet of Intelligence," with connected people (and their senses), things, and intelligence, is a fundamental enabling paradigm of Society 5.0. Indeed, connectivity, processing, actuating, sensing, data, and "intelligence" everywhere have been identified as 6G key vision elements in [39]. The current trends indicate that machine/artificial reasoning and inferring capabilities will be embedded everywhere in our society.

10.2.4 Energy Harvesting and Transfer

Society 5.0 is a sustainable wireless society Consequently, academy and industry are putting significant efforts on developing wireless technologies that innovate to zero-energy/cost/emission. Specifically, energy harvesting (EH) and wireless power transfer (WPT) techniques are especially attractive as they allow to externally recharge batteries or avoid replacements, which may be not only costly but also impossible in hazardous environments, building structures, or the human body. Indeed, EH and WPT technologies are key technological sustainability enablers as they allow [40]: (i) wireless charging, which significantly simplifies the servicing and maintenance of wireless devices, while increasing their durability thanks to contact-free feature, and (ii) enhanced energy efficiency and network-wide reduction of emissions footprint, including the e-waste processing mitigation.

Notice that WPT and EH processes can be distinguished in the sense that the former is initiated at an energy transmitting (ET) node, while the latter occurs at an energy receiving node. Figure 10.7 illustrates the broad application horizon of these technologies. In the following, we briefly discuss these two technologies.

10.2.4.1 Energy Harvesting

There are many energy sources from which it is possible to harvest energy, e.g. light (artificial and natural), motion, heat, wind, sound, vibration, and electromagnetic radiation [41, 42]. Each energy source has unique features in terms of the energy provision scale, required hardware complexity, and form-factor of the corresponding EH circuit, and other specific advantages/disadvantages [40]. A specific device and network deployment must be equipped with the proper EH solution according to such energy sources' features. For instance, it is not possible for a typical base station (BS) to re-charge using RF signals that mostly carry very

Figure 10.7   Typical use cases of EH and WPT technologies.

low energy, while that may be possible in case of an ultra-low power sensor. Solar or wind energy sources seem more appealing for charging a BS, but not a tiny sensor which cannot be equipped with a typically large add-on material required for heat/light/wind EH.

In general, the ambient energy sources may be highly unreliable and/or scarce due to their variability and dependency on climate parameters and human factors. Therefore, accurate mathematical models to describe and predict the ambient energy availability and adopt proper resource scheduling/planning strategies, are crucial [43]. This, and the development of more compact, miniaturized, and efficient EH circuits, will receive significant research effort in the coming years.

10.2.4.2   Wireless Power Transfer

Temporal/geographical/environmental circumstances may limit the service guarantees of ambient EH solutions, making them inappropriate (at least as standalone) for many use cases with QoS requirements [40, 41]. In such cases, dedicated energy sources, i.e. ETs, may be required, thus motivating WPT solutions. Some examples of near-field WPT are those based on inductive and magnetic resonance phenomena, while laser power beaming and RF sources enable far-field WPT.

WPT solutions that are widely available in the market are based on nonradiative (far-field) techniques. They are designed for short range charging, have limited multiuser support, and still confine the maneuverability and mobility of the device(s) being charged, which limits their scalability, i.e. their ability to support the IoT massive growth trends [40]. Alternatively, far-field WPT methods are more flexible and facilitate scalability, which makes them attractive for many use cases enabling a truly wireless society. However, they are much less efficient and are subject to strict regulations to prevent harmful radiation to humans and animals [44]. Meanwhile, notice that WPT and EH will appear naturally combined with wireless information transmission (WIT),[5] thus, a holistic optimization of these processes is crucial. All these issues will receive much more research attention in the coming years.

10.3 Security in Society 5.0

Society 5.0 is foreseen as a cyber–physical integration toward a human-centric society. Information and intelligence are going to be the life of Society 5.0 and these may drive toward cyber resilience. This requires to strengthen the security policy deployment linked with the industrial policies to enhance measure for critical information infrastructure protection and harmonize with the international standards.

The security researchers should move forward from cyber security to cyber resilience which will be able to withstand and recover from the hazards and threats when a cyber-attack occurs. This includes the anticipation or the prediction of cyber-attacks before they occur and be prepare for such hazards.

When the interplay between the machines and humans are inevitable while collecting data from wearable devices and IoT to create the next-generation wisdom. Ransomware attacks can target these IoT devices and have their new attack forms such as R4IoT (Ransomware for IoT). Ransomware is a type of malware that take the control of such IoT devices and prohibit the users to access them until a ransom is paid.

The success and the value of society 5.0 are mostly depending on humans and data. When human centric data are becoming the most critical factor to the operations of society 5.0, the assurance of their privacy and the personal data protection are getting more and more important. In that case both information security and data protection are highly regarded. When analyzing the encrypted forms of data, anonymous analysis is applicable with the technology of searchable encryption will enable access to data without compromising privacy.

In terms of energy-harvesting aspects of 6G networks, again security plays an important role. The fixed and security solutions and related configurations

---

5 Either in a wireless-powered or EH-enabled communication network, or simultaneous wireless information and power transfer (SWIPT) setups [40, 41].

may no longer meet the diverse requirements of 6G networks. Mainly, this is because 6G needs higher QoS, energy efficiency, and message safety and at the same time, the network threats are getting advanced. Obviously, the security protection mechanisms bring extra-energy consumption, and when there are fixed high-level security configuration, the devices will quickly run out of battery. Therefore, fixed security configurations are always low-energy efficient. Unless there are cognitive security mechanisms which are adjusted automatically with the energy existence of the devices, this issue is not going to be solved. To go on par with the widely adopted energy-harvesting techniques and the key features of envisioned 6G networks, the future security configurations need be more cognitive and adaptive to the energy availability and network threats. Both aspects have to be considered before applying certain security protection mechanisms. The concept of introducing Green Security is one such initiative taken for this purpose.

References

1 N. Mahmood, O. López, O. Park, I. Moerman, K. Mikhaylov, E. Mercier, A. Munari, F. Clazzer, S. Böcker, and H. Bartz *(Eds.)*, "White paper on critical and massive machine type communication towards 6G [white paper]," *6G Research Visions*, vol. 11, 2020. [Online]. Available: http://urn.fi/urn:isbn: 9789526226781.

2 M. Fukuyama, "Society 5.0: Aiming for a new human-centered society," *Japan Spotlight*, vol. 27, no. Society 5.0, pp. 47–50, 2018.

3 M. Maier, "6G as if people mattered: From Industry 4.0 toward Society 5.0," in *International Conference on Computer Communications and Networks (ICCCN)*. IEEE, 2021, pp. 1–10.

4 C. Narvaez Rojas, G. A. Alomia Pe nafiel, D. F. Loaiza Buitrago, and C. A. Tavera Romero, "Society 5.0: A Japanese concept for a superintelligent society," *Sustainability*, vol. 13, no. 12, p. 6567, 2021.

5 A. G. Frank, L. S. Dalenogare, and N. F. Ayala, "Industry 4.0 technologies: Implementation patterns in manufacturing companies," *International Journal of Production Economics*, vol. 210, pp. 15–26, 2019.

6 B. Meindl, N. F. Ayala, J. Mendonça, and A. G. Frank, "The four smarts of Industry 4.0: Evolution of ten years of research and future perspectives," *Technological Forecasting and Social Change*, vol. 168, p. 120784, 2021.

7 L. Georgios, S. Kerstin, and A. Theofylaktos, "Internet of Things in the context of Industry 4.0: An overview," *International Journal of Entrepreneurial Knowledge*, vol. 7, no. 1, http://dspace.vsp.cz/handle/ijek/103?show=full 2019.

8 C. Yu, X. Xu, and Y. Lu, "Computer-integrated manufacturing, cyber-physical systems and cloud manufacturing–concepts and relationships," *Manufacturing Letters*, vol. 6, pp. 5–9, 2015.

9 Y. Lu, "Industry 4.0: A survey on technologies, applications and open research issues," *Journal of Industrial Information Integration*, vol. 6, pp. 1–10, 2017.

10 P. O. Antonino, F. Schnicke, Z. Zhang, and T. Kuhn, "Blueprints for architecture drivers and architecture solutions for Industry 4.0 shopfloor applications," in *Proceedings of the 13th European Conference on Software Architecture-Volume 2*, 2019, pp. 261–268.

11 E. Y. Nakagawa, P. O. Antonino, F. Schnicke, R. Capilla, T. Kuhn, and P. Liggesmeyer, "Industry 4.0 reference architectures: State of the art and future trends," *Computers & Industrial Engineering*, vol. 156, p. 107241, 2021.

12 D. G. Victor, "Recovering sustainable development," *Foreign Affairs*, vol. 85, p. 91, 2006.

13 A. Gilchrist, *Industry 4.0: The Industrial Internet of Things*. Springer, 2016.

14 United Nations, "Resolution adopted by the general assembly on transforming our world: The 2030 agenda for sustainable development (a/res/70/1)_2015," October 2015. [Online]. Available: https://www.un.org/ga/search/view_doc.asp?symbol=A/RES/70/1&Lang=E.

15 I. Gustiana, W. Wahyuni, and N. Hasti, "Society 5.0: Optimization of socio-technical system in poverty reduction," *IOP Conference Series: Materials Science and Engineering*, vol. 662, no. 2, p. 022019, 2019.

16 M. Kinoshita, "Japan on the new industrial revolution (NIR): Direction and its global implication," 2019.

17 H. Gill, "From vision to reality: Cyber-physical systems," in *HCSS National Workshop on New Research Directions for High Confidence Transportation CPS: Automotive, Aviation, and Rail*. Austin, Texas, USA, 2008, pp. 1–29.

18 Y. Wang, M. C. Vuran, and S. Goddard, "Cyber-physical systems in industrial process control," *ACM Sigbed Review*, vol. 5, no. 1, pp. 1–2, 2008.

19 Z. Liu, D.-s. Yang, D. Wen, W.-m. Zhang, and W. Mao, "Cyber-physical-social systems for command and control," *IEEE Intelligent Systems*, vol. 26, no. 4, pp. 92–96, 2011.

20 Y. Ren, M. Tomko, F. D. Salim, J. Chan, and M. Sanderson, "Understanding the predictability of user demographics from cyber-physical-social behaviours in indoor retail spaces," *EPJ Data Science*, vol. 7, no. 1, pp. 1–21, 2018.

21 M. E. Gladden, "Who will be the members of Society 5.0? Towards an anthropology of technologically posthumanized future societies," *Social Sciences*, vol. 8, no. 5, p. 148, 2019.

22 F. Khan, R. L. Kumar, S. Kadry, Y. Nam, and M. N. Meqdad, "Cyber physical systems: A smart city perspective," *International Journal of Electrical and Computer Engineering*, vol. 11, no. 4, p. 3609, 2021.

23 S. K. Khaitan and J. D. McCalley, "Design techniques and applications of cyberphysical systems: A survey," *IEEE Systems Journal*, vol. 9, no. 2, pp. 350–365, 2014.

24 X. Li, R. Lu, X. Liang, X. Shen, J. Chen, and X. Lin, "Smart community: An Internet of Things application," *IEEE Communications Magazine*, vol. 49, no. 11, pp. 68–75, 2011.

25 M. A. Uusitalo, P. Rugeland, M. R. Boldi, E. C. Strinati, P. Demestichas, M. Ericson, G. P. Fettweis, M. C. Filippou, A. Gati, M.-H. Hamon et al., "6G vision, value, use cases and technologies from European 6G Flagship project Hexa-X," *IEEE Access*, vol. 9, pp. 160004–160020, 2021.

26 R. Li, D. Trossen, V. Sarian, M. Essa, A. Clemm, N. Wang, Y. Zhang, A. Borodin, K. Makhijani, M. Toy et al., "Network 2030 - a blueprint of technology, applications, and market drivers towards the year 2030 and beyond [White Paper]," *Network 2030 Vision, ITU-T*, 2019.

27 A. Avizienis, J.-C. Laprie, B. Randell, and C. Landwehr, "Basic concepts and taxonomy of dependable and secure computing," *IEEE Transactions on Dependable and Secure Computing*, vol. 1, no. 1, pp. 11–33, 2004.

28 O. L. A. López, N. Mahmood, M. Shehab, H. Alves, O. M. Rosabal, L. Marata, and M. Latva-aho, "Statistical tools and methodologies for designing/analyzing URLLC - a tutorial," *submitted to Proceedings of the IEEE*, 2022.

29 N. H. Mahmood, H. Alves, O. A. López, M. Shehab, D. P. M. Osorio, and M. Latva-aho, "Six Key Enablers for Machine Type Communication in 6G," *arXiv preprint arXiv:1903.05406*, 2019.

30 F. Rodriguez, I. Ahmad, J. Huusko, and K. Seppänen, "Towards dependable 6G networks," *TechRxiv*, 2022.

31 N. Rajatheva, I. Atzeni, E. Björnson, A. Bourdoux, S. Buzzi, J.-B. Doré, S. Erkucuk, M. Fuentes, K. Guan, Y. Hu et al., "White paper on broadband connectivity in 6G [white paper]," *6G Research Visions*, vol. 10, 2020. [Online]. Available: http://urn.fi/urn:isbn:9789526226798.

32 N. Batra, H. Holma, H. Viswanathan, T. Wild, P. Baracca, G. Berardinelly, G. Kunzmann, V. Ziegler, M. Montag, P. Merz et al., "Envisioning a 6G future [white paper]," *Nokia Bell Labs*, 2022. [Online]. Available: https://www.bell-labs.com/research-innovation/what-is-6g/6g-technologies/.

33 Z. Qin, X. Tao, J. Lu, and G. Y. Li, "Semantic Communications: Principles and Challenges," *arXiv preprint arXiv:2201.01389*, 2021.

34 Q. Lan, D. Wen, Z. Zhang, Q. Zeng, X. Chen, P. Popovski, and K. Huang, "What is semantic communication? A view on conveying meaning in the era of machine intelligence," *Journal of Communications and Information Networks*, vol. 6, no. 4, pp. 336–371, 2021.

35 H. Seo, J. Park, M. Bennis, and M. Debbah, "Semantics-Native Communication with Contextual Reasoning," *arXiv preprint arXiv:2108.05681*, 2021.

36 G. Zhao, M. A. Imran, Z. Pang, Z. Chen, and L. Li, "Toward real-time control in future wireless networks: Communication-control co-design," *IEEE Communications Magazine*, vol. 57, no. 2, pp. 138–144, 2018.

37 C. de Lima, D. Belot, R. Berkvens, A. Bourdoux, A. Dardari, M. Guillaud, M. Isomursu, E. Lohan, Y. Miao, A. Barreto et al. *(Eds.)*, "6G white paper on localization and sensing [white paper]," *6G Research Visions*, vol. 12, 2020. [Online]. Available: http://urn.fi/urn:isbn:9789526226743.

38 T. Wild, V. Braun, and H. Viswanathan, "Joint design of communication and sensing for beyond 5G and 6G systems," *IEEE Access*, vol. 9, pp. 30 845–30 857, 2021.

39 M. Matinmikko-Blue, S. Aalto, M. I. Asghar, H. Berndt, Y. Chen, S. Dixit, R. Jurva, P. Karppinen, M. Kekkonen, M. Kinnula et al., "White Paper on 6G drivers and the UN SDGs," *arXiv preprint arXiv:2004.14695*, 2020.

40 O. L. A. López, H. Alves, R. D. Souza, S. Montejo-Sánchez, E. M. G. Fernández, and M. Latva-Aho, "Massive wireless energy transfer: Enabling sustainable IoT toward 6G era," *IEEE Internet of Things Journal*, vol. 8, no. 11, pp. 8816–8835, 2021.

41 H. Alves and O. A. Lopez, "Wireless RF energy transfer in the massive IoT era: Towards sustainable zero-energy networks," 2021.

42 J. Huang, Y. Zhou, Z. Ning, and H. Gharavi, "Wireless power transfer and energy harvesting: Current status and future prospects," *IEEE Wireless Communications*, vol. 26, no. 4, pp. 163–169, 2019.

43 O. L. López, H. Alves, R. D. Souza, S. Montejo-Sánchez, E. M. G. Fernández, and M. Latva-Aho, "Massive wireless energy transfer: Enabling sustainable IoT toward 6G era," *IEEE Internet of Things Journal*, vol. 8, no. 11, pp. 8816–8835, 2021.

44 O. L. A. López, D. Kumar, R. D. Souza, P. Popovski, A. Tölli, and M. Latva-aho, "Massive MIMO with radio stripes for indoor wireless energy transfer," *IEEE Transactions on Wireless Communications*, vol. 21, no. 9, pp. 7088–7104, 2022.

11

6G-Enabled Internet of Vehicles

This chapter focuses on the technological evolution of Intelligent Transport Systems which pave the way to vehicular networks with highly dynamic and complex features with stringent requirements on ultra-low latency, high reliability, and massive connectivity. 6G wireless systems will cooperate with the Internet of Vehicle (IoV, technologies) to address such main challenges of modern transportation while being in line with the goals of a sustainable society. After reading this chapter, you should be able to:

- Gain an overview of key technologies and use cases of IoV with a tendency of being used in 6G networks.
- Identify the major security considerations, threat landscape, and counter measures in IoV.

11.1 Overview of V2X Communication and IoV

The IoV has emerged as a new paradigm driven by the innovations in vehicular communications. In the IoV concept, vehicles are equipped with sensors, control and computing units, communication, storage, and learning capabilities, which allow the integration of smart vehicles with the Internet, transport infrastructure, and other road users via vehicle-to-everything (V2X) communications [1, 2]. During long time, the only V2X solution was the dedicated short-range communication (DSRC), which is based on the IEEE 802.11.

In 2017, an advanced technology that relies on the capabilities of 4G, 5G, and future 6G cellular networks was incorporated by the 3rd Generation Partnership Project (3GPP), the so-called "cellular-enabled V2X or C-V2X," which can provide significantly higher system performance, higher spectral efficiency, higher range,

Security and Privacy Vision in 6G: A Comprehensive Guide, First Edition.
Pawani Porambage and Madhusanka Liyanage.

reliability, and security, thus enabling higher levels of safety to more road users than alternative technologies. C-V2X employs two complementary transmission modes to enable a very broad range of driving safety features. These modes are the short-range direct communications (C-V2X Direct) and the long-range network communications (C-V2X Mobile Communications). C-V2X Direct comprises short-range communication between vehicles (V2V), between vehicles and infrastructure (V2I), and vehicles and pedestrians (V2P). In the latter, C-V2X employs the conventional mobile network into the vehicle-to-network (V2N) communication to enable the vehicle to receive information about road conditions and traffic in the area, beyond the driver's line-of-sight (LoS) [3] (Figure 11.1).

In this regard, IoV technologies are expected to address the main challenges of modern transportation, and, at the same time, being in line with the goals of a sustainable society. It is expected then, by 2025, connected cars could save 11,000 lives and lead to 260,000 fewer accidents, while avoiding 400,000 tons of CO_2 emissions and saving 280 million hours of driving every year [3].

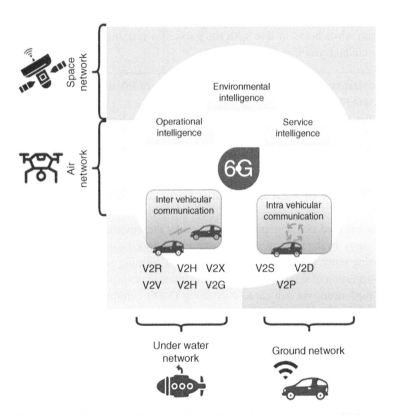

Figure 11.1 Overview of IoV with the intelligent transport systems of 6G.

The evolution to IoV will rely on the efforts from different sectors including automobile, transportation, wireless communications and networking, robotics, as well as regulation organizations. In this sense, 6G plays a pivotal role on attaining the ambitious goals for IoV by satisfying the more rigorous key performance indicators (KPIs) that were partially fulfilled by 5G for vehicle communications. Indeed, it is expected that 5G use cases categories will evolve to Further-enhanced Mobile Broadband (FeMBB), Mobile BroadBand and Low-Latency (MBBLL), ultra-massive Machine-Type Communication (umMTC), and massive Low-Latency Machine-Type communication (mLLMT) with extreme requirements such as data rates over 1 Tbps, end-to-end delays lower than 0.1 ms, network availability and reliability beyond 99.99999%, extreme connection density of over 10^7 devices/km^2, and spectrum efficiency over five times that of 5G while supporting extreme mobility [4]. Additionally, 6G is also targeting higher-frequency bands (i.e. THz), thus allowing a more precise sensing and positioning resolution and enhanced beamforming directionality and data throughput. Indeed, 6G will be a self-learning intelligent network by leveraging artificial intelligence (AI) to deal with the expected complexity of networks and network management [5].

11.2 IoV Use Cases

As mentioned in the technical specification TS 22.186 of Release 16 [6], there are five main use-case categories of cellular-enabled vehicle to everything (C-V2X) which can be also mapped with IoV.

Remote driving: This covers a remote driver or an application which is able to operate a remote vehicle when passengers cannot drive themselves or the remote vehicle is located in dangerous zones. It also considers driving based on cloud computing for predictable applications, e.g. public transportation.

Advanced driving: It contemplates for the cases of semiautomated or fully automated driving for longer inter-vehicle distance. By relying on the exchanging of data among vehicles or roadside units (RSUs) in the proximity, they are able to coordinate their trajectories or maneuvers, thus achieving safer traveling, collision avoidance, and enhanced traffic efficiency.

Vehicles platooning: This stands for the applications where vehicles are capable of dynamically assemble a group traveling together. For that purpose, there is a leading vehicle that sends periodic messages to the others to perform platoon operations. Through these operations, the distance among vehicles can be significantly reduced (in the order of sub second when distance is translated to time).

Extended sensors: They allow vehicles to improve their perception of the environment by overcoming the limitations of their sensors, as raw or processed data from sensors can be exchanged among RSUs, vehicles, devices of pedestrians, and V2X application servers.

Vehicle QoS support: It allows an IoV application to be notified of possible changes on the quality-of-service (QoS) before a change occurs, thus, the application can adjust to the conditions of 3GPP system. It is also possible for the 3GPP system to adapt the QoS according to the application's necessities.

11.3 Connected Autonomous Vehicles (CAV)

Nearly 50 leading automotive and technological companies are heavily investing in autonomous vehicle technology. The world moves forward to experience truly autonomous, reliable, safe, and commercially viable driver-less cars in near future [7]. With the advent of connected autonomous vehicles (CAV) technologies, a new service ecosystem will emerge such as driver-less taxi and driver-less public transport [8, 9].

The security issues in complex CAV ecosystem can be categorized into three categories as vehicle level, CAV supply chain, and data collecting. The vehicle-level attacks can happen by hijacking vehicle sensors, V2X communications, and taking over physical controls. Similar to unmanned aerial vehicles (UAVs), autonomous nature without human involvement will lead to possibility of physical hijacking. However, autonomous vehicles have more advanced capabilities than UAVs. Therefore, emergency security measures can be integrated within a car, for instance automatic stop of car during a terrorist attack is possible. 6G network can analyze the situation and deliver the emergency signals to vehicles.

Moreover, new types of cyber-attacks due to V2X communications in CAV ecosystem are possible. Advance CAVs have communication link with the car manufacturer, so they can constantly monitor and make instant transmission of software-related patches to mitigate any foreseen troubles over the air. However, vulnerabilities in the communication channels or forging the data downloaded from manufacturer cloud services can compromise the safety and security of the vehicles and its passengers.

The CAV ecosystem has a complex supply chain with different third-party service providers such as communication service providers (CSPs), road side equipment (RSE), and cloud service providers and regulators. Enabling common standard of security requirements and enabling the inter-operability is challenging. Privacy issue may arise when CAVs collect data about the travel routes, control sensor data, and also about their owners and passengers. Such data become a honeypot for malicious attackers. According to the National Institute of Standards and Technology (NIST), CAV security framework should target on providing device security, data security, and individuals' privacy.

Specially, when public transport modes such as trains, flights, and buses are used, protection of individual privacy while delivering 6G services such as extended reality (XR), holographic telepresence will be challenging. Therefore, 6G security framework for CAVs has to consider security convergence by combining of physical security and cybersecurity along with the concept of Privacy by Design.

11.4   Unmanned Aerial Vehicles in Future IoV

UAVs and unmanned aircraft (UA) operations are representing a special class of vehicular traffic in the 6G era. Although they are in operation with the 5G systems, novel regulatory frameworks, and new UA applications are rapidly appearing for new market needs and to manage complex air operations in highly dense urban environments.

Since 5G, UAVs are getting popular for use in various application domains. With the support of 6G- and AI-based services, UAV technologies will be used in new use cases such as passenger taxi, automated logistics, and military operations [10, 11]. Due to the limited available resources (i.e. processing and power) and latency critical applications in UAVs they should use lightweight security mechanisms which should satisfy the low latency requirements. Moreover, factors such as high scalability, diversity of devices, and high mobility have to be considered while developing the security mechanisms for UAVs. Since 6G will support AI- and Edge-AI based UAV functions such as collision avoidance, path planning, route optimization, and swarm control, it is important to deploy mechanism to mitigate AI related attacks as well. Specially, protected integrity of control data is a vital requirement for proper operation. Due to the unmanned nature of UAVs, they are highly vulnerable for physical attacks. An adversary can physically capture the UAVs by jamming control signal or use physical equipment, then steal the important data contained within the UAVs. Moreover, UAVs will have advanced computational and communication capabilities compared to other smart devices. Thus, a swarm of drones can be used to perform organized attacks. Such attacks can be range from cyber-attacks to physical terrorist attacks [12, 13].

11.5   Security Landscape for IoV

In this section, we provide an overview on the security threats that could hinder the benefits of IoV applications joint with the security requirements of V2X communications for the safe implementation of future IoV.

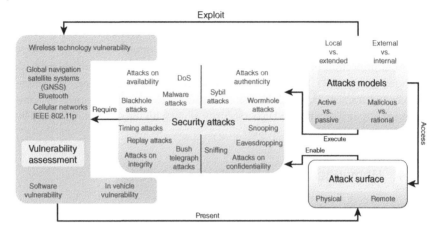

Figure 11.2 IoV security threat taxonomy.

11.5.1 Security Threats

The fast evolving of IoV may also encounter advanced and more intelligent security attacks that can create serious issues to the entire transportation system and users [14]. In [15], authors provide an extensive survey on security landscape of intelligent transport systems. Similar to their explanation, the security threats related to IoV are becoming more and more critical as human lives can be placed at risk. As a summary, we present the security threat taxonomy in Figure 11.2. Basically, the attacker models will exploit the vulnerabilities in vehicular systems by accessing physical or remote attack surfaces to execute security attacks. The attacks may create on RSUs or other physical attack surfaces such as vehicular external interfaces on board units (OBUs) via electronic control units (ECUs) [16].

There can be different attacker models in an IoV system which may intrude from outside or be internal to the system. The local attackers may target only the close-by vehicular systems, whereas the extended attacks may perform attacks with a broader scope, which is irrespective of the locality. The active attackers may inject malicious packets to block the vehicular networks causing denial-of-service (DoS), Sybil, and blackhole attacks. For instance, with sybil attacks, by falsifying or stealing multiple identities of legitimate users, the attackers may control a fraction of the network. The passive attackers will monitor the network traffic and launch eavesdropping attacks to extract useful information to create future attacks. Malicious attackers may damage the network without considering the further consequences, whereas rational attackers target specific users who can be owner of a vehicle or the passengers.

Security vulnerabilities may occur in the vehicular system itself or with wireless technologies or raise software issues. Security attacks require to have vulnerabilities to be present through the attack surfaces to enable attacks. In IoV, there are many wireless technologies in use including bluetooth, IEEE 802.11p, cellular networks, and global navigation satellite system (GNSS). Other than this, in 6G, IoV may also incorporate novel wireless technologies such as visible light communication and quantum communication. Certainly, the security vulnerabilities related to such technologies may also create a direct impact on the IoV security. In Figure 11.2, we highlight the most common types of security attacks that can occur in IoV with attacks on availability, authenticity, confidentiality, and integrity.

11.5.2 Security Requirements

The envisioned ubiquitous connectivity of vehicles in future IoV demands robust security mechanisms to prevent unauthorized access to vehicles and leakage of sensible information once the incidence and impact of security breaches is immense. At the same time, security protocols must be implemented with low communication overhead due to time constraint and low computation complexity, as well as the timeliness of authentication management should be ensured [17, 18]. According to the International Telecommunication Union (ITU) within its Telecommunication Standardization Sector (ITU-T) [19], the security requirements for V2X communication can be described as follows:

Confidentiality: It should not be possible for a unauthorized entity to reveal the messages among vehicles, between vehicles and infrastructure, vehicles and devices, and vehicles and pedestrians. Also, the unauthorized entity shouldn't be able to analyze the identification of a person through personally identifiable information (PII).

Integrity: Messages sent to or from entities in the V2X communication should be protected from unauthorized modification and deletion.

Availability: It should be possible for an entity on the V2X communication to send messages in real-time, thus low-overhead and lightweight security solutions are required.

Nonrepudiation: It should not be possible for an entity to deny that it has already sent a message. This requirement can be implemented using digital signatures in vehicular communication system.

Authenticity: OBUs, RSUs, vehicles, and nomadic devices should be able to provide proof of being an authorized owner of a legitimate identification. In group communications, the vehicle should be able to prove that is a legitimate member of the group. This requirement is called attribute authentication.

Accountability: It should be possible for an entity to detect and prevent any misbehavior of OBUs or vehicle sensors by checking their data.

Authorization: It should be defined access control and authorization for different entities.

References

1 H. Zhou, W. Xu, J. Chen, and W. Wang, "Evolutionary V2X technologies toward the Internet of Vehicles: Challenges and opportunities," *Proceedings of the IEEE*, vol. 108, no. 2, pp. 308–323, 2020.

2 D. Garcia-Roger, E. E. González, D. Martín-Sacristán, and J. F. Monserrat, "V2X support in 3GPP specifications: From 4G to 5G and beyond," *IEEE Access*, vol. 8, pp. 190946–190963, 2020.

3 GSM™,"Global System for Mobile Communications Association," [Accessed on 29.03.2021]. [Online]. Available: https://www.gsma.com/.

4 W. Saad, M. Bennis, and M. Chen, "A vision of 6G wireless systems: Applications, trends, technologies, and open research problems," *IEEE Network*, vol. 34, no. 3, pp. 134–142, 2020.

5 C. D. Alwis, A. Kalla, Q.-V. Pham, P. Kumar, K. Dev, W.-J. Hwang, and M. Liyanage, "Survey on 6G frontiers: Trends, applications, requirements, technologies and future research," *IEEE Open Journal of the Communications Society*, vol. 2, pp. 836–886, 2021.

6 3GPP TS 22.186 version 16.2.0 Release 16, "5G: Service requirements for enhanced V2X scenarios," ETSI, Technical Specification, 2020. [Online]. Available: https://www.etsi.org/deliver/etsi_ts/122100_122199/122186/16.02 .00_60/ts_122186v160200p.pdf.

7 C. Insights, "40+ corporations working on autonomous vehicles," 2019.

8 J. He, K. Yang, and H.-H. Chen, "6G Cellular Networks and Connected Autonomous Vehicles," *arXiv preprint arXiv:2010.00972*, 2020.

9 E. Peltonen, M. Bennis, M. Capobianco, M. Debbah, A. Ding, F. Gil-Casti neira, M. Jurmu, T. Karvonen, M. Kelanti, A. Kliks et al., "6G White Paper on Edge Intelligence," *arXiv preprint arXiv:2004.14850*, 2020.

10 B. Deebak and F. Al-Turjman, "Drone of IoT in 6G wireless communications: Technology, challenges, and future aspects," in *Unmanned Aerial Vehicles in Smart Cities*. Springer, 2020, pp. 153–165.

11 H. Menouar, I. Guvenc, K. Akkaya, A. S. Uluagac, A. Kadri, and A. Tuncer, "UAV-enabled intelligent transportation systems for the smart city: Applications and challenges," *IEEE Communications Magazine*, vol. 55, no. 3, pp. 22–28, 2017.

12 P. B. Johnston and A. K. Sarbahi, "The impact of US drone strikes on terrorism in Pakistan," *International Studies Quarterly*, vol. 60, no. 2, pp. 203–219, 2016.

13 J. O'Malley, "The no drone zone," *Engineering & Technology*, vol. 14, no. 2, pp. 34–38, 2019.

14 V. Sharma, I. You, and N. Guizani, "Security of 5G-V2X: Technologies, standardization, and research directions," *IEEE Network*, vol. 34, no. 5, pp. 306–314, 2020.

15 A. Lamssaggad, N. Benamar, A. S. Hafid, and M. Msahli, "A survey on the current security landscape of intelligent transportation systems," *IEEE Access*, vol. 9, pp. 9180–9208, 2021.

16 M. Hasan, S. Mohan, T. Shimizu, and H. Lu, "Securing vehicle-to-everything (V2X) communication platforms," *IEEE Transactions on Intelligent Vehicles*, vol. 5, no. 4, pp. 693–713, 2020.

17 E. B. Hamida, H. Noura, and W. Znaidi, "Security of cooperative intelligent transport systems: Standards, threats analysis and cryptographic countermeasures," *Electronics*, vol. 4, no. 3, pp. 380–423, 2015. [Online]. Available: https://www.mdpi.com/2079-9292/4/3/380.

18 5G-PPP, "5G automotive vision, *White Paper*," 5G-PPP, Tech. Rep., 2015.

19 ITU-T X.1372, "Security guidelines for vehicle-to-everything (V2X) communication," ITU, Technical Specification, 2020. [Online]. Available: https://www.itu.int/rec/T-REC-X.1372-202003-I.

12

Smart Grid 2.0 Security*

* With additional contribution from Charithri Yapa, University of Sri Jayewardenepura Sri Lanka, and Chamitha De Alwis, University of Bedfordshire, United Kingdom

This chapter provides comprehensive details of the security aspect of Smart Grid 2.0. After reading this chapter, you should be able to:

- Understand the taxonomy of Smart Grid 2.0.
- Understand security attacks related to Smart Grid 2.0.
- Understand privacy objectives of Smart Grid 2.0.
- Understand the standardization initiatives for Smart Grid 2.0 security and privacy.

12.1 Introduction

Smart Grid 2.0 (SG 2.0) is the technology embedded, novel electricity distribution scheme, which facilitates bi-directional flow of energy and information. The importance of developing an architecture that enables seamless integration of Distributed Energy Resources (DERs) to cater to the rising electricity demand was identified, which was further stimulated by the breakthrough advancements such as Internet of Things (IoT). This paved way for utilizing renewable energy in the form of Distributed Generators (DG) including roof-top solar Photo-Voltaic (PV) installations, wind generation, and microhydro projects to cater the loads in the vicinity. This helps to minimize the transmission losses and reduce the supply–demand gap to achieve a reliable power supply. SG 2.0 aims at autonomous grid operations with minimal involvement of an intermediary such as the Distribution System Operator (DSO). Applications of SG 2.0 include Peer-to-Peer (P2P) trading for excess energy generation.

Security and Privacy Vision in 6G: A Comprehensive Guide, First Edition.
Pawani Porambage and Madhusanka Liyanage.

12.2 Evolution of SG 2.0

First generation of smart girds are capable of bidirectional energy routing and communication between the DSO and the consumer, which is enabled through the smart meters installed in the consumer premises [1]. This was easily integrated with the existing top-down infrastructure of the electricity grid by upgrading the conventional meters with smart meters, which are IoT devices [2–4].

However, the electricity grid paradigm started to transform, with the interest being shifted towards renewable energy integration to overcome issues related to the conventional power generation technologies. Large-scale as well as roof-top solar PV installations and wind power plants came into existence. Additionally, consumers influenced by sustainable practices started preferring electricity generated using renewable resources, which resulted in a smaller carbon footprint. These novel trends have created new requirements such as developing a P2P network where energy trading is made possible between any two nodes without the intervention of the DSO, ensuring a reliable power supply with many intermittent generation being integrated, demand side management with consumer participation to rectify the supply–demand mismatch, and secure energy trading platform with minimal involvement of a third-party. Furthermore, increase in the use of Electric Vehicles (EVs) and their stochastic energy consumption patterns at the charging stations dispersed across a large area have resulted in an explicitly dynamic power grid [3], which require novel power management approaches.

Adding to these complexities, novel power management strategies incorporate large data sets related to electrical measurements (voltages, currents, active, and reactive power exchanged), energy trading information between nodes and among the DSO and consumers, control signals for balancing the supply and demand, power plant data for the grid integration of renewable energy, and the power flow data of the interconnected transmission and distribution network [5]. Management of information by a single entity creates vulnerabilities related to security and privacy aspect. Thus, it is evident that the change in grid paradigm requires a decentralized and delegated platform in order to share resources to overcome the security and privacy concerns arising from a single entity controlling the entire grid. This requirement of the future smart grids is facilitated through the delegated operation of IoT devices, which include sensors for advanced monitoring and smart meters for collection of energy data, facilitated by big data management and cloud computing for data analysis and efficient data management, and cybersecurity for ensuring data security and privacy [2, 6, 7].

Smart Grid 2.0 was proposed by an economic theorist, Jeremy Rifkin, in his book titled *The Third Industrial Revolution* published in 2011 [8]. This book discusses on how the technological and scientific changes could impact the economy and the energy sector is one paradigm in it. The envisaged electricity grid integrates a great diversity of energy sources and intelligent loads, while their supervision and control being handled over Internet-based protocols. This envisions converting future power grids to be self-controlled, self-optimized, and self-healing, hence autonomous. The main objective of SG 2.0 is to share energy with their relevant data, seamlessly as conveniently as sharing information over the World Wide Web [2, 6, 9].

12.3 Smart Grid 2.0

Smart Grid 2.0 (SG 2.0) refers to the next generation of electricity distribution technologies, integrating heterogeneous energy sources, which are monitored and controlled over the Internet-based P2P networks [2, 6, 10]. This facilitates the grid integration of renewable energy sources and energy storage systems, plug-and-play interfacing of EV charging, real-time monitoring/control of power grids, acquisition/management of energy data, and automation of energy balancing services in future smart grids [2, 6, 7, 11]. The first generation of smart grids enabled bi-directional flow of information through advanced Information and Communication Technologies (ICT), whereas SG 2.0 aims at the exchange of energy and monitoring data, and control information through Internet Protocols (IP) [2, 12].

Bi-directional energy flows are created through the integration of DERs at the distribution level, thereby enabling the electricity consumers to play the role of a power producer [3]. This new category of electricity producer/consumer is referred to as a "prosumer." Prosumers being dispersed across a vast geographical area maximizes the utilization of resources, while enhancing the energy security. SG 2.0 encourages the utilization of resources (i.e. solar, wind) that are available in abundance around us and fulfills the energy demand in its close vicinity [5]. This would contribute in reducing the losses that are incorporated in long distance power transmission and at the same time improve the power quality at the consumer end [13].

SG 2.0 grid architecture can be detailed as four layers namely,

1. **Physical component layer**: The *physical component layer* comprises of sensory devices, which facilitate data acquisition for real-time monitoring and decision-making process. This includes IoT devices such as smart generation technologies, smart loads, smart sensors, smart meters, Phasor Measurement Units (PMU), Remote Terminal Units (RTU), Current Transformers (CT), and Voltage Transformers (VT). The Wireless Sensor Network (WSN) communicates the sensor information. The information is required in various SG 2.0 applications incorporating financial transactions related to energy trading, grid integration of distributed generation, network, and load management.

2. **Communication and control layer**: Communication infrastructure and the Internet-based protocols facilitate sharing of the acquired data and information for real-time monitoring and control of the SG 2.0 [14]. The **communication and control layer** facilitate fast, reliable, and real-time communication among devices with minimal involvement of a third-party service provider.

3. **Application layer**: Microgeneration and large-scale power plants incorporating renewable power generation, battery storage, EVs trading with charging station through P2P energy-financial transactions, automated, efficient, reliable transmission/distribution networks facilitating bidirectional energy routing, smart buildings, and smart homes with smart meters for better load scheduling are the progressive developments of the **application layer** of future smart grids.

4. **Data analysis layer**: SG 2.0 incorporates large data sets which contain information related to instantaneous measurements of the electrical grid, energy consumption data obtained through smart meters for energy trading, and billing and control signal data for balancing the supply and demand. Data management, secure data routing, privacy-preserving, and reliable storage are thus, part and parcel of the SG 2.0 infrastructure [5, 15]. The **data analysis layer** provides assistance in cloud-centric data management, predictive analysis of trends, and underlying control of the grid.

The schematic in Figure 12.1 illustrates the basic components of each layer.

Advanced Metering Infrastructure (AMI) acquires synchronized smart meter measurements at prosumer and consumer locations, connects with the communication network, and performs data management, hence, spreads across all the above layers.

12.3.1 Comparison of Smart Grids 1.0 and 2.0

Having evolved from its predecessor, first-generation smart grid technology, the SG 2.0, shares the common objective of elevating the power system architecture to facilitate real-time operations. SG 2.0, however, incorporates a significant change in architecture which contradicts with the existing topology.

The implementations of Smart Grid 1.0 focus mainly on the smart meter communication, using the advancements in ICT. Bi-directional communication enables active supply-demand coordination through Demand Dide Integration (DSI) initiatives.

Smart Grid 2.0 facilitates further consumer participation in supply–demand matching through the contribution made in distributed power generation as a prosumer. Microgrid operation with seamless integration of RES and peak demand catering with Vehicle-to-Grid (V2G) interactions at dispersed locations are facilitated through the decentralized architecture. Distribution network

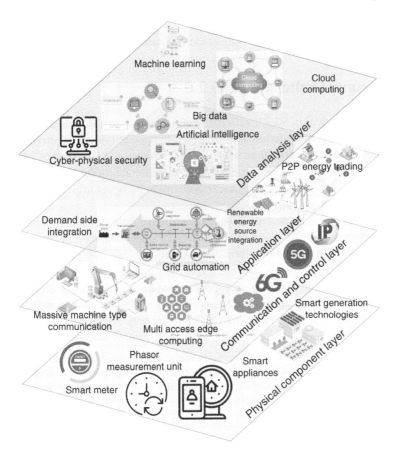

Figure 12.1   Layers of Smart Grid 2.0.

management is automated, which eliminates the roles of the trust-less intermediary stakeholders. Improved data acquisition and secure storage would enable
precise decision-making algorithms.

12.4   Role of 6G in SG 2.0

6G technology aims at autonomous operation of intelligent devices, which will
integrated to create smart cities, smart homes, smart healthcare, and smart
transportation enabled through autonomous vehicles and smart electricity distribution. The revolutionary 6G is envisaged to be facilitated by Artificial Intelligence
(AI) for intelligent, human-like decision-making, big data for data management
and Machine Learning (ML) to analyze large quotas of information. 6G is the

technology enabler of the above revolutionary advancements, which facilitate ultra-low latency information-sharing and next-generation data management.

12.5   Security Challenges of SG 2.0

Achieving the objectives of SG 2.0 with the existing grid is not a straightforward approach as the infrastructure has been designed to operate under the purview of a centralized network rather than a distributed architecture as deemed by the future grids. The rest of this section elaborates on vulnerabilities related to SG 2.0 security and privacy, which needs research attention for better utilization and secure implementation.

Grids are envisaged to be decentralized and distributed rather than centralized and vertical with the incorporation of IoT devices for state estimation of the network operation, IP for sharing of information in real-time, and cloud computing for management of energy databases. Information security, cybersecurity, and network security play a vital role in ensuring confidentiality, assuring integrity and securing the availability of information that is shared among the entities within the network thereby creating a level playing field. Security threats to which SG 2.0 is vulnerable can be categorized as physical attacks, software-related threats, network-based attacks, threats targeting control elements, encryption-related attacks, and AI- and ML-related attacks [16]. The components that are mostly targeted by the attackers include metering and data access points, processes including billing and information exchange, control elements (Supervisory Control and Data Acquisition [SCADA]) [17], and the Energy Management System (EMS) of the cyber–physical system [18]. Figure 12.2 illustrates a summary of the cyber–physical threats present in Smart Grid 2.0.

12.5.1   Physical Attacks

Physical attacks related to the SG 2.0 involve the AMI components such as smart meters, PMUs, RTUs, and the WSN. Malicious attackers might try to gain unauthorized access to smart meters and create unnecessary traffic, tamper with data transmitted through WSN, inject false data to the network, and masquerade attacks [19, 20].

- **Tampering hardware**: Smart meters and sensor nodes of the AMI could be interfered by hackers with the intention of retrieving sensitive information. This could cause physical damage to the hardware by trying to insert foreign objects to these IoT devices. Tampered meters/sensors will be used as bot by a malicious third-party to retrieve sensitive information of the users and create unnecessary traffic in the channel by overloading it with requests.

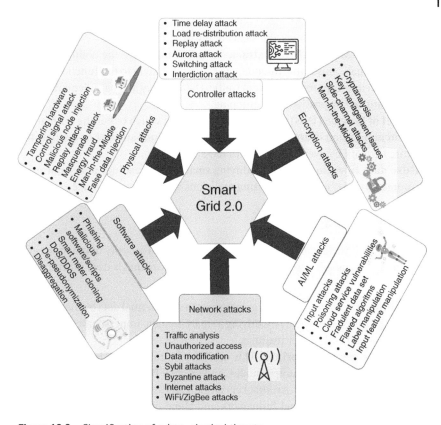

Figure 12.2 Classification of cyber–physical threats.

- **Control signal attack**: By launching such attacks, the malicious third party aims at gaining control over mission-critical devices such as Automatic Generation Control (AGC), smart inverters, Flexible AC Transmission Systems (FACTS) devices, and circuit breakers. This will create undesired effects on the smart grid with a middle-man controlling the power-related operation through compromised devices [21].
- **Malicious node injection**: Intermediate hardware installed as a middleman with the intention of modification and theft of information.
- **Replay attack**: The attacker takes control of smart sensors, captures data from them for a specified amount of time, and then replays the information on the monitoring channels while potentially introducing an exogenous signal into the system [22].
- **Masquerade**: A smart meter impersonating a different person, with the purpose of gathering information.

- **Energy fraud attack**: Attackers obtain the smart meter ID from meter spoofing to manipulate the readings and steal energy from the grid.
- **Man-in-the-Middle (MIM) attack**: A hacker may have the ability to create and inject malicious traffic intending to cause the system to malfunction, if they are able to intercept regular system traffic. Data privacy and confidentiality may be jeopardized by passive eavesdropping.
- **Denial of Service (DoS)**: This attack is twofold in which either the normal use of communication channels is inhibited by a malicious attacker or disrupts the entire network by overloading it with messages [20].
- **False data injection (FDI)**: Modifying smart meter readings and senor measurements either manually or using malicious software or deliberate errors.

12.5.2 Software Attacks

These attacks involve compromising the cyber–physical security of the smart grid through malicious software or jeopardizing the software that is used for control of the smart grid.

- **Phishing**: A malicious attacker gains access to confidential data by meter spoofing [23]. This is achieved by modifying a smart meter to impersonate some other consumer and craft meter readings.
- **Malicious software/scripts**: This includes extracts of codes or programs which could steal data, tamper nodes, and create a DoS [16, 19].
- **Cloning a smart meter**: An attacker can gain access to the smart meter's binary image of its keys, IDs, and framework and clone one. Therefore, a clone of a smart meter can change its ID with the duplicate copy, which will record lower energy consumption readings [19].
- **Denial of Service (DoS)**: Attackers could deliberately send service requests repeatedly to overwhelm the control server in the case of a centralized control architecture thereby denial of accepting authentic requests from users [23, 24].
- **Distributed Denial of Service (DDoS)**: Several attackers collaborate at the same instance to attack a single point using malicious software and algorithms [19, 23–25].
- **De-pseudonymization attacks**: Attacks violate the privacy and identification of data from smart meters [24].
- **Disaggregation attack**: In this, the attacker tries to profile the energy use patterns of the customers [24].

12.5.3 Network Attacks

Network attacks predominantly include traffic analysis attacks that gives access to confidential information, unauthorized accessing of data. This compromises the authenticity and confidentiality of the participating entities, DoS attacks, and modification of information through re-routing of the communication channels [16]:

- **Traffic analysis**: Attackers can analyze the traffic of the information being exchanged and determine the energy usage patterns and render sensitive information [18, 23].
- **Unauthorized access**: Lack of proper authentication enables attackers to access, modify, and delete data [16].
- **DoS**: Mainly targets resource consumption which includes memory and bandwidth and overwhelms the system through service requests.
- **Data modification**: A middleman can intercept the information, modify, and transmit to the receiver [24].
- **Sybil attack**: In a P2P network, a sybil attack employs a single node to manage numerous active false identities (also known as sybil identities) concurrently. By controlling the vast majority of the network's influence, this kind of attack seeks to weaken the legitimacy or power of a well-respected system. This effect is made possible by the false identities.
- **Byzantine attack**: These attacks are launched against communication networks such as cognitive radio networks and mobile ad hoc networks. Such attacks compromise the trusted routing, which affects the overall performance of the network [21].
- **Wi-Fi/ZigBee attacks**: Attacks in the AMI communication networks affecting the Wi-Fi/ZigBee networks.
- **Internet attacks**: These vulnerabilities compromise the software and systems in electric utilities.

12.5.4 Attacks to the Controller

Attacks on the controller aim at acquiring the authority of the physical device and acting upon the attacker's will by exploiting the ability to bypass data authentication. DoS attack, FDI, aurora attacks, pricing attacks, injection of a malicious control algorithms, and scripts which affect the stable operation of the network, unauthorized accessing of EMS in order to sabotage the control operation and data cloning [16] could be classified as attacks related to the controller.

- **FDI**: This includes modification of control signals transmitted, manually, or using malicious software or deliberate errors [20].
- **Unauthorized access**: A middleman can intercept the measurement data, control signals exchanged, and modify and transmit to the receiver [18].
- **Malicious code/algorithm/command injection**: Executing malicious scripts which would lead to a complete shutdown of the systems [16, 19].

- **Man-in-the-Middle attack**: This is a vulnerability created by the interception of a malicious third-party.
- **Replay attack**: The attacker hack into the controlling units of the smart grid and then replays the control information on the monitoring channels while potentially introducing an exogenous signal into the system [22].
- **Aurora attack**: Attackers attempts to seize operation of the smart grid by attempting to gain control over the AGC units of the power network.
- **Interdiction attack**: Manipulation of control commands with the intention of causing malfunctioning of the SG 2.0.
- **Switching attacks**: Deliberately issue malicious switching signals to steer the smart grid into degraded or unstable operation.
- **Load re-distribution attack**: Injecting false data into system with the intention of redistributing the load, which ultimately leads to unstable operation of the smart grid.
- **Time delay attacks**: Deliberate injection of time delays in the control signals with the intention of creating instabilities in the power network.

12.5.5 Encryption-Related Attacks

This includes the vulnerabilities of the SG 2.0 related to data encryption, key management, and distribution and key verification.

- **Cryptanalysis**: Attack, which compromises the confidentiality of the information that is exchanged has instilled a challenge in the envisaged smart grids.
- **Issues related to key management**: Decentralized operations of the SG 2.0, with minimal interference from a third party require utilization of session keys to authorize energy transactions. Attacks to the key management entity will compromise the security of the SG 2.0 transactions with malicious parties gaining access.
- **Man-in-the-Middle attack**: Malicious attackers tend to intrude into the symmetric and asymmetric key exchange process with the intention of gaining access to transaction data, energy usage patterns, and user identities.
- **Side channel attack**: The attacker can obtain the encryption key used for encrypting and decrypting data from the encryption devices of an IoT system by employing specific techniques (such as timing, power, fault, and electromagnetic analysis).

12.5.6 AI- and ML-Related Attacks

The capabilities of SG 2.0 are exploited by leverage AI and ML technologies for predictive data analysis and algorithm computation. The integration of these next-generation technological tools provoke the potential adversary of AI- and ML-related attacks [26, 27]. These attacks predominantly comprise of fraudulent data sets and flawed algorithms which would affect the accuracy of the outcomes obtained from the system.

- **Input attacks**: These include changing the data that is provided into the AI system in order to alter the system's output and achieve the attacker's objectives [28].
- **Poisoning attacks**: Attackers tamper with the AI system's creation process to alter it so that it malfunctions in their preferred way. Poison attacks can be launched through data poisoning, algorithm poisoning, and model poisoning. Poisoned AI system, poisons the data [28].
- **Cloud services vulnerabilities**: Attacks to the cloud services that are utilized in ML algorithm development and AI implementations will lead to modification of data and process [26].
- **Fraudulent data set**: Injection of false data or modification/manipulation of the sample set used for algorithm generation [26].
- **Flawed algorithms**: Use of inaccurate processes for the computation of algorithms, which would result in inaccurate results [27].
- **Label manipulation**: This indicates modifying the most vulnerable training labels to gain knowledge of the target model [26].
- **Input feature manipulation**: In these types of attacks, the attackers have the ability to manipulate the label as well as the input features of the training data set [26].

12.5.7 Stability and Reliability of Power Supply

Future electricity grids are to rely more on distributed generation including renewable energy sources, energy storage systems and dynamic loads such as EVs. The former is intermittent by nature while the latter is difficult to predict as it is difficult to predict where and when the electricity consumption would take place with the charging stations being dispersed across an area and EVs being mobilized constantly. Dynamic nature of both generation and loads pose a threat to the stability of the grid [5, 9].

Further, the future everyday applications are envisaged to be fully automated and smart, which will be heavily dependent on the electricity supply. Thus, a reliable power supply will be a top priority. Securing a reliable and stable power supply with the stochastic nature of the energy sources demands for real-time monitoring, fast decision-making capabilities, and efficient communication protocols.

However, SG 2.0 security plays a vital role in ensuring that the process is not compromised with unauthorized parties gaining control. In such a scenario where attackers are governing the control of the smart grid, actions including DG interconnection, generation distribution, pricing mechanism, and demand side management strategies will be biased and handled in an ambiguous manner. This will jeopardize the integrity of the smart grid operations and pose a threat on stability of the power supply. Further, attackers could delay the control signals, which will lead to an asynchronous system, raising concerns on the reliability of the electricity supply to the consumer.

12.5.8 Secure and Transparent Energy Trading Among Prosumers and Consumers

SG 2.0 promotes domestic level, small-scale power production, and trading energy in the close proximity, creating a microgrid. Energy trading needs to be permitted between two nodes of the network without the involvement of a third party, which increases losses and the cost. This requires a secure communication channel between the nodes while ensuring nonrepudiation of the transaction records. Double spending and energy theft have been identified as two major concerns related to the smart grid architecture, even under the purview of a centralized authority. Eliminating such adversaries in a distributed operating environment will create unique challenges to the SG 2.0 operators.

In a distributed energy market, authority is delegated to an unknown entity, which will validate authenticity of the prosumer and consumer identification and accuracy of the information relate to an electricity trading transaction. Trust management in a delegated environment is a mandatory requirement, which further needs to be achieved without the involvement of a third party [5, 6]. A malicious DG being authorized to connect to the distribution network and engage in energy trading will hinder the reliability of power supply, due to unintended loss of service. Adversaries including unauthorized access, identity spoofing, and operation of malicious nodes need to be eliminated for secure operation of the envisaged SG 2.0 architecture.

Furthermore, secure transaction data management and fair trading practices need to be implemented in SG 2.0. Trading information are required to be shared for the validation purpose and to ensure impartial trading decision. This however, needs a mechanism, which allows the prover to guarantee the verifier that the transaction details are accurate, without having to share the actual values. Transaction information will be stored in a secure platform, which is immune to data modification and data loss. Information related to the past performance of the sellers and buyers can thus, be utilized in the decisions related to energy transactions, which include seller selection and determination of the market price for electricity trading.

12.5.9   Efficient and Reliable Communication Topology for Information and Control Signal Exchange

SG 2.0 relies on IP-based communication for the exchange of information and control signals for the operation and management of the grid. Utilization of efficient and reliable technologies to establish the connectivity among the nodes of the network is indispensable for a stable operation of the grid that delivers a reliable power supply to the consumers [5, 9, 12].

Adversaries related to DoS attacks, which overwhelm the communication channels with unnecessary traffic and delay the legitimate requests, byzantine threat that hinders the efficient performance of the network layer, and aurora attacks, which launch compromised control commands to the network, are to be addressed to facilitate seamless grid integration of DGs. Further, attacks that are launched on both the controller and measurement devices compromise the selection and control decisions taken in order to maintain a continuous power supply to the consumer. This, further degrades the power quality of the electricity supply received by the end users.

12.6   Privacy Issues of SG2.0

SG 2.0 operations incorporate real-time measurements obtained by IoT sensors, energy consumption patterns received through smart meters, and real-time analysis of the data collected. This data contain sensitive information related to consumers such as customer identity, location information, and daily energy consumption patterns. The increasing number of IoT devices to obtain measurements and efficient control and interconnections in terms of RES and individual prosumers, EVs and charging stations, and energy storage units makes the network more attractive to hackers. Although SG 2.0 aims at a decentralized network architecture that eliminates the risk of single point failure, the increase in the number of nodes extensively and rapidly makes the system vulnerable to privacy attacks [6, 7].

Revealing the customer identity to neighboring nodes in the P2P network will create privacy concerns as the energy consumption patterns can be matched against each individual. This will pose a threat to the neighborhood in terms of security as the time and nature of occupancy of each consumer premises can be monitored by an unauthorized, malicious node if it gains access to the network. Further, through the analysis of one's energy consumption pattern, an external party is capable of predicting the energy consumption for the next day and deliberately attempt to cause interruptions, hindering the continuous power supply to the consumer premises. Identity information of an electricity consumer as well as his daily consumption requirement should not be managed by either a centralized authority, leading to single-point failure or revealed to nodes of the P2P network. This give rise to concerns related to misuse of sensitive information, causing breech of security and privacy. Furthermore, EVs share their exact location in real-time with the neighboring charging piles, which will pose a threat to the privacy of the EV user, if the network is compromised by an unauthorized access or the physical identity is revealed.

Such instances would compromise the privacy of the consumer hence, security of the data should be ensured during information exchange (i.e. data encryption) and storage in cloud-based platforms (i.e. cyber–physical security) [5–7, 9, 12]. A pseudo-identity would serve best for this purpose, which facilitates in verifying authenticity of the user while maintaining the secrecy of the physical identity from malicious parties. Any misconduct related to the energy data would raise security concerns for all nodes in the network. However, decentralizing the grid infrastructure with the implementation of SG 2.0 eliminates the risk of single point of failure with minimal involvement of a centralized authority.

12.7 Trust Management

Increasing amounts of microgrids comprising of roof-top solar PV and wind generation, large renewable generation connecting to the traditional grid, and EVs utilizing electricity in stochastic consumption patterns will lead to rapid acceleration of the number of stakeholders the SG 2.0 has to incorporate in its operations. In the traditional grid context as well as the first generation of smart grids, a third party (i.e. DSO) will have the authority over the management of grid infrastructure, awarding of contracts, and preparation of the power purchase agreements thereby securing the ownership of the network operation.

However, in the envisaged SG 2.0 context, the number of prosumers increases and the connection requests are made on a real-time basis [6]. Authorizing grid interconnection of all these power sources would require an automated mechanism to maximize the benefits achieved by the SG 2.0 architecture [7]. Managing trust issues related to facilitating real-time connection is a mandatory requirement for securing a safe, reliable, and continuous power supply. Stakeholders dispersed in different geographical locations would require authorization to connect with the grid and exchange energy, given that a certain predefined set of conditions has been fulfilled. Malicious requests will be identified and declined thereby, eliminating the threats of sabotaging grid operations and resulting in grid instabilities.

Future microgrid will be a collection of independent prosumers who have developed a network with their individual resources. Therefore, the central authority could no longer claim the ownership of this grid infrastructure. However, trust building among prosumers and consumers prior to engaging in energy transactions in SG 2.0 is a mandatory requirement interconnection [15]. Further, to complicate this prerequisite is expected to be fulfilled without the involvement of an external party. Furthermore, trust management with the rapidly accelerating number of stakeholders will pose a challenge in the actual implementation of future smart grids.

12.8 Security and Privacy Standardization on SG 2.0

The current legal framework does not facilitate incorporation of certain prospective features of SG 2.0 into the existing power network. 1) real-time authorization for grid interconnection of RES and DERs, 2) supply/demand responsive, flexible and transparent pricing mechanism for electricity trading among prosumers and consumers 3) real-time pricing for EV charging, security, and privacy policies regarding energy information, measurement data and pricing details that are exchanged across the network are new aspects, which are overlooked in the existing standards [5]. Further with the decentralized grid infrastructure, ownership of the network will not be entrusted to a single entity, which demands for globally accepted standards for consistency and convenience of evaluation and troubleshooting [6, 7].

With the transformation of the architecture from top-down to being delegated, the security standards are to be revised to mitigate a new set of attacks, which were not observed in the traditional grid [19]. Transparent grid operations, sharing sensitive information among peers of the network, and decentralized control and operation makes the SG 2.0 vulnerable to many security and privacy breaching attempts. Hence, the protocols viewing the security and privacy aspect of the smart grid are required to be reviewed to enhance the reliability of the envisaged grid operations. This includes addressing the security and privacy adversaries related to the novel technologies used in SG 2.0 such as AI, ML, and big data, which are not incorporated in the existing standards.

Standards regulating energy trading in the SG 2.0 architecture need to be imposed with large numbers of stakeholders (prosumers, consumers, DSOs, and TSOs) entering the network. This will create an equal level playing field for all stakeholders that participate and establish a transparent energy exchange policy. Accountability of the transactions being performed across the grid in any form including energy, information and monetary terms require standard regulatory frameworks in order to eliminate the risk of theft, illicit use of data, and threat of malicious man-made and cyber-attacks. Standards governing the privacy of the shared data incorporated with the energy transactions need to be enforced to eliminate unauthorized use of data and prevent sybil attacks that prevent from fair transmission of information that is required for sustainable smart grid operations.

References

1 X. Fang, S. Misra, G. Xue, and D. Yang, "Smart grid – the new and improved power grid: A survey," *IEEE Communications Surveys & Tutorials*, vol. 14, no. 4, pp. 944–980, 2012.

2 H. Shahinzadeh, J. Moradi, G. B. Gharehpetian, H. Nafisi, and M. Abedi, "Internet of energy (IoE) in smart power systems," *2019 IEEE 5th Conference on Knowledge Based Engineering and Innovation, KBEI 2019*, pp. 627–636, 2019.

3 K. Mahmud, B. Khan, J. Ravishankar, A. Ahmadi, and P. Siano, "An internet of energy framework with distributed energy resources, prosumers and small-scale virtual power plants: An overview," *Renewable and Sustainable Energy Reviews*, vol. 127, p. 109840, 2020.

4 W. Strielkowski, D. Streimikiene, A. Fomina, and E. Semenova, "Internet of energy (IoE) and high-renewables electricity system market design," *Energies*, vol. 12, no. 24, pp. 1–17, 2019.

5 S. M. S. Hussain, F. Nadeem, M. A. Aftab, I. Ali, and T. S. Ustun, "The emerging energy internet: Architecture, benefits, challenges, and future prospects," *Electronics (Switzerland)*, vol. 8, no. 9, 1037, 2019.

6 M. B. Mollah, J. Zhao, D. Niyato, K.-Y. Lam, X. Zhang, A. M. Ghias, L. H. Koh, and L. Yang, "Blockchain for future smart grid: A comprehensive survey," *IEEE Internet of Things Journal*, vol. 8, no. 1, pp. 18–43, 2020.

7 A. Miglani, N. Kumar, V. Chamola, and S. Zeadally, "Blockchain for internet of energy management: Review, solutions, and challenges," *Computer Communications*, vol. 151, pp. 395–418, 2020.

8 J. Rifkin, *The Third Industrial Revolution_ How Lateral Power Is Transforming Energy, the Economy, and the World*. New York: Palgrave Macmillan, 2011.

9 E. Kabalci and Y. Kabalci, *Roadmap from Smart Grid to Internet of Energy Concept*. Elsevier, 2019.

10 A. Q. Huang, M. L. Crow, G. T. Heydt, J. P. Zheng, and S. J. Dale, "The future renewable electric energy delivery and management (FREEDM) system: The energy internet," *Proceedings of the IEEE*, vol. 99, no. 1, pp. 133–148, 2011.

11 A. A. G. Agung and R. Handayani, "Blockchain for smart grid," *Journal of King Saud University - Computer and Information Sciences*, vol. 34, no. 3, 2020.

12 E. Kabalci and Y. Kabalci, *Introduction to Smart Grid and Internet of Energy Systems*. Elsevier, 2019.

13 H. Wang, M. Huang, Z. Zhao, Z. Guo, Z. Wang, and M. Li, "Base station wake-up strategy in cellular networks with hybrid energy supplies for 6G networks in an IoT environment," *IEEE Internet of Things Journal*, vol. 8, no. 7, pp. 5230–5239, 2020.

14 E. Kabalci and Y. Kabalci, "Roadmap from smart grid to internet of energy concept," *From Smart Grid to Internet of Energy*. Elsevier, pp. 335–349, 2019.

15 T. Winter, "The Advantages and Challenges of the Blockchain for Smart Grids," pp. 2018–2019, 2018.

16 I. Andrea, C. Chrysostomou, and G. Hadjichristofi, "Internet of Things: Security vulnerabilities and challenges," *Proceedings - IEEE Symposium on Computers and Communications*, vol. 2016-February, pp. 180–187, 2016.

17 D. Pliatsios, P. Sarigiannidis, T. Lagkas, and A. G. Sarigiannidis, "A survey on SCADA systems: Secure protocols, incidents, threats and tactics," *IEEE Communications Surveys & Tutorials*, vol. 22, no. 3, pp. 1942–1976, 2020.

18 E. Hossain, I. Khan, F. Un-Noor, S. S. Sikander, and M. S. H. Sunny, "Application of big data and machine learning in smart grid, and associated security concerns: A review," *IEEE Access*, vol. 7, no. C, pp. 13960–13988, 2019.

19 P. Kumar, Y. Lin, G. Bai, A. Paverd, J. S. Dong, and A. Martin, "Smart grid metering networks: A survey on security, privacy and open research issues," *IEEE Communications Surveys & Tutorials*, vol. 21, no. 3, pp. 2886–2927, 2019.

20 W. Stallings, *Cryptography and Network Security Principles and Practice Seventh Edition Global Edition British Library Cataloguing-in-Publication Data*. Pearson Education, Inc., 2017.

21 H. Zhang, B. Liu, and H. Wu, "Smart grid cyber-physical attack and defense: A review," *IEEE Access*, vol. 9, pp. 29641–29659, 2021.

22 F. Liberati, E. Garone, and A. Di Giorgio, "Review of cyber-physical attacks in smart grids: A system-theoretic perspective," *Electronics*, vol. 10, no. 10, p. 1153, 2021.

23 D. Tellbach and Y. F. Li, "Cyber-attacks on smart meters in household nanogrid: Modeling, simulation and analysis," *Energies*, vol. 11, no. 2, p. 316, 2018.

24 L. Wei, L. P. Rondon, A. Moghadasi, and A. I. Sarwat, "Review of cyber-physical attacks and counter defense mechanisms for advanced metering infrastructure in smart grid," *Proceedings of the IEEE Power Engineering Society Transmission and Distribution Conference*, vol. 2018-April, 2018.

25 M. Snehi and A. Bhandari, "Vulnerability retrospection of security solutions for software-defined cyber–physical system against DDoS and IoT-DDoS attacks," *Computer Science Review*, vol. 40, p. 100371, 2021.

26 S. Qiu, Q. Liu, S. Zhou, and C. Wu, "Review of artificial intelligence adversarial attack and defense technologies," *Applied Sciences (Switzerland)*, vol. 9, no. 5, p. 909, 2019.

27 M. Zolanvari, M. A. Teixeira, L. Gupta, K. M. Khan, and R. Jain, "Machine learning-based network vulnerability analysis of industrial Internet of Things," *IEEE Internet of Things Journal*, vol. 6, no. 4, pp. 6822–6834, 2019.

28 M. Comiter, "Attacking Artificial Intelligence," *Belfer Center Paper*, vol. 8, 2019.

Part IV

Privacy in 6G Vision

13

6G Privacy*

* With additional contribution from Chamara Sandeepa, University College Dublin, Ireland

This chapter provides comprehensive details on privacy-related aspects for B5G/6G networks. After reading this chapter, you should be able to:

- Understand the taxonomy of 6G privacy.
- Understand privacy definitions related to 6G.
- Understand the set of 6G privacy objectives.
- Understand the standardization initiatives for 6G privacy preservation.

13.1 Introduction

The possibility of achieving harmony among hundreds of devices connected around a person is underway with B5G/6G and will be a common ground or truth in future. Since these devices are connected, they may communicate with nearby devices and stream data continuously. It will digitally create a very sophisticated environment that makes it virtually possible to track every user action. Therefore, B5G/6G networks will eventually lead the way to create a fully automated smart environment that may span over a vast population.

Since this seems inevitable, it is obvious that privacy is crucial in this situation. The 6G applications also drastically increase the possibility of identifying individuals, their health status, current actions, prediction of decisions, motion, interests, habits, personal beliefs, ideologies, etc. These applications may do this by analyzing the data output through sensors, smartphones, and other personal electronic devices with network connectivity. As B5G/6G will facilitate their communication, these devices will not be isolated anymore. Therefore, third parties

Security and Privacy Vision in 6G: A Comprehensive Guide, First Edition.
Pawani Porambage and Madhusanka Liyanage.

may accumulate a wide range of signals and extract this information on the data subjects/users. For example, a smart light connected wirelessly with a remote server or a smart device can increase the house's energy efficiency. However, at the same time, it can collect data on the person when a user is at home, which rooms are used often, whether there are people in the house, and can provide insights on user habits, preferences, and daily routine [1].

With the facilitation of capabilities by B5G/6G networks, new technology applications, such as extended reality, smart medical systems, autonomous driving, can be expected soon; then, there will be many protocols, algorithms, and different device brands that will increase the complexity of the network. The work in [2] shows that since B5G/6G networks will be having this increased network heterogeneity, more concerns will be raised on privacy compared to previous generations. For example, the authors claim the involvement of connected devices in every aspect of humans, such as medical implants, that could pose serious concerns of potential leaks of personal information such as health records. Specifically, this could happen if an adversary detects a privacy vulnerability in algorithms of such systems or transfer data through protocols with lesser data privacy. However, we noted that the discussions of privacy in B5G/6G and verticals are still at its early stage, though there is a significant attention on these networks.

This chapter is exclusively focused on the privacy aspects of B5G/6G networks and its verticals, within a wide range of topics including

- **Identify privacy taxonomy**: Investigate different taxonomies for privacy defined by consideration of various aspects of privacy requirements.
- **Propose a set of privacy goals for B5G/6G**: These goals should be fulfilled to ensure and enhance privacy for B5G/6G networks.
- **Discuss 6G privacy projects and standardization**: We summarize the existing privacy projects applicable for 6G and also the standardization efforts made to achieve privacy preservation and enhancement.

13.2 Privacy Taxonomy

The term "privacy" is generally known to be the assurance that individuals get the control or influence of what details related to them may be collected and stored and by whom and to whom the information may be disclosed [3]. It is the capability that a person gets to seclude the information about themselves selectively.

Understanding the categories of privacy is essential to identify potential privacy-related aspects in B5G/6G networks since the possible privacy threats may fall into these categories. The following Section 13.4 provides an overview of the taxonomy of privacy and its relation with B5G/6G networks.

Privacy consists of psychological and social background, as it is based on personal interests and their social influence. In [4], the authors identify a set of privacy functions for six types of privacy, namely solitude, isolation, anonymity, reserve, intimacy with friends, and intimacy with family. The privacy functions are autonomy, confiding, rejuvenation, contemplation, and creativity. The work presented in [5] defines another set of seven types of privacy as follows:

- **Privacy of the person**: Right to keep body functions and body characteristics (e.g. genetic codes and biometrics) private.
- **Privacy of behavior and action**: The capability to behave in public, semipublic, or one's private space preserving privacy.
- **Privacy of communication**: Avoid the interception of communications.
- **Privacy of data and image**: Ensure that users' data are not automatically available and controlled by other individuals and organizations.
- **Privacy of thoughts and feelings**: The right not to share people's thoughts or feelings or to have those thoughts or feelings revealed.
- **Privacy of location and space**: Individuals have the right to move about in public or semipublic space without being identified, tracked, or monitored.
- **Privacy of association**: People's right to associate with whomever they wish, without being observed.

13.3 Privacy in Actions on Data

Considering the users in the B5G/6G networks, their data holders, such as network components, third-party services, and potential adversaries, required that the privacy actions should be different. The work in [6] divides the taxonomy of privacy in different actions occurring to the data subject, which is generally a B5G/6G user in our context. Figure 13.1 illustrates these actions and their components. We discuss these actions and their relation with the B5G/6G networks.

13.3.1 Information Collection

Third parties can collect user information for different reasons, including personalization and advertising for marketing purposes [7]. This may happen via surveillance, through watching, listening to, or recording individuals' activities through various methods such as click-through rates [7]. Another option is interrogation through multiple means of questioning or probing for information.

This information can be collected from users directly via user devices by third-party services operated in the B5G/6G networks. The collection of information will be done through various sensor devices and personal devices such

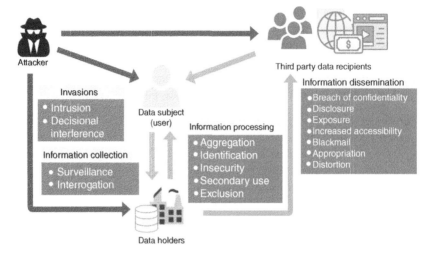

Figure 13.1 A taxonomy of privacy for interactions of the data subject with external parties. Source: Adapted from [6].

as mobile phones. A large quantity of data collection will be possible in B5G/6G compared with the current network capabilities due to the higher data rates provided in terabit speeds. Also, they will have intelligence-enabled in the collection points, which will be smart to capture only the essential information that could expose user privacy.

13.3.2 Information Processing

The use, update, and manipulation of already collected data can be considered as the information processing [6]. This can be used to get insights on the generators of data since, through processing, it may be possible to trace the origins and their links by connecting data. The work in [6] further divides the information processing into aggregation, identification, insecurity, secondary use, and exclusion.

The B5G/6G networks will be able to process data even from the edge devices with integrated intelligence. Also, from the architecture of B5G/6G defined in [8], it can be expected that there can be a tremendous processing of data in the data mining and analytics layer of B5G/6G. For this, artificial intelligence (AI) tools will be used to get insights into complex, big data, which could reveal patterns that expose user privacy. The Art. 5 [9] of the European Union (EU) data protection law General Data Protection Regulation (GDPR) defines that the personal data should be processed with principles of lawfulness, fairness, and transparency, purpose limitation, data minimization, accuracy, integrity, and confidentiality, and storage limitation.

13.3.3 Information Dissemination

This refers to the revealing of personal data or the threat of spreading of information [6]. The dissemination may include the breach of confidentiality, disclosure, exposure, an increase of accessibility, blackmail, appropriation, and distortion of information [6]. These potential privacy threats come from a third party accessing the stored information. Therefore, these issues may be mitigated by adequately keeping information with proper access rights. The next problem arises when using third parties to store information, such as cloud storage providers. They may be able to access this information due to various reasons such as poor security measures, hacking, stolen media, deliberate or unintended access by employees, accidental publishing.

The B5G/6G networks may also be vulnerable to uncontrolled information dissemination due to a lack of regulations for future data types, technologies, and loopholes in the existing laws. Also, there is currently no unified approach to solving information dissemination issues.

13.3.4 Invasion

This involves threat caused to the user, and it does not necessarily involve information [6]. The authors in [6] categorize the invasions into two types, intrusion and decisional interference. Work in [10] shows that intrusion can be detected from abnormal patterns of system usage. Ferrag et al. [11] identifies three intrusion detection methods: signature-based, anomaly-based, and hybrid systems for 4G and 5G cellular networks. The latter, decisional interference involves the governmental interference with people's decisions [6].

The invasions will be a definite threat to privacy in B5G/6G networks. The systems may not be designed by keeping privacy integrated into their systems, especially when considering edge devices and networks. It is possible due to their lack of computation capacity, which may lead to implementing only relatively weaker privacy preservation algorithms. Another possible invasion could be the attacks on AI in network management and applications in B5G/6G networks.

13.4 Privacy Types for 6G

When considering taxonomy in terms of privacy types for 6G networks, Porambage et al. [12] categorizes privacy aspects into data, personal behavior, action, image, communication, and location.

13.4.1 Data

Data privacy represents the privacy of the stored data [13]. We that observe users store their personal and sensitive data on mobile devices as mobile phones are easily accessible. Therefore, these devices need more measures to protect data to avoid leaking them to a third party without user consent. Due to the high data rates provided by future networks, an enormous volume of data will get collected at central storage within a short period. Therefore, attackers may find it easy to exploit privacy vulnerabilities in these single data nodes. Thus, privacy is much of a concern in terms of data.

13.4.2   Actions and Personal Behavior

People's actions and behavior can play a crucial role in privacy with the influence of social media. Work in [14] shows that a person's behavior is predictable using only the information provided by people around them in a social network. This includes information for a third party to jeopardize user privacy. Even when a user leaves the social network, their "shadow profile" will still be available through their friends' data. The application of IoT widens the paths to track user behavior through extended sensors in the environment. Massimo et al. [15] uses tourist data collected from IoT-enabled areas to simulate user behaviors under different contexts and scenarios and provide recommendations. In B5G/6G era, approaches such as virtual reality (VR) and smart sensing environments will gather more insights on personal behavior and may even predict human thought patterns. Therefore, user privacy protection may need to be strictly considered in such cases.

13.4.3   Image and Video

With the increase of influence from social media platforms, users tend to add their personal details via images and videos. Though the users are given the freedom to select who will be seeing the post, there is information that may consist in the images that adversely affect privacy that users may not be aware of. Also, video surveillance allows third parties to track users, and these parties may potentially use the surveillance data to compromise user privacy. The work in [16] discusses the existing image privacy protection techniques, including editing techniques: blurring, black-box, pixelation, masking, encryption, and scrambling. The authors also include other techniques such as face regions, false color, and JPEG meta-data embedding. It also shows that a balance among privacy, clarity, reversibility, security, and robustness must be maintained for public safety requirements such as identifying suspicious behavior and knowing nonsensitive information like the number of people in a specific area. Çiftçi et al. [17] presents two reversible privacy protection schemes implemented using false colors for JPEG images. The images are applied with encryption techniques to allow only authorized parties with authorization keys to reverse the false coloring. Authors in [18] propose a tool called "iPrivacy" as an automatic privacy recommendation system for image sharing from deep multitask learning. However, ultrahigh resolution and multidimensional images and videos will be very common in B5G/6G. Therefore, it will be possible to expose more unintentional details from these digital media. Thus, more safeguarding of privacy of them is essential in future.

13.4.4 Communication

Privacy in different modes of communication is another classic yet important aspect since communication with rich media has increased drastically over the development of platforms in recent years. There will be different types of communication providers to provide a wide range of services through voice, text, video, and other novel sensory modes for users in B5G/6G as options to choose. Therefore, the users will have more possibility for privacy leakages. Through IoT communication, sensor data might carry vital health signatures of individuals. The work in [19] shows that IoT platforms, with their cloud connections, pose vulnerabilities that attackers may exploit to extract useful information. It also indicates that overprivileged IoT devices and services may also collect user data themselves that may cause breaches in privacy.

13.4.5 Location

The location privacy is important for an individual since it includes details of physical locations that the person has traveled and which may also reveal the many related information including personal behavior, financial status, habits, beliefs, interests, and even political preferences [20]. They are highly valuable data for a third party and location data include moving objects, spatial coordinates, current time, and other unique features [21]. Many recent works show the importance of location privacy [20–22]. Location-based services (LBS) have ever-increasing popularity, and they consist of components: a positioning system, users, networks, LBS server content provider, and a location privacy server [22]. These components are interconnected together to provide the LBS. The authors show that an adversary can attack any of the components in the network, the location privacy server, the LBS server, or from the user side. Primault et al. [20] provides many existing Location Privacy Protection Mechanisms (LPPM) considering privacy, utility, and performance aspects. In addition, IoT systems are also vulnerable to location-based privacy issues [21], since they may include components that can sense location. Future B5G/6G are expected to be incorporated with gadget-free environment and fully automated travel and transport options, where mandatory location privacy preservation will be needed.

13.5 6G Privacy Goals

As it is clear that 5G and B5G networks can handle vast amounts of data, it may potentially be subjected to privacy leakages due to the complexity of this large data and wide range of services. Potential issues in privacy can go undetected in these latest technologies and new services. For example considering edge computation, with limited capability and lack of privacy protection, it could be possible for an adversary to exploit these issues on a large scale due to more-than-ever inter-connectivity of the new-generation networks. Therefore, we propose a set of privacy goals that need to be achieved to ensure these networks are trustworthy for general users. In [12], authors provide open challenges for privacy for 6G. The work in [13] defines a set of objectives for privacy in 5G networks. By extending this information, we provide our privacy goals for B5G/6G as follows:

13.5.1 Ensure of Privacy-Protected Big Data

The amount of big data is ever-increasing with recent technology that enables facilities such as digital and cloud storage, IoT, and network. Therefore, we can identify that B5G/6G networks contribute significantly to big data. Privacy should be considered in each stage: data generation, data storage, and data processing. Figure 13.2 shows these stages and the possibility for the attacker to exploit the process [23]. However, the work to enhance privacy in big data may not suffice because of the rate at which data are generated. There are existing techniques for big data privacy, including HybrEx, k-anonymity, T-closeness, and L-diversity [24]. Yet, more solutions for privacy enhancement should be given for B5G/6G networks to ensure user trust in them.

13.5.2 Privacy Guarantees for Edge Networks

In edge computing, the data can be processed close to or at the edge of the network. It improves the quality-of-service (QoS) and user experience. Especially, the

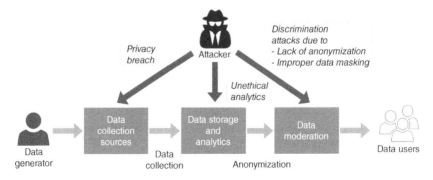

Figure 13.2 Big data system and privacy considerations on different stages of operation.

platforms like IoT providers may not consider data privacy a lot due to limited power and processing capacity.

Trusting edge devices is another challenge that future networks will face. This is due to the reason that untrusted third parties can connect their devices and services to the network [25]. Therefore, we think having guarantees for edge networks is an important factor since there is an explosive increase in the number of interconnected edge devices with the 5G [26]. It will continue in the upcoming generations.

Edge AI is another crucial area rising with the edge networks, which can be used to offload decision-making from cloud to the edge devices. This will reduce the cloud servers' load and reduce the traffic by filtering unnecessary data significantly. It is also used as a data privacy preservation technique through methods such as federated learning (FL), which will be discussed further in Section 14.3. Therefore, with this rise of this edge intelligence, the guarantees for edge privacy should also consider safeguarding the edge AI's privacy.

13.5.3 Achieving Balance Between Privacy and Performance of Services

The privacy-preserving algorithms may cause performance degradation with the implementation of methods such as encryption in the real world [27]. This can be significant for devices with limited computation capacity such as IoT, yet it is imperative to implement privacy preservation for such systems. Achieving the right balance between privacy and performance is another goal that we can propose. One of the attempts to understand the balance in the context of IoT is in [28], where authors propose a two-step approach: first, get the trade-off between the quality of collected data and performance; second, to understand how the nature of data affects adversary's ability to infer the user's private information. Today, AI-based techniques are used widely for general-purpose applications. However, with B5G/6G, they will be utilized even more, including adopting AI for B5G/6G architecture itself. The privacy-preserved AI-based decision-making may add further overhead to the operations, which need to perform at terabit speeds. Therefore, achieving the right balance between privacy and performance is important to maintain the proper QoS and data rates.

13.5.4 Standardization of Privacy in Technologies, and Applications

The technologies and applications used in B5G/6G networks evolve independently. However, when considering the real-world use of networks, we see they are interconnected. They may use their different set of protocols to communicate with each other. Therefore, we consider the proper standardization of security mechanisms as an important step to ensure privacy protection during interfacing and communicating among applications of different technologies. This could be taken in to account especially when interfacing with edge devices, as they are more vulnerable to privacy leakages. Liyanage et al. [13] also shows that harmonizing privacy services in the global context to promote a digital single market is an important objective for B5G/6G networks.

13.5.5 Balance the Interests in Privacy Protection in Global Context

For achieving standardization, it may be an ambitious goal to make a global level understanding of the importance of privacy protection, satisfying the interests in fostering privacy services at a global level. While methods such as human rights mechanisms may help, they are insufficient on their own, and stronger protective standards for cyberspace activities should be developed [29]. The work in [29] summarizes a set of actions or requirements for an effective legal regime that should be taken to ensure privacy in cyberspace:

- Define the rights of individuals and rights of data controllers.
- Provide legal means of privacy protection for individuals.
- Protect individuals from unauthorized privacy violations.
- Impose on service providers obligations to ensure lawfulness in the data processing.
- Make sure individual rights that are protected when processing data, even with incompatibilities of obligations in the legal system.
- Be easily accessible and understandable.

However, it is also essential to maintain the balance between privacy requirements and industry when imposing regulations in this case. The studies in [30] show that websites have substantially reduced their interactions with web technology vendors after GDPR is effective. Also, they mention that many firms undergo losses while major players significantly increase their market shares. This makes the market less diversified and raises the barrier of entry. The authors in [30] discuss the requirements and provide references on how organizations should consider being compliant with GDPR, using techniques such as proper distinguishing data controllers and processors, transfer of personal data with third countries or international organizations, independent supervision, and penalties. It implies a substantial effect on industries and complications in handling organizations' data with third parties.

13.5.6 Achieving Proper Utilization of Interoperability and Data Portability

Data portability is the ability to easily transfer data from one system to another without re-entering data. This can ease a consumer's life since they can avoid the redundant task of entering data in multiple applications.

We observe recent collaboration to enhance interoperability among companies, including Google, Microsoft, Twitter, and Apple. They have initiated a Data Transfer Project to allow individuals to transfer their data between online service providers [31]. It is an open-source project to extend data portability beyond a user's ability to download a copy of their data from their service provider, providing the user the ability to initiate a direct transfer of their data into and out of any participating provider. It shows that it is crucial to consider privacy principles of data minimization and transparency, with clear and concise data to transfer data among such systems. The project is implemented using adapters that convert various proprietary data formats to a small number of canonical forms or data models [32]. This helps to achieve data portability. It is easier to fulfill privacy requirements for a single type of data in each organization than different data types and multiple interfaces. Figure 13.3 shows an overview of the implementation of the project.

However, the right to data portability remains barely known among consumers and is only implemented in a fragmented manner [33].

13.5.7 Quantifying Privacy and Privacy Violations

Privacy is difficult to define as it is a subjective concept and has different levels from person to person. Even the opinion on privacy that one may have may vary over time. Due to this nature, quantification of privacy can be difficult. However, based on the context, we may provide local standards and metrics related to different types of privacy. The process may need to incorporate with other fields such as psychological aspects.

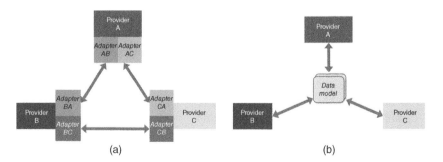

Figure 13.3 Overview of data transfer project: (a) without data model, adapters have to be written for communication among providers and (b) with data model, only a single application programming interface (API) is need to be maintained for communication with any provider. Source: Adapted from [32].

13.5.7.1 Achieving Privacy Protected AI-Driven Automated Network Management Operations

The B5G/6G networks should fulfill massive network traffic demands for billions of connected devices with better QoS. Network management functions automation is essential to make this a reality. Such an approach is called Zero-touch network and Service Management (ZSM) by ETSI, where the authors in [34] provide an architecture driven by AI for complete automation of management operations. However, it is also essential to regularly monitor the functionality of this automation for any privacy leakages since adversaries may find vulnerabilities in the AI models and decisions. Also, it is possible to attack these networks using AI to extract user information or other private details. Therefore, we suggest privacy protection is imperative when automating the network operations in B5G/6G networks.

13.5.8 Getting Explanations of AI Actions for Privacy Requirements

The users have the right to question decisions made by AI that handles their personal data. Therefore, AI used in future network operations should be explainable, and responsible entities should explain how their AI made that decision and the possible assumptions. However, many of today's machine learning algorithms put themselves in conventional black box view. Therefore, AI explainability can be considered to be one of the most important goals in terms of privacy requirements.

References

1 M. Ylianttila, R. Kantola, A. Gurtov, L. Mucchi, I. Oppermann, Z. Yan, T. H. Nguyen, F. Liu, T. Hewa, M. Liyanage et al., "6G White Paper: Research Challenges for Trust, Security and Privacy," *arXiv preprint arXiv:2004.11665*, 2020.

2 V.-L. Nguyen, P.-C. Lin, B.-C. Cheng, R.-H. Hwang, and Y.-D. Lin, "Security and privacy for 6G: A survey on prospective technologies and challenges," *IEEE Communications Surveys & Tutorials*, vol. 23, no. 4, pp. 2384–2428, 2021.

3 W. Stallings and M. P. Tahiliani, *Cryptography and Network Security: Principles and Practice*, William Stallings, vol. 6, 2014.

4 D. M. Pedersen, "Model for types of privacy by privacy functions," *Journal of Environmental Psychology*, vol. 19, no. 4, pp. 397–405, 1999.

5 R. L. Finn, D. Wright, and M. Friedewald, "Seven types of privacy," in *European Data Protection: Coming of Age*, S. Gutwirth, R. Leenes, P. de Hert, and Y. Poullet, Eds. Dordrecht: Springer, 2013, pp. 3–32.

6 D. J. Solove, "A taxonomy of privacy," *University of Pennsylvania Law Review*, vol. 154, p. 477, 2005.

7 E. Aguirre, D. Mahr, D. Grewal, K. De Ruyter, and M. Wetzels, "Unraveling the personalization paradox: The effect of information collection and trust-building strategies on online advertisement effectiveness," *Journal of Retailing*, vol. 91, no. 1, pp. 34–49, 2015.

8 C. Sandeepa, B. Siniarski, N. Kourtellis, S. Wang, and M. Liyanage, "A survey on privacy for B5G/6G: New privacy goals, challenges, and research directions," *Journal of Industrial Information Integration*, vol. 30, p. 100405, 2022.

9 "Art. 5 GDPR - Principles relating to processing of personal data," https://gdpr-info.eu/art-5-gdpr/, October 2021.

10 D. E. Denning, "An intrusion-detection model," *IEEE Transactions on Software Engineering*, SE-13, no. 2, pp. 222–232, 1987.

11 M. A. Ferrag, L. Maglaras, A. Argyriou, D. Kosmanos, and H. Janicke, "Security for 4G and 5G cellular networks: A survey of existing authentication and privacy-preserving schemes," *Journal of Network and Computer Applications*, vol. 101, pp. 55–82, 2018.

12 P. Porambage, G. Gür, D. P. M. Osorio, M. Liyanage, A. Gurtov, and M. Ylianttila, "The roadmap to 6G security and privacy," *IEEE Open Journal of the Communications Society*, vol. 2, pp. 1094–1122, 2021.

13 M. Liyanage, J. Salo, A. Braeken, T. Kumar, S. Seneviratne, and M. Ylianttila, "5G privacy: Scenarios and solutions," in *2018 IEEE 5G World Forum (5GWF)*. IEEE, 2018, pp. 197–203.

14 D. Garcia, "Privacy beyond the individual," *Nature Human Behaviour*, vol. 3, no. 2, pp. 112–113, 2019.

15 D. Massimo, E. Not, and F. Ricci, "User behaviour analysis in a simulated IoT augmented space," in *Proceedings of the 23rd International Conference on Intelligent User Interfaces Companion*, 2018, pp. 1–2.

16 L. Rakhmawati, Wirawan, and Suwadi, "Image privacy protection techniques: A survey," in *TENCON 2018 - 2018 IEEE Region 10 Conference*. IEEE, 2018, pp. 0076–0080.

17 S. Çiftçi, A. O. Akyüz, and T. Ebrahimi, "A reliable and reversible image privacy protection based on false colors," *IEEE Transactions on Multimedia*, vol. 20, no. 1, pp. 68–81, 2017.

18 J. Yu, B. Zhang, Z. Kuang, D. Lin, and J. Fan, "iPrivacy: Image privacy protection by identifying sensitive objects via deep multi-task learning," *IEEE Transactions on Information Forensics and Security*, vol. 12, no. 5, pp. 1005–1016, 2016.

19 L. Babun, K. Denney, Z. B. Celik, P. McDaniel, and A. S. Uluagac, "A survey on IoT platforms: Communication, security, and privacy perspectives," *Computer Networks*, vol. 192, p. 108040, 2021.

20 V. Primault, A. Boutet, S. B. Mokhtar, and L. Brunie, "The long road to computational location privacy: A survey," *IEEE Communications Surveys & Tutorials*, vol. 21, no. 3, pp. 2772–2793, 2018.

21 C. Yin, J. Xi, R. Sun, and J. Wang, "Location privacy protection based on differential privacy strategy for big data in industrial Internet of Things," *IEEE Transactions on Industrial Informatics*, vol. 14, no. 8, pp. 3628–3636, 2017.

22 B. Liu, W. Zhou, T. Zhu, L. Gao, and Y. Xiang, "Location privacy and its applications: A systematic study," *IEEE Access*, vol. 6, pp. 17 606–17 624, 2018.

23 S. Yu, "Big privacy: Challenges and opportunities of privacy study in the age of big data," *IEEE Access*, vol. 4, pp. 2751–2763, 2016.

24 P. Jain, M. Gyanchandani, and N. Khare, "Big data privacy: A technological perspective and review," *Journal of Big Data*, vol. 3, no. 1, pp. 1–25, 2016.

25 M. Du, K. Wang, Y. Chen, X. Wang, and Y. Sun, "Big data privacy preserving in multi-access edge computing for heterogeneous Internet of Things," *IEEE Communications Magazine*, vol. 56, no. 8, pp. 62–67, 2018.

26 Y. Liu, M. Peng, G. Shou, Y. Chen, and S. Chen, "Toward edge intelligence: Multiaccess edge computing for 5G and Internet of Things," *IEEE Internet of Things Journal*, vol. 7, no. 8, pp. 6722–6747, 2020.

27 H. Zhang, Y. Shu, P. Cheng, and J. Chen, "Privacy and performance trade-off in cyber-physical systems," *IEEE Network*, vol. 30, no. 2, pp. 62–66, 2016.

28 R. Dong, L. J. Ratliff, A. A. Cárdenas, H. Ohlsson, and S. S. Sastry, "Quantifying the utility–privacy tradeoff in the Internet of Things," *ACM Transactions on Cyber-Physical Systems*, vol. 2, no. 2, pp. 1–28, 2018.

29 M. Rojszczak, "Does global scope guarantee effectiveness? Searching for a new legal standard for privacy protection in cyberspace," *Information & Communications Technology Law*, vol. 29, no. 1, pp. 22–44, 2020.

30 C. Peukert, S. Bechtold, M. Batikas, and T. Kretschmer, "European privacy law and global markets for data," *Available at SSRN 3560392*, 2020.

31 E. Egan, "Data portability and privacy," 2019.

32 Data Transfer Project, https://datatransferproject.dev/documentation.

33 S. Kuebler-Wachendorff, R. Luzsa, J. Kranz, S. Mager, E. Syrmoudis, S. Mayr, and J. Grossklags, "The Right to Data Portability: Conception, status quo, and future directions," *Informatik Spektrum*, vol. 44, pp. 264–272, 2021.

34 C. Benzaid and T. Taleb, "AI-driven zero touch network and service management in 5G and beyond: Challenges and research directions," *IEEE Network*, vol. 34, no. 2, pp. 186–194, 2020.

14

6G Privacy Challenges and Possible Solution*

* With additional contribution from Chamara Sandeepa, University College Dublin, Ireland

This chapter provides a comprehensive analysis on 6G privacy issues and possible solutions to mitigate them. After reading this chapter, you should be able to:

- Understand challenges that appear as barriers to reach 6G privacy preservation.
- Understand the set of solutions applicable to the proposed architecture of B5G/6G networks to mitigate the challenges.

14.1 Introduction

The faster the world is moving toward a digital reality, the higher the risk people may put their privacy, which is more precisely called digital privacy. The data are collected for many applications to improve their service performance. Such processed data or the information leakage always create huge privacy issues which require well-balanced privacy preserving techniques. When more and more end devices tend to share local data to the centralized entities, the storage and processing of this data pile with the added privacy protection mechanisms will be difficult. As 6G systems may have simultaneous connectivity up to about 1000 time greater than in 5G, privacy protection should be considered an important performance requirement and a key feature in wireless communication in the envisioned era of 6G [1]. However, in the current process of data collection and analysis, privacy protection has not received the enough attention and priority level. Therefore, there are many research opportunities for finding the correct balance between increasing data privacy and maintaining them with lower computation load which may reduce the speed and accuracy of the computation. In Figure 14.1, we describe and

Security and Privacy Vision in 6G: A Comprehensive Guide, First Edition.
Pawani Porambage and Madhusanka Liyanage.
© 2023 The Institute of Electrical and Electronics Engineers, Inc. Published 2023 by John Wiley & Sons, Inc.

Figure 14.1   Summary of 6G privacy.

illustrate a summary of 6G privacy with respect to privacy types, privacy violation, privacy protection, and related technologies.

The issue in 6G with data privacy will be more challenging when the number of smart devices are increasing and tracking every move of a person with lack of transparency about what is exactly collected. Specially, in the big data era of decentralized systems, adding privacy protection mechanisms will further increase the communication and computational costs which already show a rapid growth [2]. The current European Union's General Data Protection Regulation (GDPR) for privacy assurance should be also subject to change with the evolving 6G applications and specifications.

14.2   6G Privacy Challenges and Issues

To formulate a set of privacy challenges, we considered different potential privacy issues in 6G architecture. The Figure 14.2 shows an overview of the distribution of these challenges and issues. Here, there are several actors involved in the architecture, end-users, developers, and attackers. The end-users generate data, developers consume this generated data to derive solutions for industry or B5G/6G-related requirements, and attackers attempt to obtain this data illegally through adversarial methods by attacking either the network or the user devices. However, there is a possibility for them to attack industries in a wide variety of methods as well.

Considering the data flow in Figure 14.2, first, the end-users interact with their user devices, such as smartphones or sensors, to generate private or personal data. Industries may store this data for their requirements and it may affect

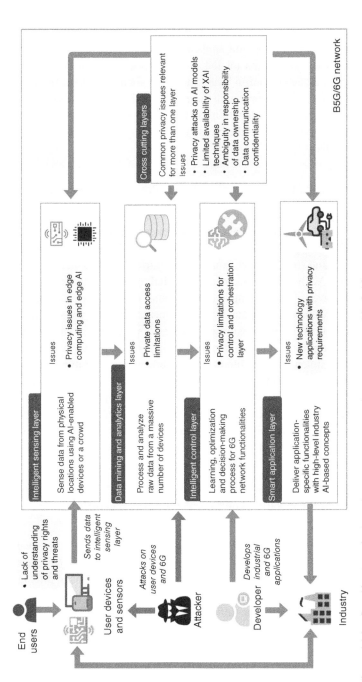

Figure 14.2 Classification of privacy issues based on architectural layers of B5G/6G.

user privacy. Then, this data are forwarded to the intelligent sensing layer in the B5G/6G enabling technology architecture. Next, the data are processed and/or forwarded to the data mining and application layer. There, they are stored and analyzed to harness more data related to user privacy. After that, the data are handled by the intelligent control layer. The smart application layer then obtains the control layer's data to interact with the external industry or services. Industries or services may forward this data to users to interact with. Several issues related to all layers are summarized as the cross-cutting layers. We discuss each of them with more details as follows:

14.2.1 Advanced 6G Applications with New Privacy Requirements

B5G/6G network capabilities provide an arena for many associated new technologies to the consumer market. One is the emergence of digital twin, which is used in many industrial applications including smart manufacturing to improve the environmental and social impact from manufacturing via virtualization of physical entities [3]. With the improvements in digital twin and extended reality, the concept of Metaverse is becoming another popular topic. Metaverse can be considered as a fully immersive virtual world. However, Metaverse incorporate many privacy requirements such as ensuring the behaviors of people are not tracked by third parties, the private and confidential data kept in the virtual environment is not exposed or stolen, and the biometric and other critical profile data used to access Metaverse is not compromised. Another example would be autonomous driving, and the advancement of vehicle-to-vehicle communication with B5G/6G. The sensitive data such as location could be tracked by third parties and the identity of the owners, their travel preferences, and their habits will be revealed easily. Therefore, we pinpoint that it is essential to discuss about potential privacy threats this eco-system of technologies bring with B5G/6G.

14.2.2 Privacy Preservation Limitations for B5G/6G Control and Orchestration Layer

After considering the associated technologies and their ecosystem, we focus on the B5G/6G architecture itself for potential privacy issues. The B5G/6G control architecture is composed of many novel features, including artificial intelligence (AI)-based zero-touch network orchestration, optimization, and management [4]. These features are essential in fulfilling the ultra-high-speed requirements of future networks and serve billions of highly interconnected devices. Intent-based networking (IBN) is also an emerging area where automated network management solutions are driven by a set of specifications called intents [5]. The intents affect the operation of network management. However, due to this, it may be of potential interest to the attackers. Closed-loop automation is another step in automating network management, where the process monitors and assesses and automatically acts in network occurrences such as congestion and faults [6]. If this monitoring process is compromised, adversaries may be able to cause various privacy-related attacks and these can go unnoticed. As the control architecture of networks evolves, it is important to consider the privacy strengths and potential weaknesses of future architectures.

14.2.3 Privacy Attacks on AI Models

As discussed in the previous two privacy issues, AI models will be used extensively for B5G/6G. For another example, zero-touch 6G networks will heavily use AI as a key component for its automated decision-making process [7]. Therefore, it is crucial to identify privacy attacks that could be possible on AI models (Figure 14.3). The systems that use AI techniques such as machine learning (ML) could be subjected to attacks named adversarial attacks, where an attacker gets details on machine learning models and the data used to train the models [8]. The solution proposed to this issue is called adversarial machine learning [8]. Another issue arises from AI models themselves. The machine learning models can learn from big data and predict patterns and trends [8]. If these trends could reveal sensitive information on individuals, then it is a privacy issue, which an AI model makes.

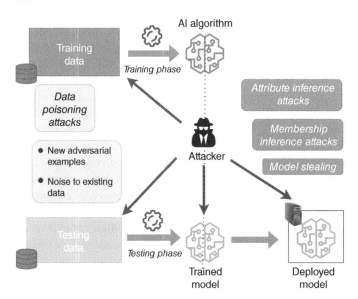

Figure 14.3 Categories of machine learning attacks in training and testing phase.

14.2.4 Privacy Requirements in Cloud Computing and Storage Environments

In the era of B5G/6G, vast collections of data will be read from many consumer products such as mobiles, computers, and sensor data in Internet of Things (IoT) devices. This makes it difficult to maintain in-house storage for many organizations as it will be practically challenging to support dedicated hardware for these data. Therefore, they will usually be outsourced to third-party cloud storage providers who maintain the data in a distributed fashion. This reduces the cost as well as improves the availability of data. However, there are some privacy issues relative to data storage in such environments. As noticed from the previous privacy issue, AI models can infer user identities and sensitive information from such massive data, which are a critical privacy leakage. Also, data processing via cloud processing environments is quite common through outsourcing computation-intensive tasks to flexible computing resources, especially, when the task takes a short time but significant computation work. The tasks such as AI model training can be done with this. This trend can be expected to increase over time. Since cloud computing is also generally distributed, there could be potential vulnerabilities in privacy leakage when processing data across multiple nodes that may have security weaknesses.

14.2.5 Privacy Issues in Edge Computing and Edge AI

Edge computing is a paradigm aimed at solving IoT and localized computing requirements by bringing computing resources close to the "edge" near the end-users. This is achieved through adding computing nodes close to user devices, thereby reducing the overhead to cloud computing [9]. The 5G brings edge computing to customers with more control on their data, but lack of consumer trust is an issue that prevents its benefits [10]. It is indubitably favorable for computing in the edge, rather than communicating directly with the cloud for resource-constrained devices, since it can save their energy usage, improve the quality-of-service (QoS), and reduce the network traffic. However, the privacy concerns here are very important to consider prior to applying it in B5G/6G (Figure 14.4). There is a frequent availability of these edge devices in many physical locations where attackers might get easy access. Also, a malicious edge computing device or a compromised device can eavesdrop or steal user data processed through it, such as sensitive health data. With the addition of many billions of more edge devices in B5G/6G, edge computing and edge AI will also increasingly face similar issues due to possible less privacy protection measures in resource constrained environments. The AI algorithms are vulnerable to all the privacy attacks discussed, such as poisoning, membership inference, and attribute inference attacks.

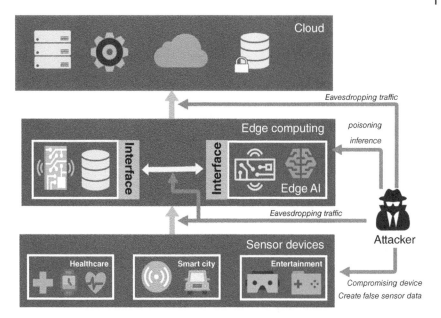

Figure 14.4 Edge computing overview and attack scenarios on sensor devices and the edge layer.

14.2.6 Cost on Privacy Enhancements

In terms of power consumption and processing, the cost of privacy can be considered inevitable when implemented in the real world. As privacy enhancement techniques such as algorithms and protocols carry the extra computation, it may affect the overall performance of any system, from cloud computing to resource-constrained devices such as sensor nodes. Also, this additional computation always results in more energy consumption on them. Even cloud computing and storage may also require privacy enhancement such that the impact of these enhancements may decrease the existing performance or operational costs of cloud services. Since we cannot fully minimize this issue, understanding the right balance between the cost and the risk is the most important aspect that can be considered here. Figure 14.5 illustrates the top six fines imposed by GDPR on various corporates in EU [11] by the end of 2021. From these high fines, it can be observed that the costs needed to pay for the lack of sufficient improvements of privacy are very high compared to the occasional development costs needed to implement these improvements. Therefore, lack of privacy improvements causes a significant risk of having to pay high levies for corporate organizations to cover up the privacy leakages or insufficient privacy measures.

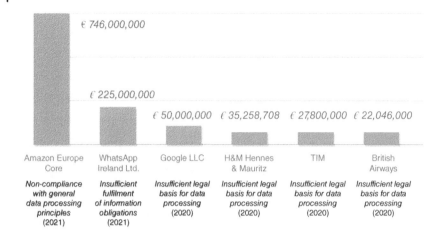

Figure 14.5 Major corporate fines from GDPR for privacy issues and breaches. Source: Adapted from [11].

14.2.7 Limited Availability of Explainable AI (XAI) Techniques

We recently observed a substantial interest in AI research, especially in the fields of deep learning, reinforcement learning, computer vision, etc. AI is commonplace in daily activities, and recently, a wide range of tools are available in web-based Machine Learning as a Service (MLaaS) platforms [12]. Despite the significant development, there is a lack of understanding about these AI models' decision-making process. These are, therefore, generally regarded as "black boxes" [13]. 6G will add intelligence as a new core element integrated to it. This will bring back the question of justifying the decision taken by the AI models, as highly sensitive private information will be transmitted through the network and may get processed by these AI models. In case of privacy violation by AI models, to investigate the roots of the issues, explainable artificial intelligence (XAI) techniques can be used. The concept of XAI, therefore, emerges again to explain and justify the underlying processes of these complex, mostly nonlinear models when making decisions, though the field itself dates back to 40 years ago [14]. We see many recent works related to XAI that attempt to describe the nature of the models in many fields, including healthcare and industrial practices [14–16]. However, the area of XAI for deep neural networks has emerged recently, and it is currently in its initial stage of development. Hence, there is a great challenge for B5G/6G in providing a reasonable explanation for the black-box approach in models such as deep neural networks.

14.2.8 Ambiguity in Responsibility of Data Ownership

With the expansion of services associated with B5G/6G, many entities will collect, process, and store personal data. Data ownership is getting complicated with this increased number of data controllers, and a problem arises when claiming the data ownership. The parties who process and store the data may be unwilling to take responsibility when a privacy leakage happens. There is also a risk that these parties may misuse data, claiming they own it. Eventually, the consumer could be the victim of any such scenario if the responsibility is not handled correctly. In B5G/6G, massive amount of big data will be collected with or without the consent of data producers. The development of multidimensional abundant sensor environment and rapid communication will facilitate this. Then, they will be processed, analyzed with edge AI and third parties who provide new technologies and services emerged with B5G/6G. Further, these data will be stored and sold to other parties. The data will be transmitted through the 6G zero-touch network, where AI models may process the data. But a question arises on who will be responsible in case of a privacy breach. Furthermore, it will be difficult to trace where the breach occurs, as there are many intermediate parties involved in the communication.

14.2.9 Data Communication Confidentiality Issues

With trillions of interconnected devices, the B5G/6G will have to fulfill scalability requirement of the network. Furthermore, with limited power of most of these devices, it is practically difficult to maintain advanced privacy protection. Furthermore, there are different modes of communications such as visible light communication (VLC), body-area communication, space internet, deep sea communication, that many incompatibilities could occur when taking privacy measures in each. During communication, there is a possibility that a third party reveals data communicated between the sender and the receiver. Ensuring confidential communication can be achieved using conventional End-to-End (E2E) data privacy solutions such as encryption, where only the intended receiver can understand the content. Since privacy is a fundamental problem, these techniques have evolved from early generations of wireless communications. However, with the involvement of several intermediate devices in B5G/6G networks, users will have to inevitably transmit their data from one device to another. Therefore, data transmission through these devices needs to have mechanisms to prevent unauthorized parties from accessing data. At the same time, upcoming networks may require to sense the type of service the users are using to provide an intelligent and differentiated QoS [17]. Furthermore, with the introduction of quantum computing, it may be possible to easily break current encryption schemes. Therefore, it is important to consider quantum-resistant cryptographic mechanisms; however, their implementation in resource constrained devices is challenging.

14.2.10 Private Data Access Limitations

Though a huge amount of data is generated each day, access to this data is limited due to privacy issues. With the B5G/6G, the data generation will be much higher as they will facilitate scalable connections for a large number of devices than the present. But, eventually, this data may get discarded without any proper use. Especially, health data generated by sensors are wasted, which may otherwise help save future lives through improved AI models for disease prediction.

Most of the current machine learning models get trained on publicly available datasets, which may not represent the present situation and do not have up-to-date information. If these models could get access to private data without breaking the privacy of data owners, we will be able to achieve a more significant leap of accuracy with improved AI models. Therefore, it will be an open issue to address and make data available without breaking privacy, especially in the era of B5G/6G that need better AI models.

14.2.11 Privacy Differences Based on Location

Most countries should enforce legal actions and agree to cooperate to make privacy regulations effective globally. However, based on locations, the definitions of privacy can be different due to many reasons such as cultural influence, religion, government regulations, rituals. Therefore, in reality, a practical challenge in solving disputes among legal entities exists.

14.2.12 Lack of Understanding of Privacy Rights and Threats in General Public

The definitions of privacy terms and rights seem far-fetched for the general public as they could be technical and require context to be understood. In the B5G/6G era, we discussed that there will be many new technologies that come with novel modes that could collect user data, such as body movements, biometrics, and even thoughts. The gadget-free communication is an emerging concept for B5G/6G where each object can sense, gather, and process the information and would be able to take context-based decisions [18]. This will enable digital services at convenience without explicit gadgets or even wearables [18]. However, this may also make ways to collect data without consent. People may lack intuition about what approaches potentially leak their data to an undesired party using these new sensing methods, and it can make privacy considerations more challenging in these circumstances. For example, though people could understand their personal information is somehow collected, they may be clueless on to which range these subtle data could be used to analyze, classify, predict, or track each aspect of their lives. With the future networks, this will be increasing even further as the complexity of the interconnectivity increases. Users may get some privacy terms and conditions when using services, but, in practice, they may not go through the privacy agreements they make when signing up for third-party applications and provide their personal and sensitive information. This creates an opportunity for third parties to collect this information and misuse them. Also, due to the uneven distribution of privacy rights, legal backgrounds, education levels, cultural influence, and many other reasons, people may have varying concerns about their privacy, making the future situation even more complicated.

14.2.13 Difficulty in Defining Levels and Indicators for Privacy

The term "privacy" is a subjective concept, which varies based on numerous reasons such as personal views, culture, and geographic locations. Therefore, it is highly challenging to define to what extent users need privacy guarantees or how much privacy is violated. However, such quantification of privacy can be very useful for AI in the decision-making process, as it simplifies which actions to be taken to ensure the privacy level. It will also then define which steps should be taken on privacy violations quickly. Also, the entities that describe and quantify privacy should justify their quantification. Having precise levels for privacy may inherently support the explainability of AI decisions for privacy.

14.2.13.1 Proper Evaluation of Potential Privacy Leakages from Non-personal Data

Nonpersonal data are another emerging concern for the B5G/6G networks due to the rapid addition of machine-generated data from various sources such as industrial robots, network sensing, and scientific equipment. Another form of nonpersonal data is the anonymized data of individuals, which were used to be personal data [19]. The increasing amount of nonpersonal data will inherently create the requirement of ensuring privacy. Though they are not directly related to individuals, there could still be potential threats to people, especially for industries or organizations that generate these data. They might reveal sensitive individuals' or confidential organizational details if a third party carefully analyzes this data. Hence, we consider it an important goal that should be fulfilled for the next generation of wireless networks.

14.3 Privacy Solutions for 6G

As discussed in Section 14.2, the privacy issues in B5G/6G networks exist, and they have to be addressed by the period we start using the next generation of wireless networks. We propose the following approaches as solutions to address the mentioned issues.

14.3.1 Privacy-Preserving Decentralized AI

AI is one of the most impactful aspects of B5G/6G networks for automated decision-making. Many applications and features thus include AI as an essential requirement. Recently, there has been an increasing interest in decentralized learning in AI. It is an approach for an AI model to learn from different sources, which may be separated from each other.

Decentralized AI has high applicability when considering the privacy aspect. This is a potential solution for the issue of having numerous new technology applications that utilize AI to learn from user data, yet the users' privacy should not be violated. In the case of ML, the privacy of data is ensured with decentralized AI, where the original data do not move from the user device. Still, a local model can learn over time and improve a global model. Therefore, a decentralized AI approach can contribute to addressing the issue of private data access limitations since the actual data are not physically transferred from the user device.

New technology applications can therefore use this approach to improve their AI models and provide a better service. Especially with IoT and privacy requirements in sensor environments, decentralized AI is emerging as a solution.

Though using different types of decentralized AI techniques seems a great research aspect, we can see there are trade-offs of each of these methods as discussed. Therefore, it may be difficult to choose the best option for real-world implementations in practice. Also, suppose such an AI approach is changed to a different one. In that case, the upgrading costs might increase significantly since the decentralized models could be spread in millions of devices unless these devices and platforms offer a flexible softwarized approach for replacing the existing models.

14.3.2 Edge AI

The idea of edge AI is the implementation of AI algorithms on edge devices. This technique is helpful for B5G/6G networks because with the ever-increasing edge connectivity, it will create huge network traffic from billions of these small devices if a central server performs AI computations on their data. Therefore, to mitigate this issue, it is inevitable to bring AI functions to the edge by offloading intelligence from the cloud to these devices to facilitate "bringing code to data, not data to code" [20].

Having Edge AI integrated into IoT environments may help solve issues such as latency related to edge computing. For example, an AI model trained to detect attacks on edge devices can be implemented to ensure data privacy. Implementing lightweight AI may be a requirement since we have to consider costs on performance and energy on them. However, having AI running on edge devices may cause performance degradation of these resource-constrained devices. Attacks on these AI models could also cause privacy losses.

The application of AI in the edge provides many benefits, including ensuring privacy preserved intelligence, low latency for the edge devices response, and reduction of network traffic. Much recent research work is in progress to provide solutions based on AI and suitable architectures to drive the change in the direction of edge AI. This is closely connected with the decentralized AI approach we discussed, where decentralized local AI models will run on these edge nodes. Therefore, we believe edge AI can be a significant part of IoT, helping B5G/6G networks offload AI functions to the edge and providing users with a smoother experience when dealing with edge services.

14.3.3 Intelligent Management with Privacy

Software service architectures often use automated resource allocation and service management for load balancing and network management tasks. This can be applied for B5G/6G networks as well. The use of intelligence to automate resource and service management will be an essential requirement for future networks since optimum decisions have to be taken considering various dynamic factors at high speeds. Therefore, in the case of privacy, the services can add such intelligence to monitor the service allocation and status to identify potential threats that could endanger user privacy.

Using intelligence-based controlling contributes to privacy enhancements of B5G/6G control and orchestration layer architecture where the resource allocation is done using automated mechanisms. This automation can further help to bring down the privacy costs of the new wireless networks, as potential privacy threats could be identified beforehand and managed automatically. Proper privacy-enhanced controlling can contribute to fixing the privacy issues related to edge devices since possible privacy leakages from edge devices will be able to be detected from the intelligent controllers.

14.3.4 XAI for Privacy

When using intelligence-based solutions, a reasonable explanation needs to be provided using XAI for justification of AI-based decisions for B5G/6G, as discussed in Section 14.2. It is critical to privacy in future networks since users have rights,

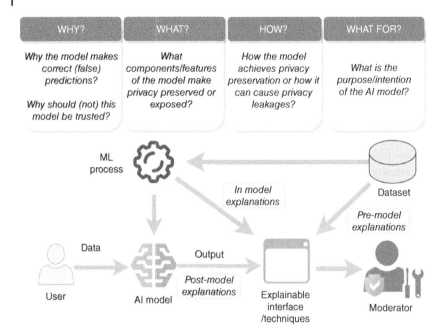

Figure 14.6 Application of XAI for privacy in B5G/6G AI models and questions to address.

and any privacy violations may cause legal actions. The decisions may depend on how explainable and reasonable the AI decision is. Figure 14.6 provides an overview of how XAI can be applied for trustworthy AI in terms of privacy for B5G/6G networks.

The application of XAI can support convincing various parties to agree upon privacy standards, which can help mitigate privacy differences between locations. We can consider that using XAI may also help define the levels for privacy violations, with explainability of actions made when taking privacy-related decisions. Further investigation of XAI techniques helps bring down the limitations of availability in explanations for AI, especially considering the issues mentioned in Section 14.2.7 on XAI itself, such as the lack of formalism and standardization of XAI techniques.

One of the main costs of XAI could be its implementation cost for XAI algorithms since it involves the engineering of XAI solutions and development costs of interfaces for presenting explainability.

14.3.5 Privacy Measures for Personally Identifiable Information

For a person, their personally identifiable information (PIIs[RHM2]) plays a key role in privacy as they directly provide a way to distinguish them easily. Therefore, the protection of these PIIs is an essential requirement for B5G/6G. For instance, information such as digital image sharing is continuously popular with new mobile phones and advanced capturing devices, which may expose the PIIs of users. Therefore, it is suitable to consider privacy awareness in these devices and services when processing them. Also, privacy concerns are rapidly escalating, considering the increasing popularity of biometric authentications, geolocation, and ever-growing personal data exposure in social media. New technologies will generate new modes of PII, such as holographic imaging and brain–computer interface (BCI) wave patterns, which need attention when considering privacy.

Maintaining proper measures for PIIs, therefore, help to address the previously discussed issues in Section 14.2, for new technologies and applications for privacy requirements. Also, it may contribute to solving private data access limitations where the data cannot be accessed due to lack of PII. Having enough privacy measures for PII can only hide crucial information on users, which may help organizations release other noncritical information publicly, thus contributing to improvements in up-to-date big data sets for AI model training.

14.3.6 Blockchain-Based Solutions

Blockchain is a decentralized and distributed public ledger technology in a peer-to-peer network [21]. In simple terms, it is a list of linked records, or blocks, which are connected with links that make it challenging to change any block after creation [22]. The blocks that store required transactions are timestamped and encrypted [22]. As hash values for blocks are unique, fraud can be effectively prevented, since modifications to a block in the chain change the hash value immediately. A new block can be added to the chain if most nodes in the network agree. We do this via a consensus method to verify the legitimacy of transactions in a block. It also verifies the validity of the block itself, Nofer et al. [23]. Even some or all blocks of a blockchain in the network are changed, it is virtually impossible to tamper in practice. These transactions are duplicated and distributed across the entire network of computer systems.

Though the implementations of blockchain have existed since 2008 with the famous Bitcoin concept [24], it has attracted great interest in the research community in recent years because of its decentralized and secured nature [21]. The application of blockchains on privacy has also made significant progress in research. Therefore, potential solutions exist, such that we can get their adoptions for B5G/6G networks.

14.3.7 Lightweight and Quantum Resistant Encryption Mechanisms

Encryption is a technique to prevent unauthorized parties from accessing information by applying the mathematical function and converting it to an incomprehensible format. There will be a key to decrypt or recover the original information during that process. This will only be converted back by an entity with this valid key, which is only revealed by the party who performed the encryption. Encryption is often used in communication in networks to ensure data privacy primarily; since if a third party receives the data, it will not be helpful as they are impossible to understand.

Since there is an apparent computational cost in encryption, it will be incorporated with a delay. As per the limitations of hardware and cost considerations such as computation capacity in network infrastructures such as IoT or cloud computing environments, there are proposed solutions for enhancing privacy through encryption mechanisms through less complex computations. It may help achieve better preservation of privacy and, at the same time, increase overall performance and quality of experience for the end-user. Therefore, B5G/6G networks need to get empowered by lightweight yet powerful encryption mechanisms. Similarly, the quantum resistant encryption techniques are necessary to ensure the data privacy is preserved in quantum computing environments facilitated by B5G/6G as the current approaches of encryption fail in them. For example, the work in [25] shows quantum algorithms can be used for factoring and the computation of discrete logarithms efficiently, making the public key cryptography techniques used today insecure.

Lightweight encryption methods support the fulfillment of privacy requirements for edge devices. E2E data communication confidentiality can be preserved during the transmission of data, which are vital for future networks. Also, it significantly supports in bringing down the costs in privacy enhancements, processing requirements, and energy consumption aspects. Having a full-fledged cryptographic environment in embedded applications is not possible due to constraints such as power dissipation, area, and cost [26]. Lightweight encryption may significantly reduce the workload on cloud computing environments as well. Quantum-resistant algorithms also play a significant role in ensuring communication privacy among multiple entities using the applications of quantum computing in B5G/6G.

14.3.8 Homomorphic Encryption

To fulfill the privacy requirements of data in B5G/6G environments, safe encryption techniques are needed. However, the drawback of most encryption techniques is that they will render the data useless for any useful task without decryption. With the emerging need for preserving privacy, meanwhile, having the possibility of analyzing data in B5G/6G, homomorphic encryption provides a promising solution. It allows mathematical operations on data to be carried out on a ciphertext, which is an encrypted form of the input data/the plain text, instead of on the actual data itself [27]. It consists of three [27] forms:

- **Somewhat homomorphic**: Supports mathematical operations for addition and multiplication, yet limited to a certain number of operations due to noise accumulation.
- **Partially homomorphic**: Supports any number of operations, yet only limited to a specific type of operation.
- **Fully homomorphic**: Support both addition and multiplication operation for any number of times.

Even though the first fully homomorphic scheme was introduced in 2009, improvements are continuously carried out to make it practical to use in every platform [28]. Figure 14.7 illustrates the basic overview of homomorphic encryption.

Homomorphic encryption helps solve the private data access issues since private organizations can allow their data to be encrypted and outsourced to commercial cloud environments for processing, while encrypted. Therefore, we can consider it as an existing solution for this privacy issue in B5G/6G network services.

14.3.9   Privacy-Preserving Data Publishing Techniques

Data can be modified for privacy during storage and usage in big data AI models in B5G/6G. Different data sanitation methods are available to remove or add noise to user data, including sensitive personal information or PII. The essential facts to consider here are protecting user privacy; meanwhile, ensuring that the original data are not mutated too much for publishing. Not limited to data publishing, privacy-preserving techniques help enhance privacy in other data-related usages, such as storing data in an unsecured environment. The actual privacy of owners of those data will not be affected even if the data are exposed. Therefore, discovering

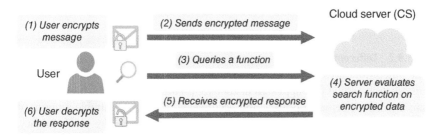

Figure 14.7   Process flow of homomorphic encryption.

more techniques on privacy-preserving data publishing may help B5G/6G networks to improve their guarantees on privacy for their users' data.

These data modification techniques can directly help in private data access issues, in which up-to-date data for AI model training and data science applications are lacking. Since third parties will not reveal the users' identity, more data can be published by private or public repositories, given that privacy is guaranteed through these methods. It also helps reduce the ambiguity of privacy responsibility in data ownership since owning data by any party may not cause privacy issues for data contributors.

There are many methods of privacy preservation for data in the associated work. Two approaches for data modification are shown by Binjubeir et al. [29]: (i) data perturbation techniques and (ii) anonymization. There are several anonymization techniques currently available as shown in [30], based on:

- **Data nature**: Tabular, data sets, graphical data
- **Anonymization approaches**: Generalization, suppression, perturbation, permutation
- **Objectives of anonymization**: k-Anonymity, L-diversity, objectives based on presumptions on attacker's knowledge

The works [30, 31] classify anonymization techniques based on (i) syntactic approaches and (ii) differential privacy. We also describe them briefly as follows.

14.3.9.1 Syntactic Anonymization

The syntactic anonymization modifies the input data set to achieve generalization. As shown in [32], most of the syntactic models are based on generalizing table entries. Goswami and Madan [30] shows several techniques for syntactic anonymization, namely (i) k-anonymity, which is the warranty that in a series of k groups, the probability of identifying that person is less than $1/k$, (ii) L-diversity, which makes the maximum probability of recognizing the sensitive user information to be $1/L$ for L different groups, and (iii) T-closeness, where the dissemination distinction between sensitive data and its values within groups does not exceed a value T.

14.3.9.2 Differential Privacy

The differential privacy is a relatively recent introduction of methods to privacy preservation, which first appeared in 2006 [33]. This is a technique used to maintain a trade-off between privacy and accuracy by adding a desirable amount of noise to data [33]. The differential privacy techniques saves the user's privacy, meanwhile, having no additional cost, and the accuracy further increases with further increase of the number of samples. Therefore, this is well applicable for the scenario of big data. In differential privacy, as data gets subjected to noise addition, its accuracy might get reduced, especially, when the data set is small. Loss of actual user information in anonymization techniques is another issue that comes as a cost for enhancing privacy.

14.3.10 Privacy by Design and Privacy by Default

With the expanding network capabilities of B5G/6G from technologies that capture and use personal data, general users will get subjected to an increased risk of privacy leakages in their lives. Therefore, privacy by design is an important aspect that should be considered in every step of the designing process life cycle of B5G/6G services, and their capability should be evaluated. Moreover, by default, these services or products should safeguard privacy requirements, even without any manual input from users. It is taking actions to protect beforehand, not after a privacy breach happened [34]. This means that, rather than having to come up with complicated and time-consuming "patches" later on, it is necessary to detect and assess potential data protection issues when creating new technology and to incorporate privacy protections into the overall design [35].

Privacy by design can help address the new technology applications with privacy requirements since designing applications focusing on privacy beforehand will eliminate most privacy threats. It also helps maintain that privacy is ensured as the general public may pay less attention to privacy rights and threats.

As the self-decision-making capability of AI increases, it will get the opportunity to impact on the B5G/6G network and end users. Therefore, the AI design process should be done with privacy concerns as a key requirement. There is a concern that if the autonomy of AI increases, it may be difficult to maintain transparency. The European Committee on Civil Liberties, Justice, and Home Affairs has given high-level indications for future regulations to stress the responsibility of AI designers and developers. They should ensure that the AI products are safe, secure, fit for purpose, and follow the procedures for data processing compliant with existing legislation, confidentiality, anonymity, fair treatment, and due process.

Considering the design of associated technologies with B5G/6G, we can provide a set of guidelines to ensure privacy in designing. For instance, the work in [36] presents an architectural view for data protection by design for an e-health application, which is compatible with the EU GDPR. Work presented by European Union Agency for Network and Information Security in [37] and [38] discusses some privacy by design strategies as follows:

- Minimize the amount of personal data as possible
- Hide personal data from plain view
- Separate personal data in a distributed fashion
- Processing of personal data should be done at the highest level of aggregation

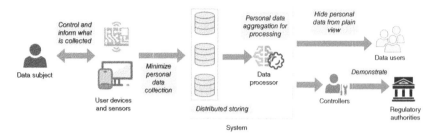

Figure 14.8 Privacy by design strategies implementation for privacy enhancement.

- Inform data subjects with transparency
- Data subjects should be supported for the processing of personal data
- Enforce privacy policy compatible with legal requirements
- Demonstrate compliance with privacy policy into force and any applicable legal requirements

Figure 14.8 illustrates how these design strategies could be applied.

Having privacy enabled in design will automatically provide enhancements of security aspects. In fact, Onik et al. [39] shows that privacy by design provides two levels of security for IoT systems: (i) check for data fitness to the context when collecting information and (ii) capability of a system to assess the scope of data sharing to the internet and other associated risks. Similarly, in [40], the authors develop an IoT environment to help developers to engage with data protection at design phase.

14.3.11 Regulation of Government, Industry, and Consumer

Many parties in B5G/6G networks handle user data, and they could be linked with the government authorities as a mediator to oversee the actions of this data by the industry. Also, governments directly collect user data with the aid of wireless communication. For instance, using online apps by the government for contact tracing in the COVID-19 pandemic [41] was such an initiative. These apps may collect vast amounts of user data with high privacy-related information such as user location, identity, and medical records. If such a government system is hacked, citizens will have critical information leakages. Also, industries collect and provide data to third parties, even without the consent of users [42]. Therefore, the involvement of regulations helps ensure user privacy since all parties who utilize privacy-related data will require to adhere to them.

Regulations can support solving privacy issues related to a lack of understanding of privacy rights and threats in the general public since imposing regulations will inherently safeguard user data. Users may not require a complex understanding of their privacy requirements. Also, it may help to solve privacy differences based on location if these regulations can tackle the general privacy requirements despite the physical location. Thus, imposing these regulations should help solve the issue of legal disputes among entities.

We can identify the different types of regulatory approaches in the literature [43] to address the issues such as

- Government regulation
- Industry self-regulation
- Consumer or market regulation

Industry and government regulations influence macrolevel general privacy concerns. Their work categorizes privacy concerns to different themes regarding general privacy concerns such as consumer attributes and privacy, macroenvironment surrounding consumer privacy, and technology-mediated E-Commerce privacy. The specific privacy concerns include vendor-related attributes and consumer privacy, consumer–vendor interaction, and consumer–vendor trust relationship. The work [43] finally presents possible future directions for consumer privacy: through technical mediation of E-commerce privacy, individual-centered situational privacy determinants, and having a decision-environment in situational privacy.

From a consumer perspective, consumers are generally aware of having privacy rights but have insufficient knowledge and resources to exercise these rights properly [44]. Consumers rely on to access privacy violations based on the moral values of trust, transparency, control, and access. However, the consumers generally have limited knowledge on security and privacy and rely on laws and safety mechanisms. Therefore, introducing regulations for privacy will ultimately mitigate users from breaching their privacy.

However, these regulations may not be applicable everywhere, yet they can cause a high extra cost in implementation in only a specific region. Also, many industries may suffer financial losses if these regulations are very strictly imposed.

14.3.12 Other Solutions

In addition to the mentioned privacy solutions, we found other solutions that are already available, which can be used for ensuring privacy issues for B5G/6G networks. We summarize them as follows:

14.3.12.1 Location Privacy Considerations

Location privacy is an important aspect to consider since it is a type of PII. The establishment of location privacy measures helps solve privacy issues related to edge computing since most of the location trackers are placed in IoT platforms and mobile devices. Associated work on the discussion of location-based privacy can be applied to ensure privacy in B5G/6G networks.

14.3.12.2 Personalized Privacy

Instead of going to a unified privacy metric, an alternative suggestion could be to rely on a personalized approach in some scenarios applicable at the consumer level since privacy preferences can differ from one individual to another. Personalized privacy is an approach to achieving privacy levels for users specialized for their requirements. Having personalized privacy may help resolve privacy issues based on different locations since no matter the location, privacy is configured based on user choices.

14.3.12.3 Fog Computing Privacy

Fog computing is an extra layer between the edge devices and the cloud server, acting as an intermediate entity for tasks such as filtering data and forwarding them to the cloud. Having such extra layer can reduce the cost for privacy enhancements since the fog layer can compute many intermediate functions. Only a fraction of data will be sent to the cloud, reducing both the cloud server's overhead and the network congestion.

Therefore, in terms of privacy, fog computing could also help to preserve the privacy of IoT and the users since they minimize the need to transmit sensitive data to the cloud for analysis [45]. However, the privacy aspects within the fog node are also important to consider since data from the edge will be directly contacting the fog layer. Also, there is a potential risk in having a compromised fog node, which the attackers may eavesdrop on or directly modify user data. We see fog computing as a promising bridge between the edge and the cloud. However, there exists a possibility of being attacked by an adversary to collect data from many edge devices, causing privacy breaches. However, fog computing is a much more active area of research as far as privacy is concerned, where associated works mention privacy enhancements can be made with the fog nodes.

References

1 M. Wang, T. Zhu, T. Zhang, J. Zhang, S. Yu, and W. Zhou, "Security and privacy in 6G networks: New areas and new challenges," *Digital Communications and Networks*, vol. 6, no. 3, pp. 281–291, 2020.

2 Y. Qu, L. Gao, T. H. Luan, Y. Xiang, S. Yu, B. Li, and G. Zheng, "Decentralized privacy using blockchain-enabled federated learning in fog computing," *IEEE Internet of Things Journal*, vol. 7, no. 6, pp. 5171–5183, 2020.

3 L. Li, B. Lei, and C. Mao, "Digital twin in smart manufacturing," *Journal of Industrial Information Integration*, vol. 26, p. 100289, 2022.

4 Z. Zhang, Y. Xiao, Z. Ma, M. Xiao, Z. Ding, X. Lei, G. K. Karagiannidis, and P. Fan, "6G wireless networks: Vision, requirements, architecture, and key technologies," *IEEE Vehicular Technology Magazine*, vol. 14, no. 3, pp. 28–41, 2019.

5 E. Zeydan and Y. Turk, "Recent advances in intent-based networking: A survey," in *2020 IEEE 91st Vehicular Technology Conference (VTC2020-Spring)*. IEEE, 2020, pp. 1–5.

6 M. Martinsson, "What is closed-loop automation?" https://www.ericsson.com/en/blog/2019/4/what-is-closed-loop-automation.

7 C. Benzaid and T. Taleb, "AI-driven zero touch network and service management in 5G and beyond: Challenges and research directions," *IEEE Network*, vol. 34, no. 2, pp. 186–194, 2020.

8 B. M. Thuraisingham, "Can AI be for good in the midst of cyber attacks and privacy violations? A position paper," in *Proceedings of the 10th ACM Conference on Data and Application Security and Privacy*, 2020, pp. 1–4.

9 W. Yu, F. Liang, X. He, W. G. Hatcher, C. Lu, J. Lin, and X. Yang, "A survey on the edge computing for the Internet of Things," *IEEE Access*, vol. 6, pp. 6900–6919, 2017.

10 P. Porambage, T. Kumar, M. Liyanage, J. Partala, L. Lovén, M. Ylianttila, and T. Seppänen, "Sec-EdgeAI: AI for edge security Vs security for edge AI," *The 1st 6G Wireless Summit, (Levi, Finland)*, 2019.

11 "GDPR Enforcement Tracker." [Online]. Available: https://www.enforcementtracker.com/.

12 M. Ribeiro, K. Grolinger, and M. A. Capretz, "MLaaS: Machine learning as a service," in *2015 IEEE 14th International Conference on Machine Learning and Applications (ICMLA)*. IEEE, 2015, pp. 896–902.

13 W. Samek and K.-R. Müller, "Towards explainable artificial intelligence," in *Explainable AI: Interpreting, Explaining and Visualizing Deep Learning, Lecture Notes in Computer Science*, W. Samek, G. Montavon, A. Vedaldi, L. Hansen, and K. R. Müller, Eds. Cham: Springer, 2019, vol. 11700, pp. 5–22.

14 F. Xu, H. Uszkoreit, Y. Du, W. Fan, D. Zhao, and J. Zhu, "Explainable AI: A brief survey on history, research areas, approaches and challenges," in *Natural Language Processing and Chinese Computing, CCF International Conference on Natural Language Processing and Chinese Computing, Lecture Notes in Computer Science*, J. Tang, M. Y. Kan, D. Zhao, S. Li, and H. Zan, Eds. Springer, 2019, vol. 11839, pp. 563–574.

15 U. Pawar, D. O'Shea, S. Rea, and R. O'Reilly, "Incorporating explainable artificial intelligence (XAI) to aid the understanding of machine learning in the healthcare domain." in *AICS*, 2020, pp. 169–180.

16 X. Li, P. Jiang, T. Chen, X. Luo, and Q. Wen, "A survey on the security of blockchain systems," *Future Generation Computer Systems*, vol. 107, pp. 841–853, 2020.

17 Huawei, "5G Security: Forward Thinking," *Huawei White paper, Tech. Rep.*, 2015.

18 T. Kumar, M. Liyanage, A. Braeken, I. Ahmad, and M. Ylianttila, "From gadget to gadget-free hyperconnected world: Conceptual analysis of user privacy challenges," in *2017 European Conference on Networks and Communications (EuCNC)*. IEEE, 2017, pp. 1–6.

19 M. Finck and F. Pallas, "They who must not be identified–distinguishing personal from non-personal data under the GDPR," *International Data Privacy Law*, 2020.

20 H. H. Kumar, V. Karthik, and M. K. Nair, "Federated k-means clustering: A novel edge AI based approach for privacy preservation," in *2020 IEEE International Conference on Cloud Computing in Emerging Markets (CCEM)*. IEEE, 2020, pp. 52–56.

21 Q. Feng, D. He, S. Zeadally, M. K. Khan, and N. Kumar, "A survey on privacy protection in blockchain system," *Journal of Network and Computer Applications*, vol. 126, pp. 45–58, 2019.

22 J. Al-Jaroodi and N. Mohamed, "Blockchain in industries: A survey," *IEEE Access*, vol. 7, pp. 36 500–36 515, 2019.

23 M. Nofer, P. Gomber, O. Hinz, and D. Schiereck, "Blockchain," *Business & Information Systems Engineering*, vol. 59, no. 3, pp. 183–187, 2017.

24 S. Nakamoto, "Bitcoin: A peer-to-peer electronic cash system," *Decentralized Business Review*, p. 21260, 2008.

25 P. W. Shor, "Algorithms for quantum computation: Discrete logarithms and factoring," in *Proceedings 35th Annual Symposium on Foundations of Computer Science*. IEEE, 1994, pp. 124–134.

26 G. Bansod, N. Raval, and N. Pisharoty, "Implementation of a new lightweight encryption design for embedded security," *IEEE Transactions on Information Forensics and Security*, vol. 10, no. 1, pp. 142–151, 2014.

27 V. Rocha and J. López, "An overview on homomorphic encryption algorithms," UNICAMP Universidade Estadual de Campinas, Tech. Rep., 2018. [Online]. Available: https://www.ic.unicamp.br/reltech/PFG/2018/PFG-18-28.pdf.

28 A. Acar, H. Aksu, A. S. Uluagac, and M. Conti, "A survey on homomorphic encryption schemes: Theory and implementation," *ACM Computing Surveys (CSUR)*, vol. 51, no. 4, pp. 1–35, 2018.

29 M. Binjubeir, A. A. Ahmed, M. A. B. Ismail, A. S. Sadiq, and M. K. Khan, "Comprehensive survey on big data privacy protection," *IEEE Access*, vol. 8, pp. 20 067–20 079, 2019.

30 P. Goswami and S. Madan, "Privacy preserving data publishing and data anonymization approaches: A review," in *2017 International Conference on Computing, Communication and Automation (ICCCA)*. IEEE, 2017, pp. 139–142.

31 X. Wang, J.-K. Chou, W. Chen, H. Guan, W. Chen, T. Lao, and K.-L. Ma, "A utility-aware visual approach for anonymizing multi-attribute tabular data," *IEEE Transactions on Visualization and Computer Graphics*, vol. 24, no. 1, pp. 351–360, 2017.

32 C. Clifton and T. Tassa, "On syntactic anonymity and differential privacy," in *2013 IEEE 29th International Conference on Data Engineering Workshops (ICDEW)*. IEEE, 2013, pp. 88–93.

33 M. U. Hassan, M. H. Rehmani, and J. Chen, "Differential privacy techniques for cyber physical systems: A survey," *IEEE Communications Surveys & Tutorials*, vol. 22, no. 1, pp. 746–789, 2019.

34 A. Cavoukian, "Privacy by design," *Identity in the Information Society*, 2009.

35 P. Schaar, "Privacy by design," *Identity in the Information Society*, vol. 3, no. 2, pp. 267–274, 2010.

36 L. Sion, P. Dewitte, D. Van Landuyt, K. Wuyts, I. Emanuilov, P. Valcke, and W. Joosen, "An architectural view for data protection by design," in *2019 IEEE International Conference on Software Architecture (ICSA)*. IEEE, 2019, pp. 11–20.

37 G. D'Acquisto, J. Domingo-Ferrer, P. Kikiras, V. Torra, Y.-A. de Montjoye, and A. Bourka, "Privacy by Design in Big Data: An Overview of Privacy Enhancing Technologies in the Era of Big Data Analytics," *arXiv preprint arXiv:1512.06000*, 2015.

38 iapp, [Online]. Available: https://iapp.org/resources/article/a-guide-to-privacy-by-design/.

39 M. M. H. Onik, K. Chul-Soo, and Y. Jinhong, "Personal data privacy challenges of the fourth industrial revolution," in *2019 21st International Conference on Advanced Communication Technology (ICACT)*. IEEE, 2019, pp. 635–638.

40 T. Lodge, A. Crabtree, and A. Brown, "IoT app development: Supporting data protection by design and default," in *Proceedings of the 2018 ACM International Joint Conference and 2018 International Symposium on Pervasive and Ubiquitous Computing and Wearable Computers*, 2018, pp. 901–910.

41 J. Li and X. Guo, "COVID-19 Contact-Tracing Apps: A Survey on the Global Deployment and Challenges," *arXiv preprint arXiv:2005.03599*, 2020.

42 H. Tuttle, "Facebook scandal raises data privacy concerns," *Risk Management*, vol. 65, no. 5, pp. 6–9, 2018.

43 M. Liyanage, J. Salo, A. Braeken, T. Kumar, S. Seneviratne, and M. Ylianttila, "5G privacy: Scenarios and solutions," in *2018 IEEE 5G World Forum (5GWF)*. IEEE, 2018, pp. 197–203.

44 L. Zhang-Kennedy and S. Chiasson, ""Whether it's moral is a whole other story": Consumer perspectives on privacy regulations and corporate data practices," in *17th Symposium on Usable Privacy and Security ({SOUPS} 2021)*, 2021, pp. 197–216.

45 A. Alrawais, A. Alhothaily, C. Hu, and X. Cheng, "Fog computing for the Internet of Things: Security and privacy issues," *IEEE Internet Computing*, vol. 21, no. 2, pp. 34–42, 2017.

15

Legal Aspects and Security Standardization

This chapter provides a comprehensive analysis on 6G privacy-issues and possible solutions to mitigate them. After reading this chapter, you should be able to:

- Understand challenges that appear as barriers to reach 6G privacy preservation.
- Understand the set of solutions applicable to the proposed architecture of B5G/6G networks to mitigate the challenges.

15.1 Legal

The next-generation 6G empowered use cases may require the creation of secure and trustworthy cyberspace with essential privacy protection and ethical considerations that can be easily accepted by all related stakeholders. Throughout this process, the key applications can be also explored with parallel business and regulation support. It will be necessary to bring together stakeholders from the industry and the public sector with academic research, standardization entities, and research community to derive requirements and develop solutions that are based on the real needs of users. There will be a scrutinized ethical and legal policy developed for intrusive applications. Therefore, all technical solutions and enabling technologies and scalable business models developed by the project for security and trust will be compatible with current regulations and laws as well, and they will be sufficient to adapt to times of crisis to their users globally.

Combining technology research with business and law proactively allows the development of novel technical solutions that address real-world problems, are globally scalable, and are adaptable to use by companies and policymakers considering the human rights and ethical requirements that are a necessity in a

Security and Privacy Vision in 6G: A Comprehensive Guide, First Edition.
Pawani Porambage and Madhusanka Liyanage.

democratic society. A multidisciplinary study with wide interaction and exchange with global and local partners from industry, academia, government, and key interest groups ensures the quality of the outcomes, benefit their recognition, and speed up dissemination by industry. More efforts should be undertaken to identify the consciousness of the legal and ethical requirements that accompany the further development of combined 5G/6G and artificial intelligence (AI) against new intrusive technologies for the well-being of society. In the further development of the applications and integration of sensing techniques bias regarding nationality, gender, race, culture, or even region should be avoided by introducing more comprehensive legal frameworks. In particular, a legal and regulatory framework for AI and their applications are still in infancy. This will require more of a global-level understating among the standardization and law enforcement agencies all over the world who are working on 6G standardization rather than a national level frameworks. Some key challenges in introducing legal frameworks to improve security and privacy in AI-enabled 6G are lack of engagement of stakeholders and legal experts, immaturity of qualified explainable metrics, and the extremely high heterogeneity in the telecommunication market and spectrum regulations.

As an example, the General Data Protection Regulation (GDPR), introduced in Europe on the 25 May 2018 is concerned with three aspects: data permission (people need to express consent to receive promotional material, i.e. marketing emails), data access (people are entitled to request companies to remove data collected from them), and data focus (companies need to justify the data they are collecting). GDPR is legally binding, and companies cannot ignore it. Under GDPR, companies need the persons' consent before installing a cookie in your computer which is used to track the behavior of the users. "The Algorithmic Accountability Act of 2019" in the United States has a similar purpose to GDPR and balances the power between the big companies and the users. Under this legislation, companies have to give a detailed description of how their automated decision systems work in a comprehensible way. In other words, companies need to justify to users the adverts they are displaying to avoid unfair biases or another kind of discrimination. There are also other law drafts like the California Privacy Rights Act and the Colorado Privacy Act, that aim at other important facets of the algorithms such as preventing companies from using patterns to manipulate individuals and undermining their ability to make free choices to avoid compulsive buying. This is very related to mental problems in the form of addictions that the users can develop by using online platforms. In short, legal frameworks provide the general lines to protect citizens from being abused by companies and all types of commercialized entities.

15.2 Security Standardization

As a critical aspect of next-generation networks and digital services, the security domain has a very active standardization and project landscape. In this section, we highlight and delineate the key standardization efforts which have a prospective impact on 6G security. Key Standards Developing Organizations (SDOs) which are relevant to 6G security are as shown in Figure 15.1.

15.2.1 ETSI

As a multipronged effort, ETSI has launched multiple Industry Specification Groups (ISG) to examine 5G component technologies, including NFV (ETSI NFV),

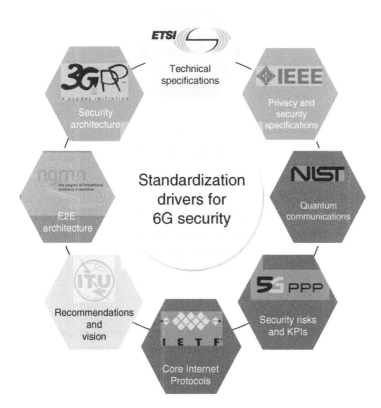

Figure 15.1 Standardization landscape relevant for prospective 6G security standards.

AI (ETSI ISG Securing Artificial Intelligence-SAI, ETSI ISG Experiential Network Intelligence – ENI) and network automation (ETSI ISG Zero touch and service management – ZSM). NFV-SEC is a WG under ISG NFV that produces industry specifications on security-related matters of NFV technology. Since 2014, the NFV SEC WG has produced multiple Group Specifications (GS) and Group Reports (GR). Work during releases 3 and 4 of ETSI NFV has increased the focus on security specifications as the scope and features of NFV platforms are expanding.

ETSI ISG ENI was also launched in 2017 to define a Cognitive Network Management architecture, using AI techniques and context-aware policies to adjust offered services based on changes in user needs, environmental conditions, and business goals. The ISG has produced a set of use cases, including network security, where the ENI system can detect various attacks and trigger a reaction by the network. Another group, ETSI ISG SAI, was formed in 2019 and aims to develop technical specifications to alleviate threats emerging from deploying AI and threats targeting AI systems originating from other AI systems and typical attack sources. This ISG has undertaken the tasks of defining AI threats, providing relevant use cases, recommending mitigation measures against such threats, and proposing possible recommendations regarding data sharing.

15.2.2 ITU-T

At a global level, International Telecommunication Union (ITU) has established the ITU-T Focus Group on ML for Future Networks (FG-ML5G) working on technical specifications for machine learning for future networks, including interfaces, network architectures, protocols, algorithms, and data formats [1]. ITU-T FG-NET2030 – Focus Group on Technologies for Network 2030 is elaborating on new drivers, requirements, and gaps to propose use cases for applications including augmented and virtual reality and holograms. The developments will also have an impact on security aspects of 6G networks [2].

15.2.3 3GPP

Similarly, 3rd Generation Partnership Project (3GPP) has already addressed the use of AI/ML in the 5G Core Service-Based Architecture (SBA) by introducing the Network Data Analytics Function. This function provides analytics and notifications to other network functions regarding the users' behavior and the network's status. 3GPP SA3 is currently working on a draft TR by identifying the security issues, requirements, and solutions regarding Network Slicing and the use of the Network Data Analytics Function in selected use cases [3].

15.2.4   NIST

Standardization of postquantum cryptographic algorithms is performed by National Institute of Standards and Technology (NIST) [4]. The ongoing work by NIST's Postquantum Cryptography Program is working to solicit candidates and then specify quantum-resistant algorithms each for digital signatures, public-key encryption, and cryptographic key-establishment. The process is now at Round 3 following the completion of the second round in July 2020. The selected algorithms will constitute the first standard developed to counter threats due to quantum decryption.

15.2.5   IETF

On the IETF front, IETF Security Automation and Continuous Monitoring (SACM) Architecture RFC defines an architecture enabling a cooperative SACM ecosystem based on entities, or components, which communicate by sharing information [5]. One or more components are consumers of information in a given flow while some are providers of information. A key component is an *orchestrator* which facilitates the automation of various functions such as configuration, coordination, and management for the SACM components. There can be also various *repositories* such as policy repositories, vulnerability definition data repositories, and security information repositories.

15.2.6   5G PPP

5G PPP has established 5G PPP Security Work Group as a joint effort on tackling 5G security risks and challenges and provide insights into 5G security and how it should be addressed [6]. It elaborates on 5G security architecture and how it fits with that of the 3GPP, access control, privacy, trust, security monitoring and management, and standardization on 5G security. Although it has a focus on 5G, the outcomes of the group have direct implications on Beyond 5G networks such as intelligent network security, security KPIs, emerging risks, threats, and countermeasures.

15.2.7   NGMN

NGMN 5G End-to-End Architecture Framework v4.3 (2020) describes the requirements in terms of network entities and functions for the capabilities of an end-to-end framework which also includes security [7]. It considers the security for the end-to-end protection of the various network features and enabling capabilities in a forward-looking 5G service paradigm.

15.2.8 IEEE

IEEE P1915.1 Standard for Software Defined Networking and Network Function Virtualisation (SDN/NFV) Security works to provide a framework to build and operate secure SDN/NFV environments. It aims for different stakeholders such as end users, network operators, and service/content providers. To this end, it specifies a security framework for SDN/NFV with related system models, analytics, and requirements [8]. Similarly, IEEE P1917.1 Standard for Software Defined Networking and Network Function Virtualisation Reliability focuses on reliability requirements and develops a framework for reliable SDN/NFV service delivery infrastructure [9].

For the quantum communications, IEEE P1913.1 (Draft) Standard for Software-Defined Quantum Communication (SDQC) defines the SDQC protocol that enables configuration of quantum endpoints in a communication network [10]. It allows dynamic creation, modification, or removal of quantum protocols or applications in a software-defined setting. This is possible with the availability of a well-defined interface to quantum communication devices, which can be reconfigured to implement a variety of protocols and measurements. The SDQC protocol functions at the application layer and communicates over TCP/Internet protocol (IP). The protocol design considers future integration with network softwarization-related standards.

References

1 ITU-T Focus Group on ML for Future Networks (FG-ML5G), "ITU FG ML5G - Unified architecture for machine learning in 5G and future networks," 2019. [Online]. Available: http://handle.itu.int/11.1002/pub/8128dfee-en.

2 International Telecommunication Union, "Focus Group on Technologies for Network 2030," 2019, [Accessed on 29.03.2021]. [Online]. Available: https://www.itu.int/en/ITU-T/focusgroups/net2030/Pages/default.aspx.

3 3GPP Technical Specification Group Service and System Aspects (TSG SA) WG3 (SA3), "3GPP SA3 - Security." [Online]. Available: https://www.3gpp.org/specifications-groups/sa-plenary/sa3-security.

4 NIST, "NIST Post-Quantum Cryptography Standardization." [Online]. Available: https://csrc.nist.gov/projects/post-quantum-cryptography/post-quantum-cryptography-standardization.

5 A. W. Montville and B. Munyan, "Security Automation and Continuous Monitoring (SACM) Architecture," Internet Engineering Task Force, Internet-Draft draft-ietf-sacm-arch-08, March 2021, work in Progress. [Online]. Available: https://datatracker.ietf.org/doc/html/draft-ietf-sacm-arch-08.

6 5G-PPP, "5G and energy," 2015, [Accessed 08.12.2020]. [Online]. Available: https://5g-ppp.eu/wp-content/uploads/2014/02/5G-PPP-White_Paper-on-Energy-Vertical-Sector.pdf.

7 "The Next Generation Mobile Networks," [Accessed on 29.03.2021]. [Online]. Available: https://www.ngmn.org/.

8 IEEE, "IEEE P1915.1 security in virtualized environments." [Online]. Available: https://site.ieee.org/p1915-1-sve/.

9 IEEE, "IEEE P1917 software defined networking and network function virtualization reliability." [Online]. Available: https://standards.ieee.org/project/1917_1.html.

10 IEEE, "IEEE P1913 software-defined quantum communication." [Online]. Available: https://standards.ieee.org/project/1913.html.

Part V

Security in 6G Technologies

16

Distributed Ledger Technologies (DLTs) and Blockchain*

* With additional contribution from Anshuman Kalla, Uka Tarsadia University, India

This chapter discusses the role of distributed ledger technologies (DLTs) and Blockchain for 6G security and privacy. After reading this chapter, you should be able to:

- Understand the basis of DLT and Blockchain in 6G networks.
- Understand the role of DLT and Blockchain in 6G network security and privacy.
- Understand the threat Landscape of Blockchain.

16.1 Introduction

Although the existing security architectures so far are able to provide a sufficient level of security; however, they are suffering from impediments such as limited scalability, over utilization of network resources (leading to increased network-level delay and congestion) and high operational cost, mainly due to the complex and static security management procedures. Moreover, they are also lacking automation to support high-speed service deployment. Thus, the current centralized architecture models in legacy mobile systems will find real difficulties to meet-up the demands of 5G/6G networks and will thus pull-down the projected capabilities. Thus, the current centralized architecture models in legacy mobile and IoT systems will struggle to scale up to meet the demands of future B5G networks.

To reconcile the pertinent issues, blockchain/DLT technologies, featured with decentralization, cryptographic techniques, and consensus-driven mechanism, can be leveraged. Also, the combination of cryptographic processes behind it can offer an intriguing alternative. Though still in infancy, blockchain technology

Security and Privacy Vision in 6G: A Comprehensive Guide, First Edition.
Pawani Porambage and Madhusanka Liyanage.
© 2023 The Institute of Electrical and Electronics Engineers, Inc. Published 2023 by John Wiley & Sons, Inc.

is turning out to be disruptive by proving its efficacy. A recent Gartner study estimates that blockchain will add US$3.1 trillion of business value by 2030. On the other hand, 5G will add business-to-business value of US$700 billion by 2030. Therefore, it is worth realizing the benefits which could be harvested with the use of blockchain for softwarized networks.

The distributed nature of blockchain allows industrial entities and various 5G/6G-enabled IoT data users to access and supply IoT data from and to peers, respectively, thereby omitting the need of centralized operations and management. Moreover, the stakeholders of the 5G ecosystem can verify the veracity of each transaction and thus brings-in accountability, auditability, along with provenance and nonrepudiation for every user.

To summarize, it is worth exploring the role of blockchain in the realm of softwarized networks along with the different use cases, opportunities, and challenges.

16.2 What Is Blockchain

Blockchain is a type of distributed ledger which is maintained in a decentralized manner by the underlying P2P network of nodes. As the name implies, blockchain comprises of series of blocks of transactions that are logically chained or linked together to form a digital ledger. Transactions occurring in a given time frame are verified and then bundled together in a unit called *block*. So each block contains a finite set of transactions. All these blocks are logically linked together in the order of their creation using cryptographic hashes. This is to say that every block stores the hash value of the previous block in the "previous block hash" field. The first block, aka genesis block, does not have any previous block. Thus, the previous block hash field of the genesis block is set to all zeros. The digital ledger holds all the past transactions in chronological order and is cryptographically sealed [1]. Further, this ledger is replicated at all the participating nodes in a P2P blockchain network. With time, the distributed ledger keeps growing because the write operations are performed in exclusively append mode [2]. From a technological viewpoint, blockchain is considered as one single technological innovation which is a unique and powerful mix of underlying concepts, techniques, and technologies [3]. For instance, blockchain uses cryptographic techniques such as public key infrastructure (PKI), hashing, digital signature, and Merkle tree. It also utilizes P2P technology for connecting nodes (aka miners) to establish a blockchain network. Furthermore, a consensus mechanism allows nodes in the blockchain network to be synchronized and to agree on the current state of the distributed ledger.

Blockchain promises to solve the issues such as exclusive peer-to-peer transactions without centralized third parties, fraudulent replication of digital asset/value or transaction (e.g. double spending), establishing trust with pseudonymity, transparent yet immutable record-keeping, provenance and auditable enabled distributed ledger, and digital signing and execution of legal agreements (between parties) in the form of softwarized contracts. The technology is also optionally driven by incentive mechanisms to attract more nodes to participate in the decentralized management of the digital ledger [4].

16.2.1 Types of Blockchain

There exist different types of blockchain and many times the clear distinction between them is missing. Figure 16.1 shows different types of blockchains, and the classification is based on (i) who is allowed to read, write, and commit transactions in blockchain (i.e. permission configuration) [3], and (ii) on what type of servers hosting is done [5]. The read permission allows a node to access and see all the past transactions, and the write permission enables nodes to create and broadcast a transaction to all the other nodes in the blockchain network. The commit permission empowers a node to update the distributed ledger [3]. Fundamentally, there are four different types of blockchain: (i) public, (ii) private, (iii) permissionless, and (iv) permissioned blockchain. If a blockchain permits any node to acquire read permission by default, and public servers are used for its hosting, then it is considered as public blockchain. On the contrary, if read permission is restricted and private hosting is used, then such a blockchain is called as private blockchain. In other words, in public blockchain anyone can read, whereas in private blockchain only the set of vetted nodes can read. Next, a blockchain is said to be permissionless, if all the nodes with read capability can also write and commit. However, in permissioned blockchain, only the authorized subset of the participating nodes (who can read) are capable to write and commit.

Based on these four basic types of blockchain, there are four combinations possible which are shown as the leaf nodes of the inverted tree structure in the Figure 16.1. *Public permissionless* blockchain are the ones where anyone can join the blockchain network and perform read and write operations. Cryptocurrencies like Bitcoin and Ethereum are examples of such blockchain. *Public permissioned* blockchain allows anyone to read, but allows only the selected few nodes to write and commit. For instance, a supply chain ledger that can be read by customers but updated by only the authorized nodes under the control of company [5] or a blockchain-based voting system. *Private permissionless* blockchain is formed by only the set of restricted nodes which can perform read, write, and commit operations. Such a type of blockchain is also called a consortium or federated blockchain. Finally, *private permissioned* blockchain enable all the nodes in the

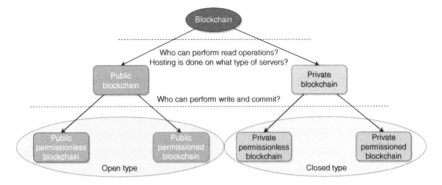

Figure 16.1 Types of blockchain [3, 5].

set of approved nodes to carry out read operation; however, the write and commit operations are exclusively performed by network operators (i.e. a subset of a set of approved nodes) [3].

16.3   What Is Smart Contracts

Yet another important concept that has extended the capability of blockchain is smart contract. The idea of smart contract existed much before the birth of blockchain technology. It was first introduced by Szabo in 1994 [6]. It can be simply defined as a computer program that encodes all the terms and conditions of an agreement between participating entities and runs on top of blockchain [7]. Any node in the blockchain network is capable of executing a smart contract either when it is called by initiating a transaction or when certain predefined conditions are met. Moreover, a smart contract can be established between untrusted and anonymous entities in the blockchain ecosystem without the need of a third party [8]. As and when the predefined encoded conditions are reached, the corresponding smart contract gets self-executed thereby empowering automation. It is worth noting that once a smart contract is deployed on the blockchain, it cannot be stopped by any entity within the system or outside the system [7].

16.4   Salient Features of Blockchain

This section briefly discusses salient features of blockchain technology. They are as follows:

- **Decentralization**: Unlike, conventional centralized technologies, blockchain technology offers decentralized modus operandi. This implies there is no central authority that owns and controls a given system completely. Rather, the system is collectively operated by a dynamic set of nodes (or miners) such that any node can create new block of transactions. Thus, blockchain decentralizes the decision-making and control of the entire system.

- **Immutability**: Once any transaction or data are stored on the blockchain in general, there is no way that it can be tampered with or altered. This is due to the distributed digital ledger, use of cryptographic techniques, and consensus-based update procedure. Thus, it is important to note that immutability also implies irrevocability and nondeletion of content from the blockchain infrastructure.

- **Transparency with pseudonymity**: Transparency implies any node in the blockchain network has access to read any content present in the digital ledger. The level of transparency is highest for public permissionless blockchain and is minimum in private permissioned blockchain. Further, nodes in the P2P blockchain network can just see transactions happening between different digital addresses (account IDs); however, the information about which ID belongs to whom (i.e. an individual or an organization) is absent. Thus, blockchain allows transparency with pseudonymity.

- **Fast processing and lower cost**: Conventional (centralized) systems involve multiple intermediaries and trusted third parties. Each of them charges a processing fee and incurs a processing delay. This collectively increases the overall cost and inflates the total delay in confirmation of transactions. Blockchain removes intermediaries and third parties, thereby resulting in fast processing and a reduction in cost per transaction.

- **Consensus-based decision-making**: To add new content or to update the current blockchain, consensus needs to be established among the majority of the nodes in the blockchain network. Thus, blockchain dethrones the monopoly and embraces democracy in the system. The algorithm required to establish the overall agreement in blockchain infrastructure is called a consensus algorithm. A variety of consensus algorithms are available, for example Proof-of-Work (PoW), Proof-of-Stake (PoS), Practical Byzantine Faulty Tolerance (PBFT), Delegated Proof-of-State (DPoS), Proof-of-Burn (PoB), and Proof-of-Activity (PoA). Based on the suitability and a given use case, one must be used.

- **Automation driven by smart contracts**: Smart contract enhances the capabilities of blockchain technology and have been found to be versatile. In particular, smart contract helps in exercising tight access control and automating blockchainized systems. At any point in time, when the predefined conditions embedded in a smart contract are met, the smart contract gets self-executed and performs the required tasks without any human interventions. This results in achieving automation among multiple untrusting stakeholders in a system.

16.5 Key Security Challenges Which Blockchain Can Solve

In future 6G networks, many security challenges may encounter due to extremely large network of heterogeneous networks, and the expected extremely higher reliability requirements. In order to thwart more sophisticated adversaries, advanced and intelligent security mechanisms will be needed [9–11]. Following are a few such security challenges identified in the envisioned 6G networks which can be solved by using blockchain.

- Confidentiality and integrity will be challenging since the future 6G infrastructure may portray enormous threat surfaces with wireless connectivity and the massive data volume generated in the network.
- Uninterrupted service accessibility will be another challenge because the broader threat surface and massive connectivity will enhance the risk of Distributed Denial of Service (DDoS) attacks [12, 13].
- The authentication and access control mechanisms need to be scaled-up to meet the diversification of 6G tenants, which are resource-intensive and may create bottlenecks in associated services. In particular, the centralized way of access control poses significant challenges in the design of future networks [14].
- Auditing will be another challenging security perspective due to the assessment requirements of a huge number of tenants (e.g. managing network slices among multiple tenants.).
- With the emergence of ubiquitous intelligence in 6G, artificial intelligence (AI)/ machine learning (ML)-based security attacks may also incur in 6G networks. In contrast, AI can be used as a tool for detecting, predicting, and mitigating security attacks. Therefore, the deployment of proactive security mechanisms and detection of zero-day attacks will be vital security challenges in 6G [15, 16].

16.5.1 Role of Blockchain

Blockchain is emerging as an intriguing solution to implement security, accountability, surveillance, and governance of mobile networks. Blockchain-based 6G networks may resist the eavesdropping and hijacking threats using the blockchain properties such as immutability, transparency, nonrepudiation, and provenance. Blockchain provides decentralized means of trust among untrusted users because it can withstand Sybil attacks [17–19]. Notably, in a Sybil attack, an attacker creates a large number of fake identities in a peer-to-peer network-based system and gains higher influence because of the sheer size of fake identities. By using efficient consensus protocols, such as PoW, blockchain ensures that only genuine users control the system [20, 21]. Also, as only the participating nodes can view or add new transactions, such a combination will mitigate the chance of data modification and man-in-the-middle attacks. In addition, authentication, access control, and accountability are key security aspects that can be addressed by deploying well-designed smart contracts on top of the blockchain.

Furthermore, as stated in [22], the use of blockchain allows (i) secure training of AI/ML models based on the immutable data stored on blockchain typically in heterogeneous network scenario, (ii) trust-building by transparently storing data related to the decision-making process in AI/ML based systems, and (iii) trustless collaborative optimization of ML models by secure exchange of model parameters. For AI/ML functions, it is vital to have trusted execution for security. Smart contracts provide a distributed mechanism for that purpose. For instance, the developed smart contract in [23] is used to detect network anomalies and malicious traffic, thus providing an analysis of network traffic between industrial IoT with a trained deep learning model in a secure distributed environment. El Azzaoui et al. [24] proposed a framework "Block5GIntell," where the support that blockchain can provide to AI for building secure 5G networks is discussed. Since that will have AI at its core, blockchain has a significant role to play for 6G networks.

From a different perceptive, 6G business ecosystem will be composed of numerous stakeholders and business players, all working together to achieve common goals. Security will be of paramount importance to build trust among these players. In summary, Figure 16.2 provides an illustration of possible security attacks and challenges in 6G networks along with the security features which are offered by blockchain. The circles indicate two key components of a 6G network, namely distributed AI and intelligent radio, which have the closest relationship with the applicability of blockchain for the security landscape. The possible types of

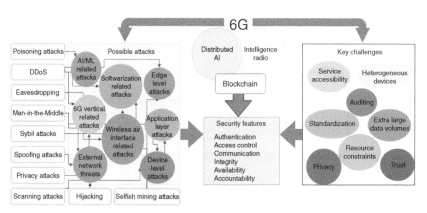

Figure 16.2 Possible security attacks and challenges in 6G networks along with the security features which may accompany with blockchain.

attacks for which blockchain based solutions can be adopted in the envisioned 6G networks are listed on the left side rectangle in the figure. The attacks may occur at different levels such as edge level, application level, device level, wireless air interfaces, external network, or different 6G verticals. Moreover, the security threats may be exacerbated with network softwarization and the use of AI/ML techniques.

Also, blockchain can be used for authentication, authorization, and key management purposes in the communication networks [25–27]. Having a common communication channel in the form of blockchain will introduce trust among multiple stakeholders in 6G network to cooperate on a common platform even at unexpected failures of the networks [28].

16.6 Key Privacy Challenges Which Blockchain Can Solve

Given that a significantly higher level of heterogeneity will characterize the 6G ecosystem (in terms of users, devices, applications, services, and stakeholders), privacy for such a complex ecosystem will be equally important, if not more, than security [29]. Preserving privacy while collecting, aggregating, and processing data in many application domains of 6G, such as healthcare, environmental protection, smart city, and social survey, will be essential [30]. Moreover, with every aspect of human lives getting connected, using of AI-enabled applications, both data privacy and user privacy must be protected. In addition, the novel 3D networking paradigm will allow the proliferation of flying connected devices. Despite the myriad of practical applications of unmanned aerial vehicles (UAVs), there are privacy concerns such as identity, trajectory (location), and mission privacy [31]. Thus, privacy will be an essential requirement for 6G networks, and compliance to privacy laws like General Data Protection Regulation (GDPR) [32] need to be ensured.

16.6.1 Key Challenges

Privacy preservation will be a significant consideration in the complex 6G network ecosystem. Following are the some of the key privacy challenges which can be solved by using blockchain.

- The unauthorized access to virtualized networking resources can impose a high risk of connected users' privacy [29]. Therefore, secure access control is one key challenge that arises for privacy preservation in vast data sets in 6G networks [33].
- With the emergence of massive-scale sensing systems (e.g. using UAVs) with cameras and sensors continuously collect data from surroundings, there can

be many instances that may cause violation of data privacy. For example, crowd-monitoring drones that collect information on large gatherings may impose threats on individual's privacy.

- Due to the enormous amount of data at centralized points, such as data aggregation servers, data providers, and service providers, there can be the risk of privacy leakage [34].

- Since AI/ML are key enabling technologies of 6G, while collecting individual personal related data for training models, the risk of revealing data may create privacy attacks [15, 35].

16.6.2   Role of Blockchain

Privacy is another technical aspect that can be met with the use of blockchain in 6G networks [34]. Having a common communication channel in the form of blockchain, can allow network users to be identified by pseudo names instead of direct personal identities [17]. Manogaran et al. [29] proposed a blockchain-based integrated security measure (dubbed as BISM) for 6G communication environment. In the proposed measure, virtualized resource access control and user privacy-preserving techniques are modeled with low time complexity and memory consumption. The authors considered the connected users and the virtualized resources, and the access control decisions are taken by maintaining the states of the resources using a Q-learning procedure. The performance of privacy preserving is measured by the user and resource level with respect to time complexity and memory consumption. However, the drawback is that the authors do not present the deployment of their proposed solution in a real blockchain platform.

As stated in [36], the use of blockchain for 6G-enabled UAV communication is significantly efficient compared to the current communication networks. It may provide higher privacy and transparency during data acquisition and air traffic management while allowing fast uploading of data with higher bandwidth and lower latency among UAV applications. In [30], the authors proposed a blockchain-based privacy-aware distribution collection-oriented strategy for data aggregation. In their solution, privacy protection is achieved by improving blockchain with a new block header structure to decompose high sensitive tasks and actors into different groups. However, they do not provide the exact details on how the block header modification can be performed for data aggregation in a trustless network.

Fan et al. [37] proposed a privacy preservation scheme that grants complete data ownership based on blockchain for content-centric 5G networks. The clearly defined access rules govern who gets to access the user's data and when, i.e. time. Moreover, the adopted key management and the use of encrypted data strengthen privacy. The work considers the mutual trust between the users and the content providers and exploits a public blockchain. In addition to that, as stated in [38], blockchain technology can be applied together with differential privacy and homomorphic encryption techniques for enabling privacy preservation in 6G. Privacy preservation using ML is an emerging research area. In [39], authors design a blockchain-based architecture for secure data sharing among distributed entities by incorporating privacy-preserved federated learning. More specifically, federated learning is used to build and share data models instead of sharing the actual data, and differential privacy is used to protect data privacy.

16.7 Threat Landscape of Blockchain

Due to the foreseen alliance of blockchain and 6G, the security vulnerabilities of blockchain and smart contracts may also implicitly impact the 6G networks [40]. Most of these attacks occurred due to the reasons such as software programming errors, restrictions in the programming languages, and security loopholes in network connectivity [41]. Moreover, these security issues can be occurred in both public and private blockchain platforms. They lead to complications such as loss of accuracy, financial losses in terms of cryptocurrency, and reduced availability of the system. Some of the critical security attacks in blockchain and smart contract systems are listed below (Figure 16.3).

Majority attack/51% attack: If malicious users capture the 51% or more nodes in the blockchain, they could take over the control of the blockchain. In a majority attack, the attackers could alter the transaction history and prevent the confirmation of new legitimate transactions from confirming [42]. Blockchain systems which use majority voting consensus [43] are usually vulnerable for majority attacks.

Double-spending attacks: The spending of the cryptographic token is a key feature of most the blockchain platforms [44]. However, there is a risk that a user can spend a single token multiple times [45] due to lack of physical notes. Such attacks are called the double-spending attacks [46] and blockchain systems should have a mechanism to prevent such double-spending attacks.

Re-entrancy attack: The re-entrancy vulnerability can occur when a smart contract invokes another smart contract iterative. Here, the secondary smart contract which was invoked can be malicious. For instance such an attack was performed to hack Decentralized Autonomous Organization (DAO) in 2016 [47]. An anonymous hacker stole USD50M worth Ethers.

Sybil attacks: Here, an attacker or a group of attackers are trying to hijack the blockchain peer network by conceiving fake identities [48]. The blockchain systems which have minimal and automated member addition systems are typically prone to Sybil attacks [49].

Figure 16.3 Key security vulnerabilities of blockchainized 6G services.

Privacy leakages: Blockchains and smart contracts are vulnerable to several privacy threats such as leakage of transaction data privacy [50], leakage of smart contract logic privacy [51], leakage of user privacy [52], and privacy leakages during execution of smart contracts [53]. Some of the blockchain nodes may follow the strict privacy roles and support too much transparency which may lead to reveal some sensitive information such as trade secrets and pricing information [50]. Moreover, business logic of the organization is required to be incorporated in the blockchain. The sensitive business logic information such as commissions and bonuses may need to be included in smart contracts, and these information can be revealed to the competitors [51].

Other attacks: Apart from the above, blockchains and smart contracts are vulnerable to several other security threats such as destroyable contracts [54], exception disorder [55], call stack vulnerability [56], bad randomness [57], underflow/Overflow errors [58, 59], broken authentication [60], broken access control [61], security misconfiguration [62], and unbounded computational power-intensive operations [63].

16.8 Possible Solutions to Secure 6G Blockchains

Obviously, when the DLT/blockchain solutions are adopted in 6G networks, they should always comply with possible mechanisms to mitigate the above security attacks. However, the deployment of some of the security mechanisms can be momentous in the public blockchains than in the private blockchains. For instance, the debugging or any correction of smart contracts might be a cumbersome process [64] since the smart contracts are adopted by all the nodes in a blockchain network. Since the smart contracts are playing a vital role in DLT/blockchain systems to enable the automation, ensuring the accuracy of the smart contract is necessary. Moreover, the proper validation of correct functionality of the smart contract is required before deploying it in thousands of blockchain nodes. The accurate functionality of smart contacts can be checked by *identifying semantic flaws* [65, 66], using *security check tools* [67–69] and performing *formal verification* [70–73].

Moreover, proper access control and authentication mechanisms should be utilized to identify the malicious bots and AI-agent-based blockchain nodes. Such mechanisms can prevent the majority and Sybil attacks. The additional privacy preservation mechanisms such as privacy by design [74, 75] and trusted execution environment (TEE) [76, 77] can be integrating to prevent privacy leakages in blockchain-based 6G services [78, 79].

Moreover, blockchain/DLT support different architecture types such as (i) public, (ii) private, (iii) consortium, and (iv) hybrid blockchain [80]. The impact of above security attacks naturally vary for different architectures. For example, the 51% attacks are highly impacting on public blockchains. In such cases, consortium or private blockchains can be suitable for certain 6G services (e.g. spectrum management, roaming) which has less number of miners [81]. Therefore, selecting the proper blockchain/DLT type according to the 6G application and services can eliminate the impact of certain attacks.

References

1 Y. Lu, "The blockchain: State-of-the-art and research challenges," *Journal of Industrial Information Integration*, vol. 15, pp. 80–90, 2019.

2 R. Zhang, R. Xue, and L. Liu, "Security and privacy on blockchain," *ACM Computing Surveys (CSUR)*, vol. 52, no. 3, pp. 1–34, 2019.

3 G. Hileman and M. Rauchs, "Global cryptocurrency benchmarking study," *Cambridge Centre for Alternative Finance*, vol. 33, pp. 33–113, 2017.

4 Y. He, H. Li, X. Cheng, Y. Liu, C. Yang, and L. Sun, "A blockchain based truthful incentive mechanism for distributed P2P applications," *IEEE Access*, vol. 6, pp. 27 324–27 335, 2018.

5 B. Carson, G. Romanelli, P. Walsh, and A. Zhumaev, "Blockchain beyond the hype: What is the strategic business value," *McKinsey & Company*, pp. 1–13, 2018.

6 N. Szabo, "Formalizing and securing relationships on public networks," *First Monday*, 1997. [Online]. Available: https://firstmonday.org/ojs/index.php/fm/article/download/548/469.

7 M. Alharby, A. Aldweesh, and A. van Moorsel, "Blockchain-based smart contracts: A systematic mapping study of academic research (2018)," in *2018 International Conference on Cloud Computing, Big Data and Blockchain (ICCBB)*. IEEE, 2018, pp. 1–6.

8 F. Casino, T. K. Dasaklis, and C. Patsakis, "A systematic literature review of blockchain-based applications: Current status, classification and open issues," *Telematics and Informatics*, vol. 36, pp. 55–81, 2019.

9 S. Yrjölä, "How could blockchain transform 6G towards open ecosystemic business models?" in *2020 IEEE International Conference on Communications Workshops (ICC Workshops)*. IEEE, 2020, pp. 1–6.

10 P. Porambage, G. Gür, D. P. M. Osorio, M. Liyanage, and M. Ylianttila, "6G security challenges and potential solutions," in *2021 Joint European Conference on Networks and Communications (EuCNC) and 6G Summit*. IEEE, 2021, pp. 1–6.

11 P. Porambage, G. Gür, D. P. M. Osorio, M. Liyanage, A. Gurtov, and M. Ylianttila, "The roadmap to 6G security and privacy," *IEEE Open Journal of the Communications Society*, vol. 2, pp. 1094–1122, 2021.

12 I. Ahmad, S. Shahabuddin, T. Kumar, J. Okwuibe, A. Gurtov, and M. Ylianttila, "Security for 5G and beyond," *IEEE Communications Surveys & Tutorials*, vol. 21, no. 4, pp. 3682–3722, 2019.

13 R. Kantola, "6G network needs to support embedded trust," in *Proceedings of the 14th International Conference on Availability, Reliability and Security*, 2019, pp. 1–5.

14 M. Wang, T. Zhu, T. Zhang, J. Zhang, S. Yu, and W. Zhou, "Security and privacy in 6G networks: New areas and new challenges," *Digital Communications and Networks*, vol. 6, no. 3, pp. 281–291, 2020.

15 S. Tanwar, Q. Bhatia, P. Patel, A. Kumari, P. K. Singh, and W.-C. Hong, "Machine learning adoption in blockchain-based smart applications: The challenges, and a way forward," *IEEE Access*, vol. 8, pp. 474–488, 2019.

16 Y. Siriwardhana, P. Porambage, M. Liyanage, and M. Ylianttila, "AI and 6G security: Opportunities and challenges," in *Proceedings of the IEEE Joint European Conference on Networks and Communications & 6G Summit (EuCNC/6G Summit)*, 2021, pp. 1–6.

17 K. Gai, Y. Wu, L. Zhu, L. Xu, and Y. Zhang, "Permissioned blockchain and edge computing empowered privacy-preserving smart grid networks," *IEEE Internet of Things Journal*, vol. 6, no. 5, pp. 7992–8004, 2019.

18 X. Liang, J. Zhao, S. Shetty, J. Liu, and D. Li, "Integrating blockchain for data sharing and collaboration in mobile healthcare applications," in *2017 IEEE 28th Annual International Symposium on Personal, Indoor, and Mobile Radio Communications (PIMRC)*, 2017, pp. 1–5.

19 A. S. Musleh, G. Yao, and S. M. Muyeen, "Blockchain applications in smart grid-review and frameworks," *IEEE Access*, vol. 7, pp. 86 746–86 757, 2019.

20 S. Kim, Y. Kwon, and S. Cho, "A survey of scalability solutions on blockchain," in *2018 International Conference on Information and Communication Technology Convergence (ICTC)*, 2018, pp. 1204–1207.

21 W. Yang, E. Aghasian, S. Garg, D. Herbert, L. Disiuta, and B. Kang, "A survey on blockchain-based internet service architecture: Requirements, challenges, trends, and future," *IEEE Access*, vol. 7, pp. 75 845–75 872, 2019.

22 Y. Liu, F. R. Yu, X. Li, H. Ji, and V. C. Leung, "Blockchain and machine learning for communications and networking systems," *IEEE Communications Surveys & Tutorials*, vol. 22, no. 2, pp. 1392–1431, 2020.

23 K. Demertzis, L. Iliadis, N. Tziritas, and P. Kikiras, "Anomaly detection via blockchained deep learning smart contracts in industry 4.0," *Neural Computing and Applications*, vol. 32, no. 23, pp. 17 361–17 378, 2020.

24 A. El Azzaoui, S. K. Singh, Y. Pan, and J. H. Park, "Block5GIntell: Blockchain for AI-enabled 5G networks," *IEEE Access*, vol. 8, pp. 145 918–145 935, 2020.

25 Z. Haddad, M. M. Fouda, M. Mahmoud, and M. Abdallah, "Blockchain-based authentication for 5G networks," in *2020 IEEE International Conference on Informatics, IoT, and Enabling Technologies (ICIoT)*. IEEE, 2020, pp. 189–194.

26 S. Kiyomoto, A. Basu, M. S. Rahman, and S. Ruj, "On blockchain-based authorization architecture for beyond-5G mobile services," in 2017 12th International Conference for Internet Technology and Secured Transactions (ICITST). IEEE, 2017, pp. 136–141.

27 X. Ling, J. Wang, T. Bouchoucha, B. C. Levy, and Z. Ding, "Blockchain radio access network (B-RAN): Towards decentralized secure radio access paradigm," *IEEE Access*, vol. 7, pp. 9714–9723, 2019.

28 H. Zhang, J. Liu, H. Zhao, P. Wang, and N. Kato, "Blockchain-based trust management for internet of vehicles," *IEEE Transactions on Emerging Topics in Computing*, vol. 9, no. 3, pp. 1397–1409, 2020.

29 G. Manogaran, B. S. Rawal, V. Saravanan, P. M. Kumar, O. S. Martínez, R. G. Crespo, C. E. Montenegro-Marin, and S. Krishnamoorthy, "Blockchain based integrated security measure for reliable service delegation in 6G communication environment," *Computer Communications*, vol. 161, pp. 248–256, 2020.

30 H. Lin, S. Garg, J. Hu, G. Kaddoum, M. Peng, and M. S. Hossain, "A blockchain-based secure data aggregation strategy using 6G-enabled NIB for industrial applications," *IEEE Transactions on Industrial Informatics*, vol. 17, no. 10, pp. 7204–7212, 2020.

31 Y. Wu, H.-N. Dai, H. Wang, and K.-K. R. Choo, "Blockchain-Based Privacy Preservation for 5G-Enabled Drone Communications," *arXiv preprint arXiv:2009.03164*, 2020.

32 P. Voigt and A. Von dem Bussche, "The EU general data protection regulation (GDPR)," *A Practical Guide*, 1st Ed., Cham: Springer International Publishing, 2017.

33 T. Alladi, V. Chamola, N. Sahu, and M. Guizani, "Applications of blockchain in unmanned aerial vehicles: A review," *Vehicular Communications*, vol. 23, p. 100249, 2020.

34 T. Nguyen, N. Tran, L. Loven, J. Partala, M.-T. Kechadi, and S. Pirttikangas, "Privacy-aware blockchain innovation for 6G: Challenges and opportunities," in *2020 2nd 6G Wireless Summit (6G SUMMIT)*. IEEE, 2020, pp. 1–5.

35 Y. Sun, J. Liu, J. Wang, Y. Cao, and N. Kato, "When machine learning meets privacy in 6G: A survey," *IEEE Communications Surveys & Tutorials*, vol. 22, no. 4, pp. 2694–2724, 2020.

36 S. Aggarwal, N. Kumar, and S. Tanwar, "Blockchain-envisioned UAV communication using 6G networks: Open issues, use cases, and future directions," *IEEE Internet of Things Journal*, vol. 8, no. 7, pp. 5416–5441, 2020.

37 K. Fan, Y. Ren, Y. Wang, H. Li, and Y. Yang, "Blockchain-based efficient privacy preserving and data sharing scheme of content-centric network in 5G," *IET Communications*, vol. 12, no. 5, pp. 527–532, 2017.

38 K. Gai, Y. Wu, L. Zhu, Z. Zhang, and M. Qiu, "Differential privacy-based blockchain for industrial internet-of-things," *IEEE Transactions on Industrial Informatics*, vol. 16, no. 6, pp. 4156–4165, 2019.

39 Y. Lu, X. Huang, Y. Dai, S. Maharjan, and Y. Zhang, "Blockchain and federated learning for privacy-preserved data sharing in industrial IoT," *IEEE Transactions on Industrial Informatics*, vol. 16, no. 6, pp. 4177–4186, 2019.

40 J. Liu and Z. Liu, "A survey on security verification of blockchain smart contracts," *IEEE Access*, vol. 7, pp. 77 894–77 904, 2019.

41 T. M. Hewa, Y. Hu, M. Liyanage, S.S. Kanhare, and M. Ylianttila, "Survey on blockchain-based smart contracts: Technical aspects and future research," *IEEE Access*, vol. 9, pp. 87643–87662, 2021.

42 S. Dey, "Securing majority-attack in blockchain using machine learning and algorithmic game theory: A proof of work," in *2018 10th Computer Science and Electronic Engineering (CEEC)*. IEEE, 2018, pp. 7–10.

43 J. Moubarak, E. Filiol, and M. Chamoun, "On blockchain security and relevant attacks," in *2018 IEEE Middle East and North Africa Communications Conference (MENACOMM)*. IEEE, 2018, pp. 1–6.

44 D. Efanov and P. Roschin, "The all-pervasiveness of the blockchain technology," *Procedia Computer Science*, vol. 123, pp. 116–121, 2018.

45 U. W. Chohan, "The double spending problem and cryptocurrencies," *Available at SSRN 3090174*, 2017.

46 S. Zhang and J.-H. Lee, "Double-spending with a sybil attack in the bitcoin decentralized network," *IEEE Transactions on Industrial Informatics*, vol. 15, no. 10, pp. 5715–5722, 2019.

47 M. I. Mehar, C. L. Shier, A. Giambattista, E. Gong, G. Fletcher, R. Sanayhie, H. M. Kim, and M. Laskowski, "Understanding a revolutionary and flawed grand experiment in blockchain: The DAO attack," *Journal of Cases on Information Technology (JCIT)*, vol. 21, no. 1, pp. 19–32, 2019.

48 P. Otte, M. de Vos, and J. Pouwelse, "TrustChain: A sybil-resistant scalable blockchain," *Future Generation Computer Systems*, vol. 107, pp. 770–780, 2020.

49 Y. Cai and D. Zhu, "Fraud detections for online businesses: A perspective from blockchain technology," *Financial Innovation*, vol. 2, no. 1, p. 20, 2016.

50 Q. Feng, D. He, S. Zeadally, M. K. Khan, and N. Kumar, "A survey on privacy protection in blockchain system," *Journal of Network and Computer Applications*, vol. 126, pp. 45–58, 2019.

51 B. Bünz, S. Agrawal, M. Zamani, and D. Boneh, "Zether: Towards privacy in a smart contract world," in *Financial Cryptography and Data Security, International Conference on Financial Cryptography and Data Security*, Lecture Notes in Computer Science, J. Bonneau and N. Heninger, Eds. Cham: Springer, 2020, vol. 12059, pp. 423–443.

52 A. Dorri, M. Steger, S. S. Kanhere, and R. Jurdak, "Blockchain: A distributed solution to automotive security and privacy," *IEEE Communications Magazine*, vol. 55, no. 12, pp. 119–125, 2017.

53 Z. Bao, Q. Wang, W. Shi, L. Wang, H. Lei, and B. Chen, "When blockchain meets SGX: An overview, challenges, and open issues," *IEEE Access*, vol. 8, pp. 170404–170420, 2020.

54 A. Groce, J. Feist, G. Grieco, and M. Colburn, "What are the actual flaws in important smart contracts (and how can we find them)?" in *Financial Cryptography and Data Security, International Conference on Financial Cryptography and Data Security*, Lecture Notes in Computer Science. Cham: Springer, 2020, vol. 12059, pp. 634–653.

55 C. Liu, J. Gao, Y. Li, H. Wang, and Z. Chen, "Studying gas exceptions in blockchain-based cloud applications," *Journal of Cloud Computing*, vol. 9, no. 1, pp. 1–25, 2020.

56 X. Li, P. Jiang, T. Chen, X. Luo, and Q. Wen, "A survey on the security of blockchain systems," *Future Generation Computer Systems*, vol. 107, pp. 841–853, 2020.

57 K. Chatterjee, A. K. Goharshady, and A. Pourdamghani, "Probabilistic smart contracts: Secure randomness on the blockchain," in *2019 IEEE International Conference on Blockchain and Cryptocurrency (ICBC)*. IEEE, 2019, pp. 403–412.

58 C. G. Harris, "The risks and challenges of implementing ethereum smart contracts," in *2019 IEEE International Conference on Blockchain and Cryptocurrency (ICBC)*. IEEE, 2019, pp. 104–107.

59 S. Kim and S. Ryu, "Analysis of blockchain smart contracts: Techniques and insights," in *2020 IEEE Secure Development (SecDev)*. IEEE, 2020, pp. 65–73.

60 H. Poston, "Mapping the OWASP top ten to blockchain," *Procedia Computer Science*, vol. 177, pp. 613–617, 2020.

61 G. Karame and S. Capkun, "Blockchain security and privacy," *IEEE Security & Privacy*, vol. 16, no. 04, pp. 11–12, 2018.

62 F. H. Pohrmen, R. K. Das, and G. Saha, "Blockchain-based security aspects in heterogeneous Internet-of-Things networks: A survey," *Transactions on Emerging Telecommunications Technologies*, vol. 30, no. 10, p. e3741, 2019.

63 A. Singh, R. M. Parizi, Q. Zhang, K.-K. R. Choo, and A. Dehghantanha, "Blockchain smart contracts formalization: Approaches and challenges to address vulnerabilities," *Computers & Security*, vol. 88, p. 101654, 2020.

64 Y. Zhang, S. Ma, J. Li, K. Li, S. Nepal, and D. Gu, "SMARTSHIELD: Automatic smart contract protection made easy," in *2020 IEEE 27th International Conference on Software Analysis, Evolution and Reengineering (SANER)*. IEEE, 2020, pp. 23–34.

65 N. Atzei, M. Bartoletti, and T. Cimoli, "A survey of attacks on ethereum smart contracts (SOK)," in *Principles of Security and Trust, International Conference on Principles of Security and Trust*, Lecture Notes in Computer Science, M. Maffei and M. Ryan, Eds. Berlin, Heidelberg: Springer-Verlag, 2017, vol. 10204, pp. 164–186.

66 M. Wohrer and U. Zdun, "Smart contracts: Security patterns in the ethereum ecosystem and solidity," in *2018 International Workshop on Blockchain Oriented Software Engineering (IWBOSE)*, March 2018, pp. 2–8.

67 C. Liu, H. Liu, Z. Cao, Z. Chen, B. Chen, and B. Roscoe, "ReGuard: Finding reentrancy bugs in smart contracts," in *Proceedings of the 40th International Conference on Software Engineering: Companion Proceedings*. ACM, 2018, pp. 65–68.

68 B. Jiang, Y. Liu, and W. Chan, "Contractfuzzer: Fuzzing smart contracts for vulnerability detection," in *Proceedings of the 33rd ACM/IEEE International Conference on Automated Software Engineering*. ACM, 2018, pp. 259–269.

69 L. Brent, A. Jurisevic, M. Kong, E. Liu, F. Gauthier, V. Gramoli, R. Holz, and B. Scholz, "Vandal: A Scalable Security Analysis Framework for Smart Contracts," *arXiv preprint arXiv:1809.03981*, 2018.

70 K. Bhargavan, A. Delignat-Lavaud, C. Fournet, A. Gollamudi, G. Gonthier, N. Kobeissi, A. Rastogi, T. Sibut-Pinote, N. Swamy, and S. Zanella-Béguelin, "Short paper: Formal verification of smart contracts," in *Proceedings of the 11th ACM Workshop on Programming Languages and Analysis for Security (PLAS), in conjunction with ACM CCS*, 2016, pp. 91–96.

71 T. Abdellatif and K.-L. Brousmiche, "Formal verification of smart contracts based on users and blockchain behaviors models," in *2018 9th IFIP International Conference on New Technologies, Mobility and Security (NTMS)*. IEEE, 2018, pp. 1–5.

72 Z. Nehai, P.-Y. Piriou, and F. Daumas, "Model-checking of smart contracts," in *IEEE International Conference on Blockchain*, 2018, pp. 980–987.

73 E. Albert, P. Gordillo, B. Livshits, A. Rubio, and I. Sergey, "ETHIR: A framework for high-level analysis of ethereum bytecode," in *Automated Technology for Verification and Analysis, International Symposium on Automated Technology for Verification and Analysis*, Lecture Notes in Computer Science, S. Lahiri and C. Wang, Eds. Cham: Springer, 2018, vol. 11138, pp. 513–520.

74 P. Schaar, "Privacy by design," *Identity in the Information Society*, vol. 3, no. 2, pp. 267–274, 2010.

75 A. Cavoukian, "Privacy by design," *Identity in the Information Society*, 2009.

76 R. Cheng, F. Zhang, J. Kos, W. He, N. Hynes, N. Johnson, A. Juels, A. Miller, and D. Song, "Ekiden: A platform for confidentiality-preserving, trustworthy, and performant smart contracts," in *IEEE European Symposium on Security and Privacy*, 2019.

77 R. Yuan, Y.-B. Xia, H.-B. Chen, B.-Y. Zang, and J. Xie, "ShadowEth: Private smart contract on public blockchain," *Journal of Computer Science and Technology*, vol. 33, no. 3, pp. 542–556, 2018.

78 R. Gupta, S. Tanwar, F. Al-Turjman, P. Italiya, A. Nauman, and S. W. Kim, "Smart contract privacy protection using AI in cyber-physical systems: Tools, techniques and challenges," *IEEE Access*, vol. 8, pp. 24 746–24 772, 2020.

79 N. Kapsoulis, A. Psychas, G. Palaiokrassas, A. Marinakis, A. Litke, and T. Varvarigou, "Know your customer (KYC) implementation with smart contracts on a privacy-oriented decentralized architecture," *Future Internet*, vol. 12, no. 2, p. 41, 2020.

80 M. Niranjanamurthy, B. Nithya, and S. Jagannatha, "Analysis of blockchain technology: pros, cons and SWOT," *Cluster Computing*, vol. 22, no. 6, pp. 14 743–14 757, 2019.

81 N. Weerasinghe, T. Hewa, M. Liyanage, S. S. Kanhere, and M. Ylianttila, "A novel blockchain-as-a-service (BaaS) platform for local 5G operators," *IEEE Open Journal of the Communications Society*, vol. 2, pp. 575–601, 2021.

17

AI/ML for 6G Security*

* With additional contribution from Yushan Siriwardhana, Center for Wireless Communications, University of Oulu, Finland

When the next-generation networks in the 6G era are closely coupled with intelligent network orchestration and management, the role of artificial intelligence (AI) is immense. In this chapter, we give some insights about how to consolidate and solidify the role of AI in securing 6G networks with more research directions.

- Gain an overview of the role of AI/machine learning (ML) in 6G security vision.
- Identify the use of AI for identification and mitigation of 6G architecture and 6G architecture and technologies.

17.1 Overview of 6G Intelligence

While 5G is well known for network cloudification with microservice-based architecture, the next-generation networks or the 6G era is closely coupled with intelligent network orchestration and management. Hence, the role of AI is immense in the envisioned 6G paradigm. As shown in Figure 17.1, AI is expected to play different roles to facilitate the functions running in 6G architecture. However, the alliance between 6G and AI may also be a double-edged sword in many cases as AI's applicability for protecting or infringing security and privacy. In particular, the end-to-end automation of future networks demands proactive threats discovery, application of mitigation intelligent techniques and making sure the achievement of self-sustaining networks in 6G. Therefore, to consolidate and solidify the role of AI and explainable artificial intelligence (XAI) in securing 6G networks, this chapter presents how AI can be leveraged in 6G security, possible challenges, and solutions.

Security and Privacy Vision in 6G: A Comprehensive Guide, First Edition.
Pawani Porambage and Madhusanka Liyanage.
© 2023 The Institute of Electrical and Electronics Engineers, Inc. Published 2023 by John Wiley & Sons, Inc.

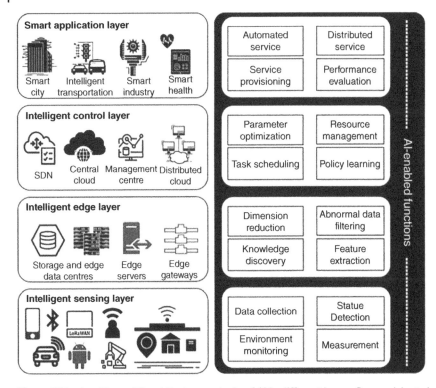

Figure 17.1 Intelligent 6G architecture and role of AI in different layers. Source: Adapted from [1]; Google LLC.

The evolution of 6G application domains calls for an innovative network architecture beyond current network designs [2]. An open and distributed reference framework for 6G architectural building blocks defined by Nokia Bell Labs comprises four major interworking components [3]. These are platform, functional, specialized, and orchestration, covering the physical layer to the service layer with the following distinguishing features. The "het-cloud" is a heterogeneous cloud environment that eases the creation, placement, and scaling of dynamic cloud services. 6G will take the network softwarization/cloudification into network intelligentization, revolutionizing wireless networks from connected things to "connected intelligence" [4, 5]. Hence, AI becomes an integral part of the network, which plays a crucial role. The distributed heterogeneous networks require ubiquitous AI services to ensure the fulfillment of 6G goals. Intelligent wireless communications, closed-loop optimization of networks, and big data analytics for 6G emphasize the use of AI in diverse aspects of 6G networks.

Beyond 2030, wireless applications will demand much higher data rates (up to 1 Tb/s), extremely low end-to-end latency (<1 ms), extremely high end-to-end reliability (99.99999%) [4, 5]. Moreover, 6G networks will comprise a collection of heterogeneous dense networks embedded with connected intelligence and utilize hyper-connected cloudification. Service provision for extreme requirements with complex 6G networks requires sophisticated security mechanisms. The security systems designed for 5G using the concepts of software-defined networking (SDN) and network function virtualization (NFV) should be further improved to cater to the security demands in 6G [6]. The end-to-end automation of future networks demands proactive threats discovery, intelligent mitigation techniques, and self-sustaining networks in 6G. Hence, the end-to-end security design leveraging AI techniques is essential to autonomously identify and respond to potential threats based on network anomalies rather than cryptographic methods. This chapter discusses the role of AI in the security provision of 6G networks.

17.2   AI for 6G Security

6G relies on AI/ML to enable fully autonomous intelligent networks. Therefore, attacks on AI systems, especially ML systems, will affect 6G. Poisoning attacks, data injection, data manipulation, logic corruption, model evasion, model inversion, model extraction, and membership inference attacks are potential security threats against ML systems. The collection of more features allows AI systems to perform better. Attacks on collected data, and the unintended use of private data, lead to privacy issues as the data processing is usually not visible to the users. This section presents AI's use in pre-6G security, security of 6G architectures, security of 6G technologies, and AI for 6G privacy (Figure 17.2).

17.3   Use of AI to Identify/Mitigate Pre-6G Security Issues

Multilayered intrusion detection and prevention using deep reinforcement learning (RL) and deep neural networks (DNN) is viable in SDN/NFV-enabled networks [7]. They effectively defend against IP spoofing attack, flow table overloading attack, Distributed Denial of Service (DDoS) attack, control plane saturation attack, and host location hijacking attack compared to several conventional approaches. ML approaches, such as Decision Trees and Random Forest, also proved useful for detecting DDoS attacks in SDN environments due to their short processing time and accuracy, respectively [8]. ML-based adaptive security approaches are effective against attacks on SDN/NFV as the 6G networks expect dynamic placement of virtual functions on-demand. The attacks also evolve using

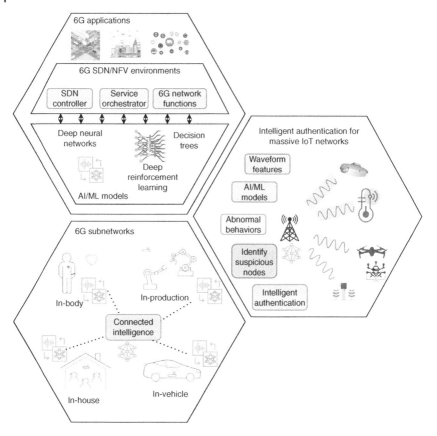

Figure 17.2 AI for 6G. Source: MicroOne/Adobe Stock; apinan/Adobe Stock.

AI techniques to learn vulnerabilities in a vastly distributed network. Hence, rule-based detection systems are ineffective.

On-device resource limitations, the difficulty of key management in massive scale heterogeneous networks, and the vast amount of device data generation make the conventional authentication/authorization systems insufficient for adequate security provision in large-scale Internet of Things (IoT). Anomaly-based intrusion detection systems in Industrial Internet of Things (IIoT) [9] detect malicious packets based on their behavior. These learning-based detection systems utilize various features of the data as the input, therefore, suitable for detecting zero-day attacks. The use of communication link attributes and user behaviors with ML for authentication and authorization [10, 11] is a better approach in future networks for resource-constrained devices. In that way, devices do not utilize their limited resources to provide additional complex security.

The subnetworks in 6G, which can be considered an expansion of local 5G networks beyond vertical domains, can benefit from learning-based security techniques within the subnetwork and between different subnetworks. ML-based algorithms deployed at the perimeter can capture the behavior of other subnetworks and detect malicious traffic from those subnetworks. Massive data transfer from one subnetworks to another might be of no use as these networks usually standalone. A subnetwork can share only the learned security intelligence to another for communication efficiency [12]. A second subnetwork can use the shared intelligence, feed it into its ML models, determine the malicious traffic of other networks, and apply dynamic policies.

17.4   AI to Mitigate Security Issues of 6G Architecture

6G will primarily depend on edge intelligence compared to present centralized cloud-based AI systems (Figure 17.3). The distributed nature enables the execution of edge-based federated learning for network security in the massive device and data regime [13], ensuring communication efficiency. 6G architecture envisages connected intelligence and uses AI at different levels in network hierarchy [1]. AI at the tiny cell level has the potential to block Denial of Service (DoS) attacks on cloud servers at the lowest level. The multiconnectivity of a device in a mesh network allows several base stations to evaluate the behavior of a device using AI classification algorithms, and collectively decide the authenticity using weighted average schemes as in [14]. The behavior-based approaches reduce the overhead caused by frequent key exchanges as the tiny cells and multiple access technologies cause frequent handovers. Different levels of authorization are possible for subnetwork level and wide area network level with federated learning. The trust score learned within the subnetwork level can be shared outside only when external

Figure 17.3   AI in 6G architecture.

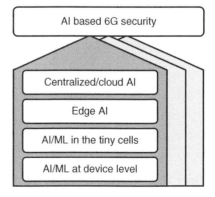

communication is needed. Learning-based intrusion detection approaches [7, 8] can be good candidates to prevent attacks on CPMS and UPMS as the edge already possess the data for intelligent service provision. Frameworks like zero touch and service management (ZSM) are equipped with domain analytics and domain intelligence services for zero-touch management of networks, predominantly based on AI. AI model assessment, AI engine for application programming interface (API) security are key security functional components that enhance the security of ZSM reference architecture [15].

17.5 AI to Mitigate Security Issues of 6G Technologies

AI and ML will definitely play integral roles in securing different types of 6G technologies as well as the use cases (Figure 17.4).

This may vary from identifying network-level anomalies or detecting/predicting security attacks to mitigating the attack propagation and recovering the networks back to normal operations. Under different chapters on each 6G technology, we have discussed the use of AI for mitigating the related security issues. For instance,

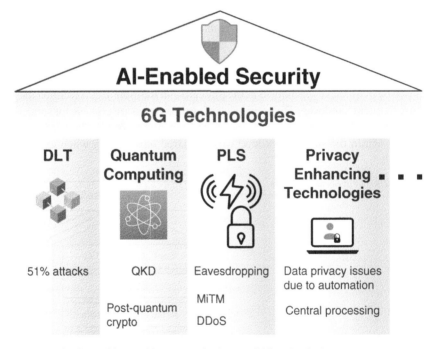

Figure 17.4 Role of AI to mitigate security issues of 6G technologies.

the predictive analytics using AI can predict attacks such as 51% attacks on blockchain before the attack occurs. A quantum computer may threaten asymmetric key cryptography. However, they can provide exponential speed-ups for AI/ML algorithms to perform tasks much faster and realize previously impossible tasks. Hence, quantum ML for network security is a potential defense technique against quantum computer-based attacks [16]. Intelligent beamforming techniques based on RL provide optimal beamforming policy against the eavesdropper attacks in visible light communication (VLC) systems [17]. Jamming resembles DoS attacks; therefore, anomaly-based detection systems equipped with AI is a possible solution to detect jamming attacks. AI-based authentication and authorization systems are also suitable for preventing node compromise attacks [10].

17.6   Security Issues in AI

6G achieves connected intelligence via AI-enabled functions, especially with ML systems that are subjected to security threats. Poisoning attacks influence the learning phase of a ML system, which leads the model to learn inaccurately. For example, data injection, data manipulation, and logic corruption are some of the poisoning attacks. Evasion attacks try to avoid the model during the inference phase using carefully crafted adversarial examples. Model extraction, model inversion, and membership inference are API-based attacks on ML models [18].

Potential countermeasures such as adversarial ML and moving target defense can create resilient AI systems. Input validation and robust learning against poisoning attacks, adversarial training and defensive distillation against evasion attacks, and differential privacy and homomorphic encryption against API-based attacks are other defense mechanisms. The balance between the increased defense and performance degradation is a design challenge with these defense mechanisms [18].

17.7   Using AI to Attack 6G

With the ability to make network-wide intelligent decisions with distributed edge-based architecture, AI itself can uncover the patterns within a large volume of data at different levels (intelligent radio, edge, and cloud). Hence, AI-based mechanisms have the potential to uncover vulnerabilities of the network. For example, AI can learn the most vulnerable IoT devices, convert them into bots, and initiate DDoS attacks [18] against a critical node.

The countermeasure for AI-based intelligent attacks is also the implementation of more intelligent defense systems. They can be empowered by AI itself using distributed intelligence. Moving target defense techniques is a proactive measure that introduces dynamicity to the network [19] and weakens the learning process of AI-enabled attackers. Quantum ML could also be used to design advanced defense techniques to resist AI-based attacks [16].

References

1 K. B. Letaief, W. Chen, Y. Shi, J. Zhang, and Y.-J. A. Zhang, "The roadmap to 6G: AI empowered wireless networks," *IEEE Communications Magazine*, vol. 57, no. 8, pp. 84–90, 2019.

2 M. Giordani, M. Polese, M. Mezzavilla, S. Rangan, and M. Zorzi, "Toward 6G networks: Use cases and technologies," *IEEE Communications Magazine*, vol. 58, no. 3, pp. 55–61, 2020.

3 V. Ziegler, H. Viswanathan, H. Flinck, M. Hoffmann, V. Räisänen, and K. Hätönen, "6G architecture to connect the worlds," *IEEE Access*, vol. 8, pp. 173 508–173 520, 2020.

4 W. Saad, M. Bennis, and M. Chen, "A vision of 6G wireless systems: Applications, trends, technologies, and open research problems," *IEEE Network*, vol. 34, no. 3, pp. 134–142, 2019.

5 C. de Alwis, A. Kalla, Q. V. Pham, P. Kumar, K. Dev, W. J. Hwang, and M. Liyanage, "Survey on 6G frontiers: Trends, applications, requirements, technologies and future research," *IEEE Open Journal of the Communications Society*, vol. 2, pp. 836–886, 2021.

6 M. Ylianttila, R. Kantola, A. Gurtov, L. Mucchi, I. Oppermann, Z. Yan, T. H. Nguyen, F. Liu, T. Hewa, M. Liyanage et al., "6G White Paper: Research Challenges for Trust, Security and Privacy," *arXiv preprint arXiv:2004.11665*, 2020.

7 I. H. Abdulqadder, S. Zhou, D. Zou, I. T. Aziz, and S. M. A. Akber, "Multi-layered intrusion detection and prevention in the SDN/NFV enabled cloud of 5G networks using AI-based defense mechanisms," *Computer Networks*, vol. 179, p. 107364, 2020.

8 R. Santos, D. Souza, W. Santo, A. Ribeiro, and E. Moreno, "Machine learning algorithms to detect DDoS attacks in SDN," *Concurrency and Computation: Practice and Experience*, vol. 32, no. 16, p. e5402, 2020.

9 M. Zolanvari, M. A. Teixeira, L. Gupta, K. M. Khan, and R. Jain, "Machine learning-based network vulnerability analysis of industrial Internet of Things," *IEEE Internet of Things Journal*, vol. 6, no. 4, pp. 6822–6834, 2019.

10 H. Fang, A. Qi, and X. Wang, "Fast authentication and progressive authorization in large-scale IoT: How to leverage AI for security enhancement," *IEEE Network*, vol. 34, no. 3, pp. 24–29, 2020.

11 H. Fang, X. Wang, and S. Tomasin, "Machine learning for intelligent authentication in 5G and beyond wireless networks," *IEEE Wireless Communications*, vol. 26, no. 5, pp. 55–61, 2019.

12 J. Wang and G. Joshi, "Cooperative SGD: A unified framework for the design and analysis of local-update SGD algorithms," *The Journal of Machine Learning Research*, vol. 22, no. 1, pp. 9709–9758, 2021.

13 C. Ma, J. Li, M. Ding, H. H. Yang, F. Shu, T. Q. S. Quek, and H. V. Poor, "On safeguarding privacy and security in the framework of federated learning," *IEEE Network*, vol. 34, no. 4, pp. 242–248, 2020.

14 Z. Chkirbene, A. Erbad, R. Hamila, A. Gouissem, A. Mohamed, M. Guizani, and M. Hamdi, "Weighted trustworthiness for ML based attacks classification," in *2020 IEEE Wireless Communications and Networking Conference (WCNC)*. IEEE, 2020, pp. 1–7.

15 C. Benzaid and T. Taleb, "ZSM security: Threat surface and best practices," *IEEE Network*, vol. 34, no. 3, pp. 124–133, 2020.

16 J. Biamonte, P. Wittek, N. Pancotti, P. Rebentrost, N. Wiebe, and S. Lloyd, "Quantum machine learning," *Nature*, vol. 549, no. 7671, pp. 195–202, 2017.

17 L. Xiao, G. Sheng, S. Liu, H. Dai, M. Peng, and J. Song, "Deep reinforcement learning-enabled secure visible light communication against eavesdropping," *IEEE Transactions on Communications*, vol. 67, no. 10, pp. 6994–7005, 2019.

18 C. Benzaid and T. Taleb, "AI-driven zero touch network and service management in 5G and beyond: Challenges and research directions," *IEEE Network*, vol. 34, no. 2, pp. 186–194, 2020.

19 J.-H. Cho, D. P. Sharma, H. Alavizadeh, S. Yoon, N. Ben-Asher, T. J. Moore, D. S. Kim, H. Lim, and F. F. Nelson, "Toward proactive, adaptive defense: A survey on moving target defense," *IEEE Communications Surveys & Tutorials*, vol. 22, no. 1, pp. 709–745, 2020.

18

Role of Explainable AI in 6G Security*

* With additional contribution from Thulitha Senevirathna, University College Dublin, Ireland

Accountability and resilience of artificial intelligence (AI)/machine learning (ML)-based 6G services is paramount now more than ever. In this chapter, we are going to discuss the potential of explainable artificial intelligence (XAI) to address the resilience of 6G services and technologies and thereby improve the accountability as a whole.

- Gain an overview of the use cases where XAI would be necessary in 6G applications.
- Gain an understanding of the role of XAI in 6G enabling technologies.
- Identify the new security issues that comes along with the addition of XAI.

18.1   What Is Explainable AI (XAI)

While the early AI systems were simple to understand, opaque decision methods such as deep neural networks (DNN) have gained popularity in recent years. Deep learning (DL) models are experimentally successful due to a combination of efficient learning algorithms and their large parametric field. DNNs are considered sophisticated black-box models since they have hundreds of layers and millions of parameters [1]. Transparency is the polar opposite of black-boxness, which is the pursuit of knowledge of how a model functions. The need for explainability among AI stakeholders is growing as black-box ML algorithms are increasingly used to make significant predictions in critical settings [2]. The risk lies in making and implementing choices that are not reasonable, lawful, or do not allow for comprehensive explanations of their actions [3]. Explanations that back up a

Security and Privacy Vision in 6G: A Comprehensive Guide, First Edition.
Pawani Porambage and Madhusanka Liyanage.

model's output are critical. For example, in medical applications, specialists need to uncover what causes are identified in the model to arrive at the forecast, which would reinforce their confidence in the diagnosis [4]. Telecommunication systems, B5G-backed autonomous cars, security, and finance are just a few other examples.

However, a better knowledge of a system may lead to its shortcomings being corrected. Interpretability as an extra design driver may enhance the implementation ability of a ML model for three main reasons according to [5]. It aids in guaranteeing objectivity in decision-making by rectifying bias in the training datasets. Second, it improves resilience by identifying possible adversarial events that may cause the forecast to alter. Finally, it will guarantee that only relevant variables are used to predict the outcomes – in other words, that the model reasoning is based on actual causation.

The literature clearly distinguishes between models that are interpretable by design and those that can be explained using external XAI techniques. XAI creates a suite of ML techniques that enable human users to understand, appropriately trust, and effectively manage the emerging generation of artificially intelligent partners [5]. This dichotomy may be thought of as the distinction between interpretable models and model interpretability methods; a more generally recognized classification is transparent models, and post-hoc explainability [6].

18.1.1 Terminologies of XAI

Transparency: A model is deemed transparent if it is understood on its own. Transparent models, by themselves, provide some degree of interpretability. Models in this domain may also be categorized according to the context in which they are interpretable, notably algorithmic transparency, decomposability, and simulatability. *Decomposability* refers to the capacity of a model to be explained in terms of its constituent components. *Simulatability* refers to a model's ability to be simulated or thought about rigorously by a person. When the model is sufficiently self-contained for a person to think and reason about it in its entirety, it can be referred to as a decomposable model with simulatability. *Algorithmic transparency* may be interpreted in a variety of ways. Prominently, it refers to the user's capacity to comprehend how the model generates any given result from its input data. The primary restriction on algorithmically transparent models is that they must be completely explorable using mathematical techniques and analysis [7]. Each of these classes includes its antecedents; for example, a simulatable model is both decomposable and algorithmically transparent. Some popular models that fall under transparent models are Linear/Logistic regression, Decision Trees, K-Nearest Neighbors, Rule-based models, General Additive Models (GAM), and Bayesian models. These models are deemed to be expressive enough to be human-understandable [5].

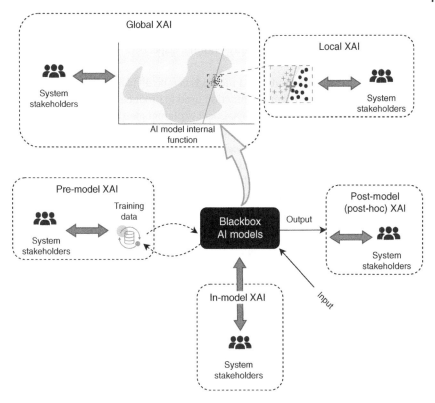

Figure 18.1 XAI taxonomy. Premodel XAI explains the training data used for building AI model (e.g. principal component analysis [PCA], t-distributed stochastic neighbor embedding [t-SNE]). In-model XAI refers to transparent AI models that are self-explanatory (e.g. decision trees, random forests). Post-hoc XAI models explain the results given by the trained AI models (e.g. LIME, SHAP).

18.1.2 Taxonomy of XAI

The XAI methods can be divided into multiple categories based on various criteria [5, 8]. The most common XAI-based taxonomy is represented in Figure 18.1. XAI methods that fall into those categories that are not necessarily exclusive for each group. According to the taxonomy, there can be methods that belong to even two or more categories.

(a) **Model-agnostic vs. model-specific**: Model agnostic methods for XAI are the ones that are not constrained by the core parts of an AI algorithm when making a prediction. They are helpful in decoding black box models' decision processes and provide good flexibility for developers to apply them to a wide variety of ML models. On the other hand, model-specific methods are bespoke for specific models and take the use of core components of an ML model to interpret the outcomes. This characteristic makes model-specific methods more suitable to identify granular aspects of ML models, but they lack flexibility.

(b) **Local vs. global methods**: XAI methods can be divided into two main categories based on the interpreter function used to produce an explanation. They are local and global explanations. Local explainers are designed to interpret a portion of the model function that contributes toward the outcome given a specific data point. The close vicinity of a ML function to that datapoint is explored when generating an explanation. On the contrary, global methods take the ML function as a whole when generating explanations for the inference. This quality generally makes these methods slow but robust, where local methods are fast but error prone.

(c) **Premodel, in-model vs. postmodel explainers**: Depending on the stage at which XAI methods are applied in the development process, there are three main categories of XAI; premodel, in-model, and postmodel. Premodel methods are used mainly used during the dataset preparation time in the model development pipeline. These methods are helpful in data analysis, feature engineering, and explaining any underlying patterns seen in data at a glance. In-model XAI methods are embedded in the ML algorithms. It includes all the transparent models such as linear regression, decision tree, random forest. In addition, in-model explanations are also generated through modifications to the existing ML model architectures using inherently transparent models. Posthoc/postmodel explanations are applied after training an ML model. It enables us to identify what the model has learned during the training process.

(d) **Surrogate vs. visualization**: The XAI method is divided into two main categories based on what is explained during the process. Surrogate model-based explainers generate explanations from an approximated model of the black-box model trained in a similar way to mimic the original model's behavior. Surrogate models are mostly inherently interpretable. Otherwise, one can use visualization techniques (e.g. heatmaps, graphs) on the original black-box model to explore its internal workings without using a representation. These XAI methods fall under the visualization category.

18.1.3   XAI Methods

There are numerous methods studied in the literature to explain black-box AI/ML models. Here, we discuss a selected set of popular XAI methods that are more established in the academic and industrial community, as shown in Figure 18.2.

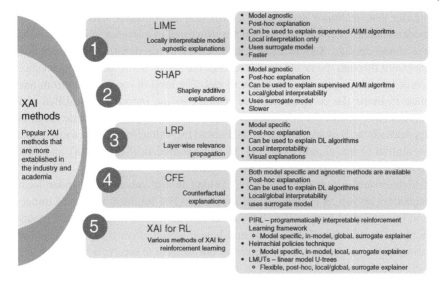

Figure 18.2 Various XAI methods are currently popular among the research and industrial community. Each method has its own strengths and weaknesses, as depicted. LIME, SHAP, LRP, CFE, PIRL, Hierarchical policies, and LMUTs are selected only to represent the categories of XAI methods.

Local interpretable model-agnostic explanations (LIME) [9] can be considered as one of the most popular model-agnostic XAI methods used to interpret model outputs in various applications. LIME provides a locally faithful explanation based on the feature importance that contributed to the output. It is achieved through a surrogate dataset obtained after sampling perturbations in the proximity of the original inputs. Then it creates a simpler interpretable model that can be used in identifying important features of a given output. Because of this local fidelity, LIME is faster and can be applied to all types of black-box models. Stemming from its popularity in the research community, application-specific variations of LIME such as OptiLIME [10] and CBR-LIME [11] are gaining popularity.

SHAP [12] is a model-agnostic interpreter that can be used with black-box models to interpret their model outcomes. SHAP uses the concept of shapely values derived from the cooperative game theory. Shapley values represent the average marginal contribution of a feature calculated over all the subsets of features with and without the said feature. Feature importance values generated here can be coalesced to obtain a global (unlike LIME) explanation of the model outcome space, which paves the way for better explanations. TreeSHAP [13] and DeepSHAP (DeepLift [14] + Shapley values) [12] are few derivatives of SHAP that are developed to fit specific models and improve the computation efficiency.

Layer-wise relevance propagation (LRP) [15] is popularly used to interpret models such as neural networks that are structured in a layered manner, making it a model-specific XAI method. LRP operates by propagating the prediction backward through the layers until it reaches the inputs while assigning relevance scores to each functional unit in the model. The level of contribution from one node to make the consequent node relevant to the output is quantified and aggregated to obtain the relevance of each layer. It is applied in a wide range of applications such as identifying model biases [16], extracting points of interest in prevention of side-channel attacks [17], audio source localization [18], and EEG pattern recognition in brain–computer interfaces [19].

Counter factual explanations (CFE) belong to the subcategory under local explainers known as example-based explainers. The intuition here is to understand what would happen if a slightly different data point is given to the model in place of the original data point and how it would affect the prediction. For example, if a model classifies a person as not suitable to receive a loan, then the CFE would provide the reason that he/she needs to have a saving of € 10 000 for the model to classify him/her as a suitable person to receive the loan. Unlike many other interpreters, the explanations are closer to human nature and provide actionable and precise recommendations. The basic idea of CFEs are model agnostic, but several variations are developed for model-specific applications [20].

XAI for reinforcement learning (RL) has been explored for agent-based RL AI systems from as early as 1994 [21]. It is not easy to deliver XAI for RL because it generally includes several judgments made over time, often seeking to offer the next action in real-time. Unlike conventional ML, RL explanations must cover a collection of acts spread across a plethora of different states that are connected in some manner. Also, the absence of an explicit training dataset can contribute to the difficulty in applying XAI techniques [22]. Some of the widely known AI techniques for RL is discussed below [23].

(a) Programmatically interpretable reinforcement learning framework (PIRL) is a global, in-model method used in place of deep reinforcement learning (DRL) [24]. In DRL, neural networks reflect the policies and are difficult to comprehend. On the other hand, PIRL policies are expressed using a high-level, human-readable programming language. However, unlike standard RL, they limit the number of target policies by using a (policy) sketch. They use a framework based on imitation learning called Neurally Directed Program Search (NDPS) to uncover these regulations.

(b) Hierarchical policies technique [25] is another in-model XAI technique used to interpret the decision process of multitask complex RL systems but locally. The core tenet of this approach is to decompose a complicated task into smaller subtasks. These smaller tasks will be accomplished with the already learned policies or learn a new skill. This model also takes the temporal connections and task priorities to improve efficiency and accuracy. The technique builds on multitask RL with modular policy design and a two-layer hierarchical policy based on minimal assumptions and limits. In [25], they assess this technique in object manipulation tasks in the Minecraft game.

(c) Linear model U-trees (LMUT) [26] is a posthoc explanation method unlike the methods mentioned above. They are flexible to be used to generate both local and global approximations of an RL model's Q-predictions [23]. LMUTs are an extension of Continuous U-Trees with the contrast in using linear models at each leaf node instead of constants, making them more interpretable and comprehensible. Because of the inherent interpretable nature of the trees, it becomes easier to generate explanations from the LMUTs as they mimic the original Q-function.

18.2 Use of XAI for 6G

Adopting XAI in 6G-enabling technologies will encourage the creation of new applications that were unattainable previously owing to the absence of account-ability in AI/ML-based systems. Figure 18.3 represents only a few of the areas that can have major commercialization potential for AI/ML with the advent of XAI. Following is a brief discussion of how XAI can impact some of those use cases. Smart cities were proposed to manage and optimize resource and energy usage. AI-based techniques will be found in many sectors of smart cities such as Intelligent transportation, cyber-security, smart transportation, electric and water system, waste management, public safety, and UAVs-assisted next-generation communication (5G and B5G). [27–29]. As shown in Figure 18.3, attacks such as DoS, phishing, and spoofing on these systems can be mitigated by ML-driven intrusion detection systems (IDS). XAI is a game changer when it comes to protect-ing these IDS systems. Therefore, current literature is starting to emphasize on use of XAI to ensure accountability before these applications are deployed in the real world. For example, Embark [30] emphasize the importance of XAI in the transi-tion stage from heritage cities to smart cities. Some of the other example can be shown as [31–33].

In smart healthcare, the use of AI has become essential in processing data obtained from personalized and real-time health monitoring services [34]. As shown in Figure 18.3, user-level information collected from wearables, smart-phones, and healthcare applications is processed with AI during health screenings and treatment plan selections, diagnostics, and emergency responses. But the lack of trust in AI systems is due to the black-box nature and lack of fail-safe mechanisms in case of security breaches. Authors of [35] have pointed out the

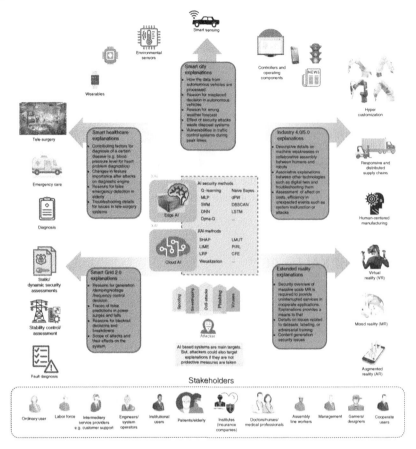

Figure 18.3 XAI is expected to create waves in the implementation of AI/ML models in many publicly accessible services that relies on 6G communication technologies. Accountability of these AI-based services are reinforced with explanations generated. In order to efficiently function, explanations must be communicated to the stakeholders in different levels of expertise in a comprehensible format. Each explanation should be coherent and unambiguous. Source: RF BSIP/Adobe Stock; xiaoliangge/Adobe Stock.

impacts of XAI in Healthcare as threefold: increased transparency, result tracking, and model improvement. If a user is classified to having a certain ailment by an AI/ML models, say high blood sugar, physicians will get a report detailing the attributes employed in arriving to the said conclusion. Some attributes can be heart rate, body temperature, and calorie consumption. With XAI, the most responsible characteristic for the result can be identified (e.g. calorie intake). After that, the medical practitioners can swiftly assess the features and propose drugs or exercises. Then again, the same feature's important values can be used to identify any intruders that might be meddling with the central ML model.

In addition to the above discussed technologies, smart grid 2.0, extended reality, Industry 4.0/5.0, Holographic telepresence are some of the other use cases where XAI can provide additional security in the 6G era.

18.3 XAI for 6G Security

Human-centric AI-powered telecommunication in the B5G era would attract the attention of various contingent parties that need assurance for trusting these systems – giving convincing evidence on how the decision-making process inside an AI model will be challenging pertaining to the technical knowledge gap and obfuscated nature of internals in the widely used AI/ML model. Especially, the security of this new technology in telecommunication has gained much attention from both malevolent and benevolent agents relevant to all the layers of the network.

Network softwarization (NS) and network function virtualization (NFV) introduced with 5G are expected to be significantly enhanced as we enter the B5G era. In the first layer of B5G architecture, data are gathered through IoT devices such as smartwatches, phones, drones, to enable real-time services in higher layers. Although it is one of the essential operations, the convictions behind data collection and security/privacy issues are highly influenced by the demography and underlying regulations. Using XAI, such differences can be addressed evenly by giving more details about how collected data are used inside AI models in the rest of the pipeline. Further, it enables system operators to identify the performance of each device more intimately with respect to the overall AI system. Radio access network (RAN), edge, core, and backhaul layers provide the infrastructure to reach higher speeds and quality of services based on enhanced virtualization techniques. The security of these layers is envisioned to be addressed through AI-/ML-based methods to accommodate the massive volumes of data. Getting automated feedback on the performance of those AI/ML systems is paramount to ensure maximum resilience by identifying false predictions and diagnosing any system issues. It will benefit system operators and stakeholders who are not technically sound in AI/ML. In E2E slicing and zero-touch network and service management (ZSM) [36, 37] security of AI/ML components are used in integral parts of the system architecture. For example, the ZSM's E2E service intelligence enables decision-making based on data collected in the domain and standard data services. An attacker may create inputs to cause the ML model to make incorrect judgments and threaten performance, financial loss, service level agreement (SLA) fulfillment, and security assurances. XAI would be highly useful during the response process to estimate the overall effect and trace back to the most basic module responsible for the anomaly. Finally, the application layer would require the most high-level explanations relevant to the end-users in B5G. Techniques like counterfactual explanations are ideal for inculcating trust and confidence in the users in the application layer.

When designing a system with explainable security, one must evaluate the 6W questions: Why, Who, What, Where, When, and How to generate security explanations. Inspired by Vigano and Magazzeni [38], Figure 18.4 depicts the flow of identifying basic building blocks to design an explainable security system. First, the apparent reason to *why* the system needs XAI must be identified. Then *to whom* and *who* create the explanation and decide the level of granularity of the content that is broadcast to each group of actors. Identifying the needs of each actor early helps to decide on *what* aspects of the system need to be explained. Here, the system designers must consider the layer of B5G architecture and fit the explanation to meet its requirements. Although the explanation is generated in one layer, it will not be the same *where* it will be accessible. Whether it will be a separate service or embedded in the system/output must be decided. It is also essential to decide *when*, explanations are needed during the process, i.e. during design, installation, maintenance, defense, etc. Finally, the nature of the explanation is decided by answering the question of *how* to interpret AI/ML model. It will lay the groundwork for choosing the correct XAI methods for high-quality explanations.

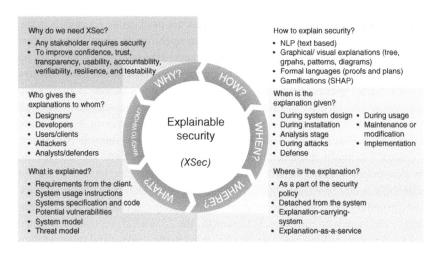

Figure 18.4 6W analysis for explainable security in 6G. The procedure shown can be used as a framework to initiate laying the groundwork when designing security aspects of explainable intelligent systems built in/on B5G network. NLP–natural language processing.

18.3.1   XAI for 6G Devices and IoT Security

The ever-changing and heterogeneous nature of the IoT systems can make the security of the IoT and 6G devices to face many security issues and challenges. In [39], access control techniques are implemented using naive Bayes and SVM algorithms to mitigate intrusions. In other literature such as [40], reinforcement-based techniques (Q-learning and Dyna-Q) were used for authentication to prevent spoofing attacks, while in [41] and [42], the authors used SVM and DNN, respectively. However, in all those algorithms, the ability to explain the outputs is lacking. Due to the nature of the application and its gravity, the outputs of these algorithms must be reliable. Hence, these black-box models need to be wrapped with some explainable technique to visualize the reasoning behind each decision made by black-box models in those systems. DQNviz proposes a visual analytics-based method to expose the blind training in four levels which make the large experience space more comprehensible to the users [43]. Such visualization helps the security operators to filter and give faster response during an attack reconciliation. However, explainability could incur various additional costs and challenges as well. Additional in-device explanation techniques would require computation power straining the already available limited processor, memory, bandwidth, and battery. Addition of more powerful devices would cause the cost-to-serve to increase. Driving trends such as zero-energy IoT envisaged in 6G [44] would require more subtle XAI techniques deployed in the IoT devices.

18.3.2   XAI for 6G RAN

The ORAN alliance has brought all the latest C/V/ORAN (Cloud, Virtual, Open RAN) technologies together to realize 6G RAN. The architecture of ORAN can be slightly modified to accommodate XAI techniques and improve the resilience of AI/ML workflow. Self-organization and intelligence-based technologies will be major characteristics in the deployment process of ORAN [45]. Therefore, the heavy automation involved in this process calls for reliable and more secure intelligence-based methods (i.e. AI/ML). IDSs in RANs are one of the abundantly researched areas in RAN security. AI-/ML-based IDS takes a prominent place due to their data-driven, self-improving capabilities. Insights on black-box ML models make those models transparent for the operators, developers, and engineers who interact with them. Interpretations obtained through XAI are a possible way in to understanding the false classifications of anomalous traffic as benign traffic or vice versa in an IDS system.

To ensure accountability, open distributed units that host those models will also require pipelines built to generate and communicate explanations. This requirement calls for more computation power and resources [46]. ML and AI models use real-time data from the RAN to monitor the RAN's health and performance. XAI techniques applied to those ML techniques will require additional time, effort, and resources. Near-real-time and nonreal-time RAN intelligence controller (RIC) will require additional computation power to host interpreters incurring nontrivial costs. Added costs are justifiable as O-RAN's security, and management capabilities are enhanced because of the obtained results.

18.3.3 XAI for 6G Edge

Edge networks are seemingly becoming more popular [47] due to their advantages in cost-effectiveness of data usage [48], privacy improvement, and bandwidth usage [49, 50] which in turn enables the implementation of novel ML applications on it [51]. Current research describes the use of AI as a facilitator of edge security including more general applications and complete architectures that rely on AI such as AI4SAFE-IoT [52]. The three-layer (network, application, and edge) architecture uses an AI engine for security across all three layers. Network layer IDS is claimed to mitigate sinkhole, DoS, rank, and local repair attacks in the proposed architecture. The security risks associated with AI can possibly be reduced by providing a layer of posthoc XAI techniques that continuously monitors false outputs. Malicious features responsible for particular attack can be highlighted for further analysis which are otherwise hidden inside black-box models. Deploying XAI methods in edge layer will enable secure analysis of the AI/ML models since it reduces the possibility of data pollution/interception ensuring overall fail-tolerant systems while making the models more transparent. Access to edge caching for generating and storing premodel explanations is necessary to ensure security in the edge and IoT layers. Thus, additional storage spaces will be required when implementing XAI in 6G edge.

18.3.4 XAI for 6G Core and Backhaul

Core network consists of communication facilities that link primary nodes providing routes to communicate between subnetworks. The backhaul network links BSs to network controllers within a coverage region, which interconnects to the core network through the core transport network. In the 6G realms, low latency and high throughput are highly emphasized aspects of backhaul and core networks. Using ML in this regard is imperative considering the gravity of the applications relying on the 6G infrastructure. There is a growing trend of using reinforcement and ML methods to backhaul and core networks. For instance, in [53], a Q-learning method is proposed for increasing the dependability of a millimeter-wave (mmW) nonline-of-sight small cell backhaul system. Adversarial attacks on such models can cause disruption in traffic management which relies on a healthy backhaul system. The need is to be explainable enough without adding a burden on latency and throughput. The multi layer perceptron (MLP) used in [54] for network optimization can be improved with various posthoc techniques of explainability such as case-based reasoning (CBR) [55] coupled into the MLP network. Whether to explain an RL model used in the core/backhaul network with a posthoc explanation technique (LMUT), or to completely replace them with an interpretable alternate model such as Programmatically Interpretable Reinforcement Learning (PILR model) is a decision that the system developers have to make considering the users and the criticality of security features involved. Deploying XAI methods in energy-efficient small cell backhauling techniques in UAV, high altitude platform stations, and satellites [56] will be highly challenging in terms of costs.

18.3.5 XAI for 6G Network Automation

Out of many 6G technologies, it is safe to say that network automation takes the lead in the greatest percentage of AI/ML used in their implementation. The ultimate automation goal in 6G is to create fully closed-loop autonomous networks through frameworks such as ZSM. Such wide use of AI/ML methods in network automation opens the door to many new possible scenarios for XAI applications. Authors of [57] have shown that a variety of ML techniques such as RNNs (long short-term memory networks [LSTMs]), support vector data descriptions (SVM-inspired technique), Q-learning, and Gaussian models would be necessary to enable functionalities such as self-configuration, self-monitoring, self-healing, and self-optimization without human involvement. Incidentally, IDS development and implementation is a salient concern to secure the full automation paradigm. For example, [58] gives an evaluation of perturbation-based posthoc XAI tools in the intrusion detection field with network traffic data. In this case, we can observe that all of these tools work quite well, with LIME and SHAP providing exceptional results. Furthermore, the work in [59, 60] proposes to use SHAP with their Multimodal DL-based Mobile Traffic Classification system (MIMETIC) to evaluate the input importance. It shows that XAI will play a key role in fostering the trustworthiness and transparency of AI applications in security for ZSM. Additional channels to communicate explanations generated about domain intelligent services and data collection services must be looped into domain analytics of each management domain (MD) in ZSM, so that any changes required in domain control and orchestration are properly executed.

18.4   New Security Issues of XAI

Although providing explainability to AI and ML solutions brings many benefits to B5G, it can be detrimental to ML models' security and the systems that embed these models. XAI may increase their vulnerability to ML attacks (i.e. attackers know how the black box model works), complicate their design (i.e. explainability must be considered in the trade-off between model performance and security), and open new attack vectors (i.e. the explanation itself can be falsified).

18.4.1   Increased Vulnerability to Adversarial ML Attacks

Many attacks that already exist target ML models: adversarial ML attacks [61]. *Membership inference*, and *model extraction* [62] attacks compromise the confidentiality of the training data and the ML model, respectively. *Model poisoning* and *model evasion* attacks (a.k.a. adversarial examples) compromise the integrity of the ML model and its predictions. A common characteristic of adversarial ML attacks is that their effectiveness increases as the attacker's knowledge about the ML model and its decision process increase. Consequently, the obfuscation of ML models' decision process, by making it a *black-box*, is an effective defense to mitigate adversarial ML attacks [63].

Explainability deobfuscates the decision process of black-box ML models, thereby revealing helpful information to an attacker, as depicted in Figure 18.5. It has been shown that the information produced by explainable ML techniques can be leveraged to design more effective black-box attacks [64]. The effectiveness of membership inference, model extraction, poisoning, and evasion attacks increases against black-box ML models augmented with explainability.

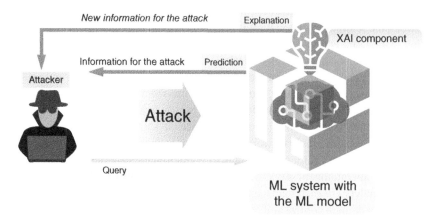

Figure 18.5   Explainability reveals new information that "white-box" and black-box ML models facilitate adversarial ML attacks against them.

The explanation provided for an ML model prediction enables an attacker to modify a sample that they want to get misclassified manually. Explanation reveals the features from a sample that are the most significant in the prediction provided by the model. Attackers can iteratively modify these features, using the feedback from the explanation, to eventually change the ML model prediction for the sample.

Defending against adversarial ML attacks is still an open issue, and there is no foolproof defense against any of these attacks. Some vulnerabilities exploited by adversarial ML attacks are even claimed to be necessary features of ML models [65]. Most defenses, like model obfuscation, only increase the effort required to perform a successful attack, but they do not fully protect the model from attacks. Explainability denies the usage of obfuscation to make ML models more resilient to attacks. This increased exposure decreases the security of all ML-based systems used in B5G.

For instance, ML-based security measures are deployed in the perimeter of 6G subnetworks for monitoring anomalous behavior within the subnetwork or coming from other subnetworks [66]. Decision tree (DT), Random Forest, DNN, clustering, ensemble methods, Gradient Boosting Machines (GBM), etc., are used to detect common network attacks, such as DDoS attacks, from traffic data [66–69]. The evasion and poisoning of these ML-based anomaly detectors are made easier if they are explainable [64]. They can cause malicious traffic to bypass the system defenses and cause exhaustion of network resources. The availability of the system resources to serve legitimate users will be constrained. As a result, many critical applications (e.g. telesurgery, smart grid stability control systems) that depend on the service layer of B5G could be affected by the exhaustion of resources. When the outermost security layer of a network fails, it leaves the internal modules exposed to a higher attack risk.

ML-based decisions are also used for intelligence services in closed-loop E2E service management. Adversarial examples against these ML models can be more easily generated if the models are explainable. These adversarial examples can lead to inaccurate predictions and choices, such as falsely predicting the future requirement of resources for an E2E service or reconfiguring the management policies [57]. Likely results can range from performance deterioration, financial loss, and loss of security guarantees.

18.4.2 Difficulty to Design Secure ML Applications

The design and implementation of ML-based systems are guided by the sole requirement of maximizing performance, i.e. high accuracy, high generalizability, and low-response time. Adding security requirements to ML-based systems introduced the first trade-off between antagonist properties: performance vs. security.

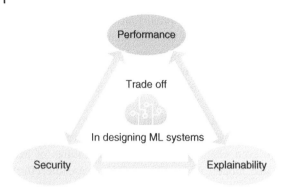

It has been shown that effective defenses against adversarial examples are adversarial training, degrading the accuracy [70] and the generalizability [71] of protected ML models. There also exists trade-offs between security properties. For instance, increasing the resilience of ML models against evasion attacks makes them more vulnerable to privacy attacks like membership inference [72].

Explainability is a new requirement adding to the existing trade-off. Three properties partly detrimental to each other need now to be fulfilled by ML systems, as illustrated in Figure 18.6: performance–security–explainability. When transparency provides explainability, all three requirements are applied to the ML model and its training algorithm. Providing explanation through transparency reduces the choice in training algorithms and models during the design of the ML system. This potentially leads to discarding the solution providing the best accuracy, security, or privacy to meet the explainability requirement.

The new requirement of achieving a performance–security–explainability trade-off makes it challenging to design well-balanced ML systems for B5G networks. B5G networks are massive scale heterogeneous networks where small form factor devices are used in many applications that collect information from the environment. This information is currently transmitted to centralized cloud-based servers for intelligent processing and decision-making. However, with the advent of IoE in B5G, there is a shift toward edge intelligence. Deploying ML models on-device enables training using federated learning and local decision-making, making communication more efficient. On the other hand, device resource limitations make it challenging to run ML models on-device. Performance becomes thus a primary requirement constrained by device resources, relegating security, and explainability to secondary places.

For example, body-sensors/fit-bits collecting vital signals to provide dietary and physical recommendations struggle to squeeze out the necessary computational power to run sophisticated cryptographic techniques on top of ML models, and they fail to provide sufficient security [66]. In such a case, running posthoc explanation techniques would burden the already exhausted computation power found in such devices. Nevertheless, these are end-devices dealing with highly sensitive health data, and it is essential to include some form of explanation to make their operations trustworthy for customers. These constraints require developers to use transparent or in-model explanations, which might not be an ideal model selection for the particular use case in terms of accuracy, robustness, or privacy.

18.4.3   New Attack Vector and Target

Post-hoc methods for XAI are new components added to ML-based systems. This new component can complement the prediction of ML models, weighing heavily on the actions of systems and humans that depend on the ML model. In some cases, the explanation itself is more important than the prediction. This is the case for AI used in applications having a societal impact, where predictions must be fair and unbiased. This is also the case for security applications like detection and response (D&R), where an explanation is used to counter and recover from detected attacks using appropriate measures.

Due to the importance of explanation, the XAI component can become the main target of an attack, as depicted in Figure 18.7. Directly attacking posthoc XAI methods can change the explanation while the prediction of the ML model remains the same, as demonstrated in [73, 74]. The ML model makes the right decision, but the dependent system or human takes a wrong course of action based on the incorrect explanation.

There is also the possibility of concealing unfair outcomes of an ML model with deceptive justifications to veil any underlying problems using XAI. It is

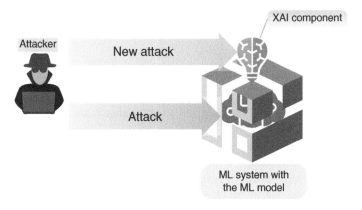

Figure 18.7   The XAI component becomes a new target and a new attack vector to compromise the whole ML-based system.

defined as *fairwashing* [75]; misapprehension that an ML model adheres to specific standards, although its actual performance significantly deviates from its explanations. Both model explanations and outcome explanations are vulnerable to this issue. It is demonstrated further that posthoc explanatory approaches depending on input perturbations, like LIME and SHAP, are unreliable and do not give definitive information regarding fairness [76]. An interpreter-only attack technique known as *scaffolding* is built based on this observation. An attacker can generate desired explanations for a given unfair ML model (which uses LIME/SHAP) by masking any biases in the model. Through this hack, a compromised XAI method enables hiding biased/unfair outcomes indicating that they are harmless/unbiased.

This threat exists for every ML application to B5G where explanation weighs equally or more than the prediction in the action it triggers. For instance, in D&R where an explanation is used to counter and recover from an attack, modifying the explanation obtained from a prediction (by a malevolent agent) will lead to fixing a nonexisting or irrelevant issue. It will not block the detected attack or prevent it from happening again in the future.

A reliable explanation also increases trust and improves the user experience for end-users of B5G networks. It fosters the adoption and usage of services provided over B5G networks. Certain decisions require user data to ensure the security and safety of the services provided, and it is also essential to provide proper explanations on how the data is used in decision-making. Critical applications such as autonomous driving are envisaged to rely on B5G networks [77]. When a system fails or crashes, the explanation for the incorrect prediction that led to it will be paramount for handling the lawsuits and other legalities that follow. Although an accurate prediction would have prevented a crash, an explanation would be of critical importance for future proofing a model. A compromised explanation can divert attention from the real issue and protect the responsible party from any consequence.

References

1 D. Castelvecchi, "Can we open the black box of AI?" *Nature News*, vol. 538, no. 7623, p. 20, 2016.

2 A. Preece, D. Harborne, D. Braines, R. Tomsett, and S. Chakraborty, "Stakeholders in Explainable AI," *arXiv preprint arXiv:1810.00184*, 2018.

3 D. Gunning and D. Aha, "DARPA's explainable artificial intelligence (XAI) program," *AI Magazine*, vol. 40, no. 2, pp. 44–58, 2019.

4 E. Tjoa and C. Guan, "A survey on explainable artificial intelligence (XAI): Toward medical XAI," *IEEE Transactions on Neural Networks and Learning Systems*, vol. 32, no. 11, pp. 4793–4813, 2020.

5 A. B. Arrieta, N. Díaz-Rodríguez, J. Del Ser, A. Bennetot, S. Tabik, A. Barbado, S. García, S. Gil-López, D. Molina, R. Benjamins et al., "Explainable artificial intelligence (XAI): Concepts, taxonomies, opportunities and challenges toward responsible AI," *Information Fusion*, vol. 58, pp. 82–115, 2020.

6 R. Guidotti, A. Monreale, S. Ruggieri, F. Turini, F. Giannotti, and D. Pedreschi, "A survey of methods for explaining black box models," *ACM Computing Surveys (CSUR)*, vol. 51, no. 5, pp. 1–42, 2018.

7 Y. Lou, R. Caruana, and J. Gehrke, "Intelligible models for classification and regression," in Proceedings of the 18th ACM SIGKDD International Conference on Knowledge Discovery and Data Mining, 2012, pp. 150–158.

8 A. Singh, S. Sengupta, and V. Lakshminarayanan, "Explainable deep learning models in medical image analysis," *Journal of Imaging*, vol. 6, no. 6, p. 52, 2020.

9 M. T. Ribeiro, S. Singh, and C. Guestrin, ""Why should i trust you?" Explaining the predictions of any classifier," in *Proceedings of the 22nd ACM SIGKDD International Conference on Knowledge Discovery and Data Mining*, 2016, pp. 1135–1144.

10 G. Visani, E. Bagli, and F. Chesani, "OptiLIME: Optimized LIME Explanations for Diagnostic Computer Algorithms," *arXiv preprint arXiv:2006.05714*, 2020.

11 J. A. Recio-García, B. Díaz-Agudo, and V. Pino-Castilla, "CBR-LIME: A case-based reasoning approach to provide specific local interpretable model-agnostic explanations," in *Case-Based Reasoning Research and Development, International Conference on Case-Based Reasoning*, Lecture Notes in Computer Science, I. Watson and R. Weber, Eds. Cham: Springer, 2020, vol. 12311, pp. 179–194.

12 S. M. Lundberg and S.-I. Lee, "A unified approach to interpreting model predictions," *Advances in Neural Information Processing Systems*, vol. 30, 2017.

13 S. M. Lundberg, G. Erion, H. Chen, A. DeGrave, J. M. Prutkin, B. Nair, R. Katz, J. Himmelfarb, N. Bansal, and S.-I. Lee, "From local explanations to global understanding with explainable AI for trees," *Nature Machine Intelligence*, vol. 2, no. 1, pp. 56–67, 2020.

14 A. Shrikumar, P. Greenside, and A. Kundaje, "Learning important features through propagating activation differences," in *International Conference on Machine Learning*. PMLR, 2017, pp. 3145–3153.

15 A. Binder, S. Bach, G. Montavon, K.-R. Müller, and W. Samek, "Layer-wise relevance propagation for deep neural network architectures," in *Information Science and Applications (ICISA) 2016*. Springer, 2016, pp. 913–922.

16 S. Lapuschkin, S. Wäldchen, A. Binder, G. Montavon, W. Samek, and K.-R. Müller, "Unmasking Clever Hans predictors and assessing what machines really learn," *Nature Communications*, vol. 10, no. 1, pp. 1–8, 2019.

17 B. Hettwer, S. Gehrer, and T. Güneysu, "Deep neural network attribution methods for leakage analysis and symmetric key recovery," in *International Conference on Selected Areas in Cryptography*. Springer, 2019, pp. 645–666.

18 L. Perotin, R. Serizel, E. Vincent, and A. Guérin, "CRNN-based multiple DoA estimation using acoustic intensity features for Ambisonics recordings," *IEEE Journal of Selected Topics in Signal Processing*, vol. 13, no. 1, pp. 22–33, 2019.

19 I. Sturm, S. Lapuschkin, W. Samek, and K.-R. Müller, "Interpretable deep neural networks for single-trial EEG classification," *Journal of Neuroscience Methods*, vol. 274, pp. 141–145, 2016.

20 S. Verma, J. Dickerson, and K. Hines, "Counterfactual Explanations for Machine Learning: Challenges Revisited," *arXiv preprint arXiv:2106.07756*, 2021.

21 W. L. Johnson, "Agents that learn to explain themselves." in *AAAI*, 1994, pp. 1257–1263.

22 L. Wells and T. Bednarz, "Explainable AI and reinforcement learning-a systematic review of current approaches and trends," *Frontiers in Artificial Intelligence*, vol. 4, p. 48, 2021.

23 E. Puiutta and E. Veith, "Explainable reinforcement learning: A survey," in *International Cross-Domain Conference for Machine Learning and Knowledge Extraction*. Springer, 2020, pp. 77–95.

24 A. Verma, V. Murali, R. Singh, P. Kohli, and S. Chaudhuri, "Programmatically interpretable reinforcement learning," in *International Conference on Machine Learning*. PMLR, 2018, pp. 5045–5054.

25 T. Shu, C. Xiong, and R. Socher, "Hierarchical and Interpretable Skill Acquisition in Multi-Task Reinforcement Learning," *arXiv preprint arXiv:1712.07294*, 2017.

26 G. Liu, O. Schulte, W. Zhu, and Q. Li, "Toward interpretable deep reinforcement learning with linear model u-trees," in *Joint European Conference on Machine Learning and Knowledge Discovery in Databases*. Springer, 2018, pp. 414–429.

27 A. Kumari, R. Gupta, and S. Tanwar, "Amalgamation of blockchain and IoT for smart cities underlying 6G communication: A comprehensive review," *Computer Communications*, vol. 172, pp. 102–118, 2021.

28 Z. Ullah, F. Al-Turjman, L. Mostarda, and R. Gagliardi, "Applications of artificial intelligence and machine learning in smart cities," *Computer Communications*, vol. 154, pp. 313–323, 2020.

29 T. M. Ho, T. D. Tran, T. T. Nguyen, S. Kazmi, L. B. Le, C. S. Hong, and L. Hanzo, "Next-Generation Wireless Solutions for the Smart Factory, Smart Vehicles, the Smart Grid and Smart Cities," *arXiv preprint arXiv:1907.10102*, 2019.

30 O. Embarak, "Explainable artificial intelligence for services exchange in smart cities," in *Explainable Artificial Intelligence for Smart Cities*. CRC Press, 2021, pp. 13–30.

31 B. P. L. Lau, S. H. Marakkalage, Y. Zhou, N. U. Hassan, C. Yuen, M. Zhang, and U.-X. Tan, "A survey of data fusion in smart city applications," *Information Fusion*, vol. 52, pp. 357–374, 2019.

32 D. Luckey, H. Fritz, D. Legatiuk, K. Dragos, and K. Smarsly, "Artificial intelligence techniques for smart city applications," in *International Conference on Computing in Civil and Building Engineering*. Springer, 2020, pp. 3–15.

33 Z. Boudanga, S. Benhadou, and J. P. Leroy, "IoT-and XAI-based smart medical waste management," in *Explainable Artificial Intelligence for Smart Cities*. CRC Press, 2021, pp. 31–46.

34 S. Tian, W. Yang, J. M. Le Grange, P. Wang, W. Huang, and Z. Ye, "Smart healthcare: Making medical care more intelligent," *Global Health Journal*, vol. 3, no. 3, pp. 62–65, 2019.

35 U. Pawar, D. O'Shea, S. Rea, and R. O'Reilly, "Explainable AI in healthcare," in *2020 International Conference on Cyber Situational Awareness, Data Analytics and Assessment (CyberSA)*. IEEE, 2020, pp. 1–2.

36 Y. Xiao and M. Krunz, "Dynamic network slicing for scalable fog computing systems with energy harvesting," *IEEE Journal on Selected Areas in Communications*, vol. 36, no. 12, pp. 2640–2654, 2018.

37 C. Benzaid and T. Taleb, "ZSM security: Threat surface and best practices," *IEEE Network*, vol. 34, no. 3, pp. 124–133, 2020.

38 L. Vigano and D. Magazzeni, "Explainable security," in *2020 IEEE European Symposium on Security and Privacy Workshops (EuroS&PW)*. IEEE, 2020, pp. 293–300.

39 M. A. Alsheikh, S. Lin, D. Niyato, and H.-P. Tan, "Machine learning in wireless sensor networks: Algorithms, strategies, and applications," *IEEE Communications Surveys & Tutorials*, vol. 16, no. 4, pp. 1996–2018, 2014.

40 L. Xiao, Y. Li, G. Han, G. Liu, and W. Zhuang, "PHY-layer spoofing detection with reinforcement learning in wireless networks," *IEEE Transactions on Vehicular Technology*, vol. 65, no. 12, pp. 10 037–10 047, 2016.

41 M. Ozay, I. Esnaola, F. T. Y. Vural, S. R. Kulkarni, and H. V. Poor, "Machine learning methods for attack detection in the smart grid," *IEEE Transactions on Neural Networks and Learning Systems*, vol. 27, no. 8, pp. 1773–1786, 2015.

42 C. Shi, J. Liu, H. Liu, and Y. Chen, "Smart user authentication through actuation of daily activities leveraging WiFi-enabled IoT," in *Proceedings of the 18th ACM International Symposium on Mobile Ad Hoc Networking and Computing*, 2017, pp. 1–10.

43 J. Wang, L. Gou, H.-W. Shen, and H. Yang, "DQNViz: A visual analytics approach to understand deep q-networks," *IEEE Transactions on Visualization and Computer Graphics*, vol. 25, no. 1, pp. 288–298, 2018.

44 C. De Alwis, A. Kalla, Q.-V. Pham, P. Kumar, K. Dev, W.-J. Hwang, and M. Liyanage, "Survey on 6G frontiers: Trends, applications, requirements, technologies and future research," *IEEE Open Journal of the Communications Society*, vol. 2, pp. 836–886, 2021.

45 L. Gavrilovska, V. Rakovic, and D. Denkovski, "From cloud RAN to open RAN," *Wireless Personal Communications*, vol. 113, no. 3, pp. 1523–1539, 2020.

46 B. Brik, K. Boutiba, and A. Ksentini, "Deep learning for B5G open radio access network: Evolution, survey, case studies, and challenges," *IEEE Open Journal of the Communications Society*, vol. 3, pp. 228–250, 2022.

47 W. Shi, J. Cao, Q. Zhang, Y. Li, and L. Xu, "Edge computing: Vision and challenges," *Internet of Things Journal*, vol. 3, no. 5, pp. 637–646, 2016.

48 D. Floyer, "The Vital Role of Edge Computing in the Internet of Things," October 2015. [Online]. Available: https://wikibon.com/the-vital-role-of-edge-computing-in-the-internet-of-things/.

49 M. S. Murshed, J. J. Carroll, N. Khan, and F. Hussain, "Resource-aware on-device deep learning for supermarket hazard detection," in *2020 19th IEEE International Conference on Machine Learning and Applications (ICMLA)*. IEEE, 2020, pp. 871–876.

50 J. Wang, Z. Feng, Z. Chen, S. George, M. Bala, P. Pillai, S.-W. Yang, and M. Satyanarayanan, "Bandwidth-efficient live video analytics for drones via edge computing," in *2018 IEEE/ACM Symposium on Edge Computing (SEC)*. IEEE, 2018, pp. 159–173.

51 M. Murshed, C. Murphy, D. Hou, N. Khan, G. Ananthanarayanan, and F. Hussain, "Machine Learning at the Network Edge: A Survey," *arXiv preprint arXiv:1908.00080*, 2019.

52 H. HaddadPajouh, R. Khayami, A. Dehghantanha, K.-K. R. Choo, and R. M. Parizi, "AI4SAFE-IoT: An AI-powered secure architecture for edge layer of Internet of Things," *Neural Computing and Applications*, vol. 32, no. 20, pp. 16 119–16 133, 2020.

53 H. Tong and T. X. Brown, "Adaptive call admission control under quality of service constraints: A reinforcement learning solution," *IEEE Journal on Selected Areas in Communications*, vol. 18, no. 2, pp. 209–221, 2000.

54 M. Abdulkadir, "Optimizing mobile backhaul using machine learning," 2019.

55 C. Nugent and P. Cunningham, "A case-based explanation system for black-box systems," *Artificial Intelligence Review*, vol. 24, no. 2, pp. 163–178, 2005.

56 B. Tezergil and E. Onur, "Wireless Backhaul in 5G and Beyond: Issues, Challenges and Opportunities," *arXiv preprint arXiv:2103.08234*, 2021.

57 C. Benzaid and T. Taleb, "AI-driven zero touch network and service management in 5G and beyond: Challenges and research directions," *IEEE Network*, vol. 34, no. 2, pp. 186–194, 2020.

58 J. Tritscher, M. Ring, D. Schlr, L. Hettinger, and A. Hotho, "Evaluation of post-hoc XAI approaches through synthetic tabular data," in *International Symposium on Methodologies for Intelligent Systems*. Springer, 2020, pp. 422–430.

59 A. Nascita, A. Montieri, G. Aceto, D. Ciuonzo, V. Persico, and A. Pescapé, "XAI meets mobile traffic classification: Understanding and improving multimodal deep learning architectures," *IEEE Transactions on Network and Service Management*, vol. 18, no. 4, pp. 4225–4246, 2021.

60 A. Nascita, A. Montieri, G. Aceto, D. Ciuonzo, V. Persico, and A. Pescapè, "Unveiling mimetic: Interpreting deep learning traffic classifiers via XAI techniques," in *2021 IEEE International Conference on Cyber Security and Resilience (CSR)*. IEEE, 2021, pp. 455–460.

61 B. Biggio and F. Roli, "Wild patterns: Ten years after the rise of adversarial machine learning," *Pattern Recognition*, vol. 84, pp. 317–331, 2018.

62 M. Juuti, S. Szyller, S. Marchal, and N. Asokan, "PRADA: Protecting against DNN model stealing attacks," in *2019 IEEE European Symposium on Security and Privacy (EuroS&P)*. IEEE, 2019, pp. 512–527.

63 N. Papernot, P. McDaniel, I. Goodfellow, S. Jha, Z. B. Celik, and A. Swami, "Practical black-box attacks against machine learning," in *Proceedings of the 2017 ACM on Asia Conference on Computer and Communications Security*, 2017, pp. 506–519.

64 A. Kuppa and N.-A. Le-Khac, "Adversarial XAI methods in cybersecurity," *IEEE Transactions on Information Forensics and Security*, vol. 16, pp. 4924–4938, 2021.

65 I. Andrew, S. Shibani, T. Dimitris, E. Logan, and T. Brandon, "Adversarial examples are not bugs they are features," in *Advances in Neural Information Processing Systems (NeurIPS)*, 2019.

66 Y. Siriwardhana, P. Porambage, M. Liyanage, and M. Ylianttila, "AI and 6G security: Opportunities and challenges," in *Proceedings of the IEEE Joint European Conference on Networks and Communications & 6G Summit (EuCNC/6G Summit)*, 2021, pp. 1–6.

67 P. K. Singh, S. K. Jha, S. K. Nandi, and S. Nandi, "ML-based approach to detect DDoS attack in V2I communication under SDN architecture," in *TENCON 2018-2018 IEEE Region 10 Conference*. IEEE, 2018, pp. 0144–0149.

68 G. C. Amaizu, C. I. Nwakanma, S. Bhardwaj, J. Lee, and D.-S. Kim, "Composite and efficient DDoS attack detection framework for B5G networks," *Computer Networks*, vol. 188, p. 107871, 2021.

69 M. Idhammad, K. Afdel, and M. Belouch, "Semi-supervised machine learning approach for DDoS detection," *Applied Intelligence*, vol. 48, no. 10, pp. 3193–3208, 2018.

70 A. Kurakin, I. Goodfellow, and S. Bengio, "Adversarial Machine Learning at Scale," *arXiv preprint arXiv:1611.01236*, 2016.

71 A. Raghunathan, S. M. Xie, F. Yang, J. C. Duchi, and P. Liang, "Adversarial Training Can Hurt Generalization," *arXiv preprint arXiv:1906.06032*, 2019.

72 L. Song, R. Shokri, and P. Mittal, "Membership inference attacks against adversarially robust deep learning models," in *2019 IEEE Security and Privacy Workshops (SPW)*. IEEE, 2019, pp. 50–56.

73 A. Kuppa and N.-A. Le-Khac, "Black box attacks on explainable artificial intelligence (XAI) methods in cyber security," in *2020 International Joint Conference on Neural Networks (IJCNN)*. IEEE, 2020, pp. 1–8.

74 A. Galli, S. Marrone, V. Moscato, and C. Sansone, "Reliability of explainable artificial intelligence in adversarial perturbation scenarios," in *International Conference on Pattern Recognition*. Springer, 2021, pp. 243–256.

75 U. Aïvodji, H. Arai, O. Fortineau, S. Gambs, S. Hara, and A. Tapp, "Fairwashing: The risk of rationalization," in *International Conference on Machine Learning*. PMLR, 2019, pp. 161–170.

76 D. Slack, S. Hilgard, E. Jia, S. Singh, and H. Lakkaraju, "Fooling lime and shap: Adversarial attacks on post hoc explanation methods," in *Proceedings of the AAAI/ACM Conference on AI, Ethics, and Society*, 2020, pp. 180–186.

77 X. Chen, S. Leng, J. He, and L. Zhou, "Deep-learning-based intelligent intervehicle distance control for 6G-enabled cooperative autonomous driving," *IEEE Internet of Things Journal*, vol. 8, no. 20, pp. 15180–15190, 2020.

19

Zero Touch Network and Service Management (ZSM) Security

This chapter presents the importance of the Zero-Touch Network and Service Management (ZSM) concept for 6G network and security issues associated with it. After reading this chapter, you should be able to:

- Understand the concept of ZSM architecture.
- Understand the security issues associated with ZSM concept.

19.1 Introduction

Network infrastructures and supporting technologies have witnessed immense growth both horizontally and also vertically. The conventional 5G cellular networks are composed of terrestrial infrastructures such as IoT and mobile devices, small cells, and macrocells. In order to support such massive connectivity ensuring global coverage, future sixth-generation (6G) wireless systems would comprise of underground, underwater, and aerial communications. In particular, an aerial radio access network consists of three main tiers, including low-altitude platforms (LAPs), high-altitude platforms (HAPs), and low-earth orbit (LEO) satellites. LAP systems usually connect to users directly and support very high-quality of services (QoS). The LEO satellite tier supports sparse-connectivity scenarios and global coverage with reasonable QoS, while HAP systems maintain a balance of LAP systems and LEO satellite communications. Along with a massive number of IoT and mobile devices, managing the network in a fully automated manner is a great challenge [1]. Although many solutions and concepts have been developed over the last decade, such as network functions virtualization (NFV), software-defined networking (SDN), multi-access edge computing (MEC), and network slicing (NA), still manual processes are required for the operation and management

Security and Privacy Vision in 6G: A Comprehensive Guide, First Edition.
Pawani Porambage and Madhusanka Liyanage.
© 2023 The Institute of Electrical and Electronics Engineers, Inc. Published 2023 by John Wiley & Sons, Inc.

of present network systems, i.e. human intervention is a must to ensure fully autonomous network and service management solutions [2–4]. These difficulties have motivated extra efforts from academic and industry communities.

19.1.1 Need of Zero-Touch Network and Service Management

The following limitations of existing network management and orchestration (MANO) solutions have motivated the adoption of the ZSM concept [5–14].

- **Network complexity**: Massive IoT connectivity, many emerging services, and new 5G/6G technologies result in extremely heterogeneous and complex mobile networks, and thus significantly increase the overall complexity of the network orchestration and management.
- **New business-oriented services**: Many new services will be available in future networks, which should be quickly implemented to meet business opportunities. Along with key-enabling technologies such as NS, NFV, and MEC, the ZSM concept allows an agile and more straightforward deployment of new services.
- **Performance improvement**: Diverse QoS requirements and the need to reduce the operational cost and improve network performance triggers robust solutions of network operation and service management.
- **Revolution for future networks**: Even 5G networks are not fully available worldwide, and numerous activities have been dedicated to the research and development of future 6G wireless systems. Many new technologies, services, applications, and IoT connections will be available, which will make the future network very complex and complicated to be efficiently managed by conventional MANO approaches [1].

The above limitations explain a strong need for the ZSM concept for the complete automation and management of future networks. In order to eliminate such limitations, enabling fully automated network operation and management solutions, the European Telecommunications Standards Institute (ETSI) ZSM group was established in December 2017.

19.1.2 Importance of ZSM for 5G and Beyond

The recent advancements in IoT technology increase the number of connected devices [15]. As the number of devices increases, there is a need to improve network infrastructure to ensure good communication or connectivity among geographically spread devices [16]. These advancements should enable real-time operations to be performed with minimal latency and improved performance. To be successful in achieving these goals, a suitable communication medium is required. 5G and beyond is the promising next-generation network that enables various enhanced capabilities such as ultra-low latency, high reliability, seamless connection, and mobility support [17]. To meet enterprise requirements, 5G is built with service-aware globally managed infrastructures, highly programmable. SDN, NFV, MEC, and NS are critical foundations for 5G and beyond [18]. New business models, such as multidomain, multiservice, and multitenancy, will emerge in 5G and beyond due to new technologies, thus bolstering new industry dynamics. The existing infrastructure will result in a complex 5G architecture in terms of operations and services [19].

Traditional network management techniques do not fulfill the new paradigm, hence, the need arises for an efficient end-to-end (E2E) automated network system capable of providing faster services to end-users [20]. The goal of automation is to drive services through an autonomous network governed by a set of high-level policies and rules. Enabled by the ZSM implementation, 5G and beyond networks can be operated independently, i.e. without human intervention [21]. Keeping the requirements in the account, ETSI developed the ZSM ISG in 2017. The ZSM objective is to create an underlying paradigm that enables fully autonomous solutions for network operation and service management of 5G and beyond networks. The ZSM comprises operational processes and tasks such as planning, delivery, deployment, provisioning, monitoring, and optimization that are executed automatically without human intervention [22].

19.2 ZSM Reference Architecture

The architecture of the ZSM was created to fully automate the network and service management in the environments with multidomains, where the operations span across the legal boundaries of the organizations [23, 24]. Cross-domain data services, multiple management domains (MDs), intra- and cross-domain integration fabrics, and an E2E service MD are all part of the system architecture, as shown in Figure 19.1.

Every MD is responsible for smart automated resource and service management within their scope. The E2E service MD is treated as an in-charge of E2E service management across various administrative domains. The differentiation of MDs and E2E service MDs encourages device modularity and helps them to grow independently. Each MD is made up of several management functions organized into logical groups through which service interfaces are exposed by management services [25].

The intradomain integration fabric is used to provide and consume resources that are local to the MD. The cross-domain integration fabric consumes resources are spread across domains. Intelligence services within E2E service MDs and

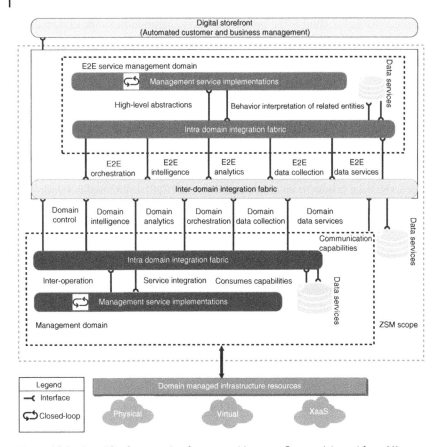

Figure 19.1   The ZSM framework reference architecture. Source: Adapted from [6].

MDs may use data in cross-domain data services to support cross-domain, and domain-level artificial intelligence (AI)-based closed-loop automation (CLA) [26].

19.2.1   Components

The components of the ZSM reference architecture are discussed in brief below.

19.2.1.1   Management Services
The fundamental building block in the ZSM architecture is its "management service" Management services provide capabilities for consumption by consumers with the help of standardized management service end-points. A management service's capabilities collectively describe its role in the organization it manages. Multiple service customers may use the same service capabilities. Several service

end-points may be allocated to one or more service capabilities. For invocation, all management services have a consistent collection of capabilities. In the case of interactions between management domains, it allows a high degree of automation and consistency. Management services that are already available can be merged to create new management services. Management resources with higher abstraction and broader reach are supported by each higher layer in the composition hierarchy. The infrastructure resources communicate with the management service producers to provide their capabilities, either directly through their management interfaces or indirectly through the consumption of other management services through their service end-points [27].

19.2.1.2 Management Functions

Management functions produce and use management services, which can be both producer and consumer of the service. If the management function produces certain capabilities, it is a service producer. On the contrary, it is a service consumer if it consumes certain management services [28].

19.2.1.3 Management Domains

The administrative responsibilities are classified by the management domains to establish "separation of concerns" in a given ZSM framework, depending on several implementation, organizational, governance, and functional constraints. It federates management services with the capabilities required to control the resources/resource-facing services in a given domain. For example, some management services are constrained by approvals when the authorized consumers consume the management domain. In contrast, others remain available to the authorized consumers, both within and outside the management domain, at all times. Management domains manage one or more entities and provide service capabilities by consuming service end-points. Sometimes, the consuming service being managed by the management domain can also potentially consume management services [29].

19.2.1.4 The E2E Service Management Domain

It is a unique management domain that offers E2E management of customer-facing services, combining resource-facing services from several management domains. However, it does not control infrastructure resources directly [30].

19.2.1.5 Integration Fabric

It facilitates communication and interoperation between management functions that include the communication between management functions, discovery, registration, and invocation of management services. It also offers management service integration, interoperation, communication capabilities, and consumes capabilities [31].

19.2.1.6 Data Services

Registered consumers can access and persist shared management data across management services using the data services. Data processing and data persistence are enabled by removing management functions to handle their data persistence [32].

19.3 Security Aspects

To ensure privacy preservation and security, E2E security management is crucial to establish clear identities. The threats to the network have increased rapidly and continuous evolution in association with the rising number of connected devices have also been observed [33–35]. The major security challenges faced by ZSM and the potential counterattack mechanisms are discussed in this section.

The security threats related to ZSM can be categorized as violation threats, deliberate threats, and accidental threats. Figure 19.2 illustrates major threats related to ZSM.

19.3.1 ML/AI-Based Attacks

Implementing machine learning (ML) approaches on network and service management results in substantial enhancements in service efficacy, performance, and time management, enabling new business models to emerge. ML/AI

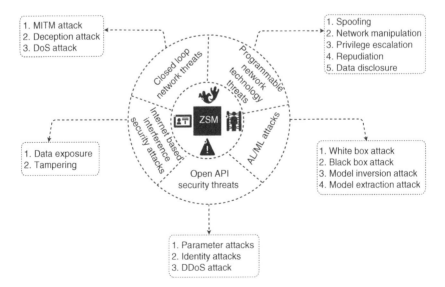

Figure 19.2 Possible threats and attacks on ZSM.

approaches are predominantly used for intelligent network management and activity capabilities. In addition, they also support Traffic management, forecast, mobility assessment, and resource distribution. Network security has recently gained immense momentum [36] because if the vulnerability around the usage of AI/ML techniques is not resolved, its use in next-generation network management may be dampened. Undoubtedly, the use of AI/ML and other data analytic technologies have initiated new attack vectors. As an example, model inversion attacks and model extraction attacks have emerged recently that target ML as a Service [37, 38]. In model inversion attacks, the training data are reconstructed from the model parameters that can be used to extract private and sensitive data. Model extraction attacks use model parameters extracted by querying the model. It has also been identified that ML approaches are susceptible to a variety of interventions [36] that exploit both the training phase (i.e. poisoning attacks) and the testing phase (i.e. evasion attacks).

Based on the attacker's knowledge, attacks on ML can be classified into two categories, the white-box and black-box attacks [39].

19.3.1.1 White-Box Attack
In case of a white box attack, the intruder has a full understanding of the classification model. The intruder is completely aware of the training algorithm and can thus exploit the training data distribution and the parameters of the entire trained model architecture [40].

19.3.1.2 Black-Box Attack
On the contrary, the attackers do not know the classification model, its algorithm, training data, and the model architecture [41]. The black box attack analyses the model's vulnerability using knowledge about its settings or previous inputs. Black box attacks can be categorized as strict black-box attacks, Nonadaptive & adaptive black-box attacks. In nonadaptive black box attack, only the distribution of the training dataset can be accessed by the attacker while in adaptive black box attack, the attacker will not have access to the distribution of the training dataset; however, the attacker will have access to the training model. In strict black-box attack, the attacker does not have access to the distribution of the training dataset and cannot modify the input query to observe the output of the model. In addition, the attacker will not be able to change the input query in order to observe the model's results.

19.3.2 Open API Security Threats

Application programming interfaces (APIs) is a technology used for incorporating web-based applications [42]. APIs are an extremely critical component of the ZSM architecture, enabling communication that interfaces between its components and services. Parameter attacks such as Script insertions, structured query language (SQL) injections, buffer overflow attacks along with man-in-the-middle (MITM) attack, identity attacks, and denial of service (DoS) are possible API-based attacks working against the ZSM system [43].

19.3.2.1 Parameter Attacks

It takes advantage of data that are transmitted through an API, which includes the query parameters, HTTP headers, uniform resource locator, and post content [44]. The following are possible API parameter attacks.

Script insertions: This type of attack takes advantage of systems that interpret the submitted parameter content as a script.

SQL injections: This is a query language-based attack in which parameters are designed to load a particular input into a database query. The query is tampered with to alter the intent of an underlying SQL template.

Buffer overflow attacks: These attacks are triggered by data beyond the intended types or ranges. It results in a system crash, thereby providing access to memory spaces.

Identity attacks: These attacks try to gain access to a target API using a list of previously compromised passwords, stolen credentials, or tokens. It feeds large quantities of random data into a framework to find vulnerabilities [45].

Denial of service attacks: These attacks overwhelm essential API resources by sending large traffic volumes from multiple sources and interrupting access to these services.

Application and data attacks: These attacks incorporate data breach, data deletion or modification, code injection, and script disruption.

19.3.3 Intent-Based Security Threats

The intent-based interfaces use network orchestration, AI, and ML to automate administrative capabilities [46]. The goal is to minimize manual associations. The potential threats related are data exposure, unusual behavior, and inappropriate configurations.

19.3.3.1 Data Exposure

Automation will expose data regarding the application's interests such as communicating with associates, advertising content, and managing traffic. Consequently, an unauthorized party intercepts such information endangering system objectives leading to the launch of other attacks.

19.3.3.2 Tampering

The intruder makes physical changes to an interface or a contact connection. As a consequence, disconnecting or modifying the physical connection occurs along with modification of the transmitted data.

19.3.4 Automated Closed-Loop Network-Based Security Threats

CLA is a continuous process that tracks, evaluates, and assesses real-time network traffic enhancing end-user Quality of Experience. External CLA capabilities are required to deal with the expanding threat in 6G technologies [47, 48]. CLA security mechanisms have the potential to automatically detect threats such as DOS, MITM attack, Deception attack, unknown threats, and quickly mitigate them using ML and AI.

19.3.4.1 MITM Attack

When an intruder intercepts messages between two parties to remotely eavesdrop or manipulate traffic, it is known as a MITM attack. It captures user credentials, personal information, and spies on users interrupting their messages, leading to data corruption.

19.3.4.2 Deception Attacks

Deception is an exchange of information between two entities – a deceiver and a target. In such attacks, the deceiver effectively convinces the target to accept an incorrect version of the truth as a fact and manipulates the target to behave in a way that benefits the deceiver.

19.3.5 Threats Due to Programmable Network Technologies

SDN and NFV technologies are used to create a programmable networking solution. When users have programmatic access to SDN, the risks increase. These threats are predominant in situations wherein the users are forced to "trust" and rely on third-party applications or standard-based solutions for network access. Also, in the absence of appropriate isolation, the control information and network element management get exploited to attacks such as Network Manipulation, address resolution protocol (ARP) Spoofing Attack, and others. Functional Virtualization needs to support an infrastructure that is independent of hardware. NFV is prone to generic virtualization threats, generic networking threats, and virtualization of technology threats [49]. Virtual network functions (VNFs) are vulnerable to design, implementation, and configuration defects, leading to inappropriate monitoring of data that misleads the service intelligence and E2E analytics in ZSM.

19.3.6 Possible Threats on ZSM Framework Architecture

The E2E service intelligence offered by ZSM facilitates decision-making and also helps in predicting capabilities. Information from data services and domain data collection services are used to make important decisions. As a result, an intruder may generate inputs to deceive the ML algorithm used by the E2E intelligence services leading to incorrect assumptions or conclusions, resulting in decreased efficiency, financial loss, and endangering of service-level agreement fulfillment and security guarantees. APIs are used extensively during the Services provisioning, governance, orchestration, and monitoring in the ZSM, making them the possible ideal target for intruders. An attacker can potentially access or interfere with ZSM's services by using insecure APIs. Data loss, theft of personal information, server compromise, and service outage are all possible outcomes of API-based attacks.

One of the core principles of the ZSM architecture is intent-based interfaces. A registered consumer may use the ZSM domain orchestration service to create, modify, and terminate domain-level network services. An attacker may try to use the orchestration service from a compromised consumer and tamper with the interfaces. ZSM architecture supports closed-loop control automation of domain data collection. An attacker may initiate a deception attack by sending a fabricated fault event to the ZSM domain data collection interface claiming VNF as faulty.

The fault event service in a domain data collection process publishes fake fault cases. It is accepted by the domain intelligence services as part of the closed-loop service at the domain level and responds to the attacker. If the attacker successfully hijacks the response or reroutes the traffic via an attacker-controlled switch, the man-in-the-middle attack is performed. The ZSM architecture is built on a foundation of the programmable network approach by integrating with SDN and NFV technologies. The attacker uses possible attacks such as tampering, spoofing, information disclosure, repudiation, DoS attack, and privilege escalation on SDN, used by the ZSM framework. The attacker can also compromise VNFs providing inaccurate monitoring data thereby misleading the analytics and intelligence services of the ZSM framework.

References

1 C. De Alwis, A. Kalla, Q.-V. Pham, P. Kumar, K. Dev, W.-J. Hwang, and M. Liyanage, "Survey on 6G frontiers: Trends, applications, requirements, technologies and future research," *IEEE Open Journal of the Communications Society*, vol. 2, pp. 836–886, 2021.

2 C. Benzaid and T. Taleb, "AI-driven zero touch network and service management in 5G and beyond: Challenges and research directions," *IEEE Network*, vol. 34, no. 2, pp. 186–194, 2020.

3 M. Bunyakitanon, X. Vasilakos, R. Nejabati, and D. Simeonidou, "End-to-end performance-based autonomous VNF placement with adopted reinforcement learning," *IEEE Transactions on Cognitive Communications and Networking*, vol. 6, no. 2, pp. 534–547, 2020.

4 D. Bega, M. Gramaglia, M. Fiore, A. Banchs, and X. Costa-Perez, "AZTEC: Anticipatory capacity allocation for zero-touch network slicing," in *IEEE INFOCOM 2020-IEEE Conference on Computer Communications*. IEEE, 2020, pp. 794–803.

5 "Zero-touch network and Service Management (ZSM); Requirements based on documented scenarios," [Accessed on 29.03.2021]. [Online]. Available: https://www.etsi.org/deliver/etsi_gs/ZSM/001_099/001/01.01.01_60/gs_ZSM001v010101p.pdf.

6 "Zero-touch network and Service Management (ZSM); Reference Architecture," [Accessed on 29.03.2021]. [Online]. Available: https://www.etsi.org/deliver/etsi_gs/ZSM/001_099/002/01.01.01_60/gs_ZSM002v010101p.pdf.

7 "Zero-touch network and Service Management (ZSM); End to end management and orchestration of network slicing," [Accessed on 29.03.2021]. [Online]. Available: https://portal.etsi.org/webapp/WorkProgram/Report_WorkItem.asp?WKI_ID=54284.

8 "Zero-touch network and Service Management (ZSM); Landscape," [Accessed on 29.03.2021]. [Online]. Available: https://www.etsi.org/deliver/etsi_gr/ZSM/001_099/004/01.01.01_60/gr_ZSM004v010101p.pdf.

9 "Zero-touch network and Service Management (ZSM); Means of Automation," [Accessed on 29.03.2021]. [Online]. Available: https://www.etsi.org/deliver/etsi_gr/ZSM/001_099/005/01.01.01_60/gr_ZSM005v010101p.pdf.

10 "Zero touch network and Service Management (ZSM); Proof of Concept Framework," [Accessed on 29.03.2021]. [Online]. Available: https://www.etsi.org/deliver/etsi_gs/ZSM/001_099/006/01.01.01_60/gs_ZSM006v010101p.pdf.

11 "Zero-touch network and Service Management (ZSM); Terminology for concepts in ZSM," [Accessed on 29.03.2021]. [Online]. Available: https://www.etsi.org/deliver/etsi_gs/ZSM/001_099/007/01.01.01_60/gs_ZSM007v010101p.pdf.

12 "Zero-touch network and Service Management (ZSM); Cross-domain E2E service lifecycle management," [Accessed on 29.03.2021]. [Online]. Available: https://portal.etsi.org/webapp/WorkProgram/Report_WorkItem.asp?WKI_ID=56825.

13 "Zero-touch network and Service Managment (ZSM); Closed-loop automation: Solutions for automation of E2E service and network management use cases," [Accessed on 29.03.2021]. [Online]. Available: https://portal.etsi.org/webapp/WorkProgram/Report_WorkItem.asp?WKI_ID=58055.

14 "Zero-touch network and Service Management (ZSM); General Security Aspects," [Accessed on 29.03.2021]. [Online]. Available: https://portal.etsi .org/webapp/WorkProgram/Report_WorkItem.asp?WKI_ID=58436.

15 K. Dev, R. K. Poluru, L. Kumar, P. K. R. Maddikunta, and S. A. Khowaja, "Optimal radius for enhanced lifetime in IoT using hybridization of rider and grey wolf optimization," *IEEE Transactions on Green Communications and Networking*, vol. 5, no. 2, pp. 635–644, 2021.

16 P. K. R. Maddikunta, T. R. Gadekallu, R. Kaluri, G. Srivastava, R. M. Parizi, and M. S. Khan, "Green communication in IoT networks using a hybrid optimization algorithm," *Computer Communications*, vol. 159, pp. 97–107, 2020.

17 A. Osseiran, F. Boccardi, V. Braun, K. Kusume, P. Marsch, M. Maternia, O. Queseth, M. Schellmann, H. Schotten, H. Taoka et al., "Scenarios for 5G mobile and wireless communications: The vision of the METIS project," *IEEE Communications Magazine*, vol. 52, no. 5, pp. 26–35, 2014.

18 X. Foukas, G. Patounas, A. Elmokashfi, and M. K. Marina, "Network slicing in 5G: Survey and challenges," *IEEE Communications Magazine*, vol. 55, no. 5, pp. 94–100, 2017.

19 Z. Zhang, Y. Xiao, Z. Ma, M. Xiao, Z. Ding, X. Lei, G. K. Karagiannidis, and P. Fan, "6G wireless networks: Vision, requirements, architecture, and key technologies," *IEEE Vehicular Technology Magazine*, vol. 14, no. 3, pp. 28–41, 2019.

20 T. Huang, W. Yang, J. Wu, J. Ma, X. Zhang, and D. Zhang, "A survey on green 6G network: Architecture and technologies," *IEEE Access*, vol. 7, pp. 175 758–175 768, 2019.

21 B. Holfeld, D. Wieruch, T. Wirth, L. Thiele, S. A. Ashraf, J. Huschke, I. Aktas, and J. Ansari, "Wireless communication for factory automation: An opportunity for LTE and 5G systems," *IEEE Communications Magazine*, vol. 54, no. 6, pp. 36–43, 2016.

22 "Zero-touch network and Service Management (ZSM); End-to-end architectural framework for network and service automation," [Accessed on 29.03.2021]. [Online]. Available: https://www.etsi.org/committee?id=1673.

23 I. Vaishnavi and L. Ciavaglia, "Challenges towards automation of live telco network management: Closed control loops," in *2020 16th International Conference on Network and Service Management (CNSM)*. IEEE, 2020, pp. 1–5.

24 H. Hantouti, N. Benamar, and T. Taleb, "Service function chaining in 5G & beyond networks: Challenges and open research issues," *IEEE Network*, vol. 34, no. 4, pp. 320–327, 2020.

25 R. Rokui, H. Yu, L. Deng, D. Allabaugh, M. Hemmati, and C. Janz, "A standards-based, model-driven solution for 5G transport slice automation and assurance," in *2020 6th IEEE Conference on Network Softwarization (NetSoft)*. IEEE, 2020, pp. 106–113.

26 I. Afolabi, M. Bagaa, W. Boumezer, and T. Taleb, "Toward a real deployment of network services orchestration and configuration convergence framework for 5G network slices," *IEEE Network*, vol. 35, no. 1, pp. 242–250, 2020.

27 E. G. ZSM, "Zero touch network and service management (ZSM) landscape, version 1.1. 1," *ETSI: Sophia Antipolis, France*, 2020.

28 Q. Duan, "Intelligent and autonomous management in cloud-native future networks-a survey on related standards from an architectural perspective," *Future Internet*, vol. 13, no. 2, p. 42, 2021.

29 A. Boudi, M. Bagaa, P. Pöyhönen, T. Taleb, and H. Flinck, "AI-based resource management in beyond 5G cloud native environment," *IEEE Network*, vol. 35, no. 2, pp. 128–135, 2021.

30 A. Muhammad, T. A. Khan, K. Abbass, and W.-C. Song, "An end-to-end intelligent network resource allocation in IoV: A machine learning approach," in *IEEE 92nd Vehicular Technology Conference (VTC2020-Fall)*. IEEE, 2020, pp. 1–5.

31 K. Samdanis and T. Taleb, "The road beyond 5G: A vision and insight of the key technologies," *IEEE Network*, vol. 34, no. 2, pp. 135–141, 2020.

32 J. Baranda, J. Mangues-Bafalluy, E. Zeydan, L. Vettori, R. Martínez, X. Li, A. Garcia-Saavedra, C. Chiasserini, C. Casetti, K. Tomakh et al., "On the integration of AI/ML-based scaling operations in the 5Growth platform," in *2020 IEEE Conference on Network Function Virtualization and Software Defined Networks (NFV-SDN)*. IEEE, 2020, pp. 105–109.

33 D. C. Le, N. Zincir-Heywood, and M. I. Heywood, "Analyzing data granularity levels for insider threat detection using machine learning," *IEEE Transactions on Network and Service Management*, vol. 17, no. 1, pp. 30–44, 2020.

34 M. Liyanage, I. Ahmad, A. B. Abro, A. Gurtov, and M. Ylianttila, *A Comprehensive Guide to 5G Security*. John Wiley & Sons, 2018.

35 M. Liyanage, A. Braeken, P. Kumar, and M. Ylianttila, *IoT Security: Advances in Authentication*. John Wiley & Sons, 2020.

36 J. Chen, X. Lin, Z. Shi, and Y. Liu, "Link prediction adversarial attack via iterative gradient attack," *IEEE Transactions on Computational Social Systems*, vol. 7, no. 4, pp. 1081–1094, 2020.

37 X. Liu, L. Xie, Y. Wang, J. Zou, J. Xiong, Z. Ying, and A. V. Vasilakos, "Privacy and security issues in deep learning: A survey," *IEEE Access*, vol. 9, pp. 4566–4593, 2021.

38 Y. Siriwardhana, P. Porambage, M. Liyanage, and M. Ylianttila, "AI and 6G security: Opportunities and challenges," in *Proceedings of the IEEE Joint European Conference on Networks and Communications & 6G Summit (EuCNC/6G Summit)*, 2021, pp. 1–6.

39 K. Ren, T. Zheng, Z. Qin, and X. Liu, "Adversarial attacks and defenses in deep learning," *Engineering*, vol. 6, no. 3, pp. 346–360, 2020.

40 Q. Liu, J. Guo, C.-K. Wen, and S. Jin, "Adversarial attack on DL-based massive MIMO CSI feedback," *Journal of Communications and Networks*, vol. 22, no. 3, pp. 230–235, 2020.

41 H. Yan, X. Li, H. Li, J. Li, W. Sun, and F. Li, "Monitoring-Based Differential Privacy Mechanism Against Query-Flooding Parameter Duplication Attack," *arXiv preprint arXiv:2011.00418*, 2020.

42 H. Gu, J. Zhang, T. Liu, M. Hu, J. Zhou, T. Wei, and M. Chen, "DIAVA: A traffic-based framework for detection of SQL injection attacks and vulnerability analysis of leaked data," *IEEE Transactions on Reliability*, vol. 69, no. 1, pp. 188–202, 2019.

43 P. Porambage, G. Gür, D. P. M. Osorio, M. Liyanage, A. Gurtov, and M. Ylianttila, "The roadmap to 6G security and privacy," *IEEE Open Journal of the Communications Society*, vol. 2, pp. 1094–1122, 2021.

44 S. Sihag and A. Tajer, "Secure estimation under causative attacks," *IEEE Transactions on Information Theory*, vol. 66, no. 8, pp. 5145–5166, 2020.

45 Y. Li, Z. Chen, H. Wang, K. Sun, and S. Jajodia, "Understanding account recovery in the wild and its security implications," *IEEE Transactions on Dependable and Secure Computing*, vol. 19, no. 1, pp. 620–634, 2020.

46 Y. Lee, R. Vilalta, R. Casellas, R. Martínez, and R. Mu noz, "Auto-scaling mechanism in the ICT converged cross stratum orchestration architecture for zero-touch service and network management," in *2018 20th International Conference on Transparent Optical Networks (ICTON)*. IEEE, 2018, pp. 1–4.

47 C. Benzaid and T. Taleb, "ZSM security: Threat surface and best practices," *IEEE Network*, vol. 34, no. 3, pp. 124–133, 2020.

48 P. Porambage, G. Gür, D. P. M. Osorio, M. Liyanage, and M. Ylianttila, "6G security challenges and potential solutions," in *2021 Joint European Conference on Networks and Communications & 6G Summit (EuCNC/6G Summit)*. IEEE, 2021, pp. 1–6.

49 S. Maaroufi and S. Pierre, "BCOOL: A novel blockchain congestion control architecture using dynamic service function chaining and machine learning for next generation vehicular networks," *IEEE Access*, vol. 9, pp. 53 096–53 122, 2021.

20

Physical Layer Security*

* By Diana P. M. Osorio, Center for Wireless Communications, University of Oulu, Finland; José D. Vega-Sánchez, Universidad de Las Américas (UDLA), Escuela Politécnica Nacional (EPN), Ecuador; Edgar E. B. Olivo, São Paulo State University (UNESP), Campus of São João da Boa Vista, Brazil; Andre N. Barreto, Nokia, France

This chapter presents the fundamentals of physical layer security (PLS) and its road-map toward 6G. After reading this chapter, you should be able to:

- Understand the fundamentals of PLS and the main techniques.
- Understand the role of PLS for security 6G networks.

20.1 Introduction

The 6G is driven by the principles of trustworthiness, sustainability, efficiency, and automation by integrating sensing capabilities. Thus, 6G should push beyond the limitations of 5G to cope with really ambitious demands in highly dynamic and heterogeneous environments with several constraints [1]. Therefore, the emergent security challenges are equally diverse, thus requiring novel mechanisms to face the new unprecedented threats raised by the pervasive use of artificial intelligence (AI) and the rapid evolution of quantum computing [1]. Then, it is expected that security protocols in 6G will be acting at all layers of the communication stack, including the traditionally disregarded physical layer. In this chapter, the fundamentals of PLS, main techniques, and the road-map toward 6G are introduced.

Security and Privacy Vision in 6G: A Comprehensive Guide, First Edition.
Pawani Porambage and Madhusanka Liyanage.

20.2 Physical Layer Security Background

20.2.1 PLS Fundamentals

Properties of wireless channels can serve as security resources, and designing the physical layer accordingly provides information-theoretic security guarantees, also known as Physical Layer Security (PLS) [2, 3]. This security paradigm can potentially offer several advantages. For instance, it is considered to be quantum secure, as no assumptions on the attacker are made; the level of security is tied to the communication properties; it allows for on-the-fly secret keys; and PLS implementations can be lightweight.

The idea behind PLS goes back to 1949, when the work of Shannon [4] unveiled the concept of secrecy transmission in his cipher system, where a transmitter (Alice) communicates with a legitimate receiver (Bob) by encrypting a message M into a codeword X via a secret key to prevent an eavesdropper (Eve) from decoding it, thus introducing the notion of perfect secrecy. Later, in 1975, Wyner [5] showed that secrecy can be reached in a discrete memoryless wiretap channel without the need of a secret key, as illustrated in Figure 20.1, where Alice encodes M into n-length codeword X^n, and Y^n and Z^n are the observations at Bob's and Eve's channels, respectively. From [5, 6], quantities of secrecy in terms of mutual information were defined as weak and strong, formulated as $\lim_{n\to\infty}(1/n)I(M;Z^n) = 0$ and $\lim_{n\to\infty}I(M;Z^n) = 0$, respectively, with $I(\cdot;\cdot)$ being the mutual information. Besides, in [5] was introduced the concept of secrecy capacity, C_S, defined as the difference between the capacities of the main and wiretap channels and given by $C_S = \max\{\log_2(1+\gamma_B) - \log_2(1+\gamma_E), 0\}$, where γ_B, and γ_E denote the received signal-to-noise-ratio (SNR) at Bob and Eve, respectively. Based on the above theory, other insightful secrecy metrics have been proposed in the literature to measure the secrecy performance of wireless systems. For instance,

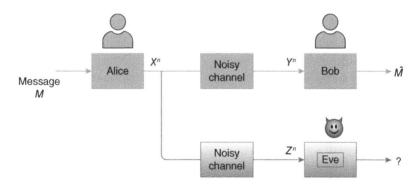

Figure 20.1 The classic wiretap channel.

in [7], the secrecy outage probability (SOP) is introduced as SOP $= \Pr\left\{C_S < R_S\right\}$, where R_S is the secrecy rate threshold and $\Pr\{\cdot\}$ stands probability. Also inspired by Wyner's ideas, the achievable secrecy capacity against eavesdropping was addressed in [8] and [9] for the broadcast and Gaussian channels, respectively. Recently, a large body of research has been motivated by PLS in fading channels to characterize upcoming wireless communications (e.g. THz bands), e.g. PLS over α–μ [10, 11], κ–μ shadowed [12], fluctuating two rays [13], and N-wave with diffuse power [14], to name a few.

20.2.2 PLS Approaches

In the following, the main approaches proposed in the literature for exploiting physical layer characteristics to protect wireless communications from different attacks against confidentiality, authenticity, and integrity are described.

20.2.2.1 Confidentiality (Edgar)

Various PLS techniques have been proposed as a means of preserving the confidentiality of sensitive information transmitted over the wireless medium by preventing illegitimate receivers from intercepting it. Next, the most widely used PLS techniques are summarized:

Artificial noise (AN): The idea behind this technique is to inject a noise-like signal, which is purposely generated to degrade the eavesdropper's channel; thus, by following a proper design, the artificial noise (AN) can be removed at the legitimate receiver so that the received signal is not compromised [15]. Particularly, in multiple-input multiple-output (MIMO) systems, the transmitter superimpose the AN signal on the information signal by properly using weighting vectors on these signals and by splitting the transmit power between them, so that AN is steered toward the eavesdropper while lying in the null space of the legitimate receive. In this case, the knowledge of the eavesdropper's channel state information (CSI) is required at the transmitter; otherwise, AN can be broadcasted isotropically [16].

Beamforming: In this technique, a beamforming vector matched to the legitimate channel is employed so that the transmitted signal is steered in the direction of the legitimate user while suppressed in the direction of the eavesdropper, thus the signal-to-noise ratio at the legitimate receiver is maximized [17]. different methods for the design of MIMO-beamforming systems aiming at improving PLS in cellular networks have been investigated. For instance, in [18], the range of secure transmissions when employing beamforming arrays at 60 GHz were estimated, assuming typical transmission and reception parameters specified for 5G cellular systems. However, although beamforming allows the transmit power to be highly concentrated in the main lobe beam, a leakage of confidential information may occur since some of the power can be lost in the minor side lobes which can be coincident with the direction of potential eavesdroppers.

Cooperative relaying/jamming: These techniques use cooperating relays to improve the performance of secure wireless communications in the presence of one or more eavesdroppers. The widely known amplify-and-forward (AF) and decode-and-forward (DF) schemes are employed to improve the secrecy rate by enhancing the channel quality between the transmitter and legitimate receiver, whereas a technique so-called cooperative jamming (CJ) exploits the assistance of relays to intentionally inject a jamming signal into the communication with the aim of degrading the eavesdropper's channel thereby improving the secrecy rate [19]. It is worthwhile to mention that the benefits of using relay-based communication techniques in the aforementioned scenarios lie on the premise that the relay is considered to be a trustworthy node. However, relays may try to leak information for its own benefit, thus becoming potential untrustworthy nodes. In this context, cooperative scenarios with untrustworthy relays have also raised attention [20, 21].

20.2.2.2 Physical Layer Authentication

The authentication process aims at verifying the identity of a device or user, thus preventing attacks where multiple devices try to use a single identity or spoofing attacks, or where a single device claims multiple identities, i.e. Sybil attacks [22]. Traditionally, authentication processes are realized above the physical layer through digital signatures and certificates. However, cryptographic-based authentication is increasingly becoming difficult to implement due to the massiveness, dynamism, and heterogeneous characteristics of scenarios beyond 5G. At the physical layer, the randomness and uncorrelated nature of the communication channel can be exploited to provide flexible and cost-effective authentication of devices, which is so-called physical layer authentication (PLA) [23]. The integration of PLA with upper-layer authentication mechanisms can be seen as a multifactor authentication framework, which is able to provide robustness for network security.

PLA schemes can be categorized into two types [24]:

- **Radio-frequency (RF) or hardware-based PLA**: The RF signals emitted by wireless transceivers possess unique features that can be extracted and used to identify devices, such as emission patterns, the I/Q offset, the power spectral density (PSD).
- **Location or channel-based PLA**: Devices at different locations will observe different channels features, specifically, when the legitimate transmitter is separated by more than a half of wavelength from the adversary, it is impossible

to both observe the same physical-layer features. Therefore, these features can be measured and exploited to authenticate devices. These features can be based on statistical channel information such as the received signal strength (RSS), or instantaneous channel information such as channel frequency response (CFR) and channel impulse response (CIR), which consider small-scale fading and provide a higher level of security.

The accurate estimation of physical-layer features may be especially challenging when considering time-varying channels with unpredictable interference conditions, mobility of devices, nonsymmetrical observations at transceivers, and measurement errors, thus reducing the reliability of PLA. To circumvent this, multiple physical-layer features can be considered to improve the authentication performance. For this purpose, machine-learning-based techniques have been successfully used to design adaptive PLA schemes based on multiple features [25, 26].

All in all, it is still dubious that PLA could effectively replace existing cryptography-based authentication mechanisms, but it is promising as a first line of defense. It can provide continuous authentication with low-latency until attacks are recognized, and upper-layer authentication is invoked [27]. Thus, more research and efforts are required to the design of robust PLA solutions that can be standalone solutions at least for some applications in 6G.

20.2.2.3 Secret Key Generation

Due to the strong assumptions about the position, propagation channel, antenna configuration, and noise characteristics of the eavesdropper, and that the absolute certainty about the secrecy can in practice never be guaranteed, wiretap coding is not expected to be widely employed in wireless cellular systems, but only in specific applications. However, there is another more recent technique in the PLS toolbox that can find wider usage in the 6G time frame. This technique is so-called secret key generation (SKG), which exploits random characteristics of the propagation channel to distill sequences of secret bits. This can be used as a secret pairing code or as symmetric encryption key that can be used with existing encryption methods, like advanced encryption standard (AES).

The idea behind it was first proposed simultaneously in [28] and [29] in 1993. Let us suppose that we have a common random source X, which can be observed by Alice, Bob, and Eve, with their observations given by X_A, X_B, and X_E. It can be shown that Alice and Bob can derive a certain number of secret bits S from this source, which is upper bounded by

$$S(X_A; X_B | X_E) \leq \min\left[I(X_A; X_B), I(X_A; X_B | X_E)\right]. \tag{20.1}$$

Thus, by considering that time-division duplexing (TDD) is employed, which is increasingly the case, then channel reciprocity holds and the observations at both ends of a given communication links, e.g. between Alice and Bob, are likely to be strongly correlated and yield a high mutual information. Also, in most propagation scenarios, the channel has a large degree of spatial decorrelation, i.e. observations of a few wavelengths away (few centimeters in higher frequencies) that are strongly uncorrelated. This means that any observation by an eavesdropper that is reasonably apart will be strongly uncorrelated, and the secret key capacity can be in the best case upper bounded by the term that's independent from X_E in (20.1). Thus, secret bits can be generated by sampling and quantizing the channel response. The eavesdropper can try to derive the channel using ray tracing methods, but this requires an enormous amount of knowledge about the environment and of processing power, such that it is ineffective [30] in practice.

However, SKG also have some drawbacks and challenges. First, the channel is never truly reciprocal, as the observations are made at different instants, the hardware is never exactly the same, and noise and interference realizations are different at Alice and Bob. This can be solved using different reconciliation methods [31], which basically employ forward error correction (FEC) to guarantee that both nodes can generate the same code word as a key. Also, some information has to be exchanged, possibly in an unprotected channel, thus some information can be leaked to Eve, and thus a hash function is usually applied to hide the information from Eve, in what is called privacy amplification [32].

Whereas the nonreciprocity issue is largely solved, some others still remain. One of the most important is that the SKG rate is strongly dependent on the channel statistical properties, and a guaranteed rate cannot be ensured [33–35]. SKG is particularly challenging in line of sight (LoS), where this has been shown that only small SKG rates can be achieved, and, in some circumstances, the eavesdropper cannot be ignored. Finally, hardware implementation architecture has to be also considered. Most commercial wireless modem chipsets do not provide CSI, and, even if they did, one would have to trust the information provided by them. Chipsets are targets of security attacks, thus for SKG be dependable; we either have to build new trustworthy chipsets, or we have to think of a different architecture, where the secret keys are generated by an independent trusted component [36].

SKG may not be the panacea to all security issues, but it is a promising security enhancement for 6G systems without the need for a centralized and vulnerable key distribution server. It is also a scalable and quick method, which is likely to be useful for highly dynamic scenarios and for low-energy massive machine-type communications (mMTC).

20.3 The Prospect of PLS in 6G

In the 6G time frame, PLS may have a great momentum with the advent of disruptive technologies that will be integral part of 6G design. Particularly, the opportunities for more secure transmissions provided by the moving toward higher frequencies in the sub-THz bands [37], the smart reconfiguring of the radio propagation provided by reconfigurable intelligent surfaces (RIS) [38], the more secure transmissions provided by the use of visible-light communication (VLC) [39], and the additional sensing functionality that can facilitate the provision of security [40] are forecasting a rapid move toward having PLS schemes as part of the future security definitions. In the following, the application scenarios in 6G as well as the prominent enabling technologies are introduced.

20.3.1 Application Scenarios of PLS in 6G

In the following is presented some of the application scenarios in 5G Advance and 6G where PLS techniques have a great potential to provide enhanced security and privacy protection, also illustrated in Figure 20.2.

mMTC+: 6G will continue the tendency from 5G to connect everything, thus it will consider a massively scalable connectivity of low-rate, low-power, low-complexity, and very light devices with native trustworthiness, mainly from the Internet of Things (IoT). It is expected that there will be stringent requirements on the cost, energy consumption, and environmental impact for the operation of these networks, as well as highly accurate localization. Therefore, cryptography-based security methods may conflict with some of the requirements of the evolution of mMTC, thus PLS approaches can provide viable alternatives for safeguarding these applications [22].

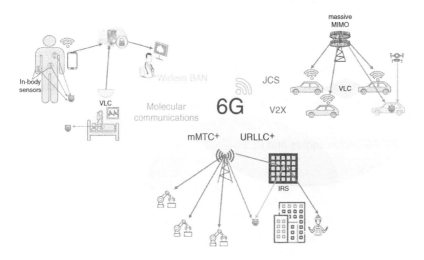

Figure 20.2 Application scenarios of PLS in 6G.

URLLC+: In 6G, applications will require real-time interactions to provide automated services in the evolution of ultrareliable low-latency communications (URLLC). Then, as PLS techniques are tailored to the communication channel and do not require key distribution or encryption/decryption, they can potentially protect these kind of applications from eavesdropping attacks without violating the strict latency requirement [41].

MC: This is a biologically inspired type of communication where chemical signals are employed to exchange information. Molecular communications (MC) may handle highly sensitive information, thus being particularly sensitive to security and privacy issues [42]. PLS techniques are potential candidates for providing security and privacy for MC that consider biological devices with limited computational power [43].

WBAN: It considers the connection of low-power, low-cost, and resource-constrained devices installed in-body or on-body. These networks require a high level of security and privacy while complying with energy and miniaturization constraints. Thus, lightweight and less-demanding solutions are critical in wireless body-area networks (WBANs), and PLS is one strong candidate to handle the confidentiality of in- and on-body devices [44].

Vehicular-to-everything (V2X) communications: It considers the communication among smart vehicles that are equipped with sensors, control and computing units, and communication, storage, and learning capabilities to be integrated to the Internet, the transport infrastructure and other road users in the so-called Internet of Vehicles (IoV). PLS techniques can potentially complement cryptography solutions by providing security with low-latency and light-signaling overhead in large-scale vehicular networks [45].

JCS: Joint communications and sensing (JCS) systems unify two key operations of the network, radar sensing and communication, in an energy, spectrum, and cost-efficient manner. However, JCS raises unique security and privacy challenges as the inclusion of information signaling into the probing waveform for target sensing makes the communication susceptible to attacks [40]. Therefore, security can be integrated to JCS systems with PLS.

20.3.2   6G Technologies and PLS

20.3.2.1   IRS

Intelligent reflecting surface (IRS) has drawn enormous attention for its potential to enable the customization of the wireless medium to improve security, coverage, and energy or spectrum efficiency [46]. An IRS is a planar structure, metasurface, or reflecting phased antenna array that comprises a large number of nearly passive reflecting/refracting elements that can be dynamically controlled [47]. Due to these features, IRS has a great potential to improve the security of 6G networks by

Figure 20.3 Basic IRS-aided wiretap wireless communication system.

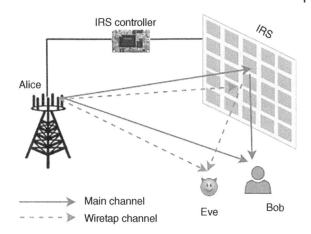

allowing a smart exploitation of a customizable radio environment. From a PLS perspective, it can be highlighted that key practical challenges associated with the IRS's implementation and deployment, namely phase-shift noise, transmit beamforming vector, spatial channel correlation, electro-magnetic interference (EMI), and signal cancelation/jamming.

By considering an IRS-assisted PLS scenario as shown in Figure 20.3, recent works have proposed different schemes to provide information-theoretic security guarantees for wireless networks. For instance, the studies in [48, 49] presented a deep-in inspection of imperfect phase-shift compensation and quantization error, and it was revealed that these hardware impairments have a detrimental impact on the secrecy performance. Also, the impact of EMI at both the legitimate and the eavesdropper sides was explored in [50] for a multiple-input single-output (MISO) system.

Moreover, the IRS has been investigated as a new degree of freedom for wireless SKG, where each element acts as an individual scatterer to increase the secret key capacity [51]. Also based on the same principle, the authors in [52] proposed a novel PLS strategy where the IRS operates as a backscatter device to create a scatter jamming signal toward the eavesdropper while the transmitter is assumed as a RF source. Finally, a hybrid relay-IRS implementation in millimeter-wave (mmWave) MIMO systems was introduced in [53] as a promising approach to overcome the restricted beamforming gain of the conventional fully passive reflecting surface, thus offering significant gains in terms of secrecy.

20.3.2.2 Unmanned Aerial Vehicles

Due to the significant advantages of 3D mobility, autonomous flight capabilities, and LoS links, the integration of unmanned aerial vehicles (UAVs) in beyond 5G and 6G has raised huge attention. This integration can be seen from two perspectives, as new aerial users of wireless cellular networks, i.e. cellular-connected UAVs or as new aerial communication platforms, such as base station (BS) and relays, in UAV-assisted wireless communications [54].

The benefits from the integration of UAVs come along with serious security risks such as the LoS-dominant airto-ground links that make them prone to jamming and eavesdropping attacks. Moreover, malicious UAVs misused by unauthorized agents can take advantage of their high mobility and flexibility to track their targets over time, thus overhearing or jamming their communications more effectively [55]. At the same time, the advantageous characteristics of UAVs can be exploited to provide security to ground communications as illustrated in Figure 20.4.

Promising solutions to safeguard networks with cellular-connected UAVs and UAV-assisted wireless communications, by relying on PLS techniques, have been intensively studied in recent years. Particularly, UAVs have been considering as facilitators to provide secrecy transmissions to ground legitimate communications by acting as *friendly jammers*, i.e. generating AN to confuse eavesdroppers. Then, optimal 3D deployment of UAVs, power allocated for jamming, as well as an efficient coverage of friendly jamming have been investigated [56–59].

The case of malicious UAVs acting as eavesdroppers have been also considered, where machine learning algorithms as deep reinforcement learning are applied

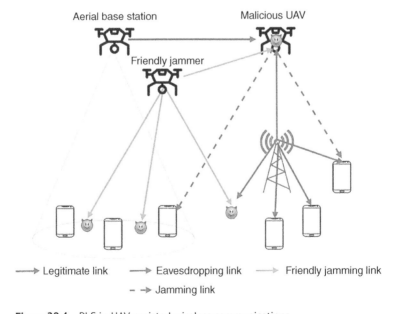

Figure 20.4 PLS in UAV-assisted wireless communications.

to optimize trajectories, power control, and user scheduling [60]. Moreover, UAVs acting as relays to improve the reliability of mobile networks may not have the same level of clearance of legitimate nodes in the network, thus being treated as potential eavesdroppers or untrusted relays. In those cases, the joint design of trajectory, velocity, and power allocation can enable enhanced secrecy performance [61].

It is expected that these approaches will continue evolving to more robust solutions in the time frame of 6G, together with the evolution of other 6G key technologies.

20.3.3 Cell-Free mMIMO

Traditionally, in transitioning from one cellular mobile technology to the next, a common practice is to consider smaller and smaller cells to increase the network's capacity. For instance, 5G deals with picocells and femtocells, and following this trend, 5G Advance networks would be expected to be managed with even smaller cells. However, without the design of advanced techniques to mitigate intercell interference, the cell's size reduction could even worsen the network's capacity instead of benefiting. In this regard, using ever-smaller cells would not be the first option for 6G networks to achieve further network densification [62].

To overcome this shortcoming, a novel 6G architecture has been conceived based on the creation of a number geographically distributed access points (APs) that jointly serve a smaller number of user equipments (UEs) instead of creating centralized cells [63]. All APs are connected via a fronthaul path to a central processing unit (CPU), which is responsible for two main tasks, namely, (i) APs cooperation and (ii) APs baseband signal processing. The multiple CPUs are all connected via backhaul links to the core network. A key advantage of this scheme is that a subset of APs associated with different CPUs can serve UEs, leading to a user-centric cellular network to solve intercell interference issues and QoS variations [63].

A critical aspect of cell-free massive multiple-input multiple-output (mMIMO) is information security since a plethora of 6G applications that will coexist in the environment are prone to eavesdropping. In this context, PLS is a candidate that can efficiently alleviate the security issues in cell-free mMIMO networks. In Figure 20.5, a basic schematic view of PLS in cell-free mMIMO networks is depicted. Departing from this setup, some research activities on secure cell-free mMIMO deployments have been investigated in the state of the art, especially against pilot-spoofing attacks. For instance, A secure transmission in a multigroup multicasting cell-free mMIMO network was investigated in [64]. Therein, the authors introduced an approach that relies on the minimum description length algorithm to detect pilot spoofing attacks. In this mechanism, pilot signals are

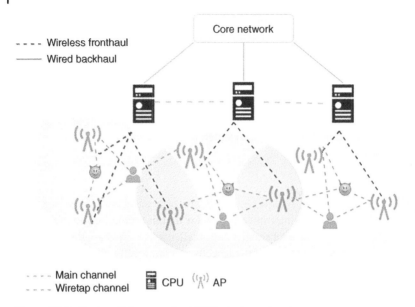

Figure 20.5 Basic cell-free massive MIMO wiretap network.

designed for active relay attack detection, which helps to reduce information leakage to the eavesdropper. Later, the researchers in [65] investigate the effect of hardware impairments at APs and UEs on PLS for cell-free mMIMO by assuming the presence of a pilot spoofing attack. The paper's takeaway was to demonstrate that the hardware-quality scaling impact vanishes as the number of APs increases in the cell-free mMIMO system, assuming the eavesdropper is equipped with perfect hardware. Recently, the authors in [66] presented a framework to provide secure transmissions on cell-free mMIMO systems by employing location technology. Specifically, this novel method is based on fingerprint positioning to help reduce the impact of active pilot attacks on user's transmission. As a forward-looking research direction, federated learning is resounding as a means to alleviate security issues for cell-free mMIMO networks in the 6G era.

20.3.4 Visible Light Communication (VLC)

VLC has been proposed as an alternative (or complementary) solution to cope with the spectrum scarcity problem and the ever-increasing demand for wider bandwidths and higher data rates required for 6G networks. Major advantages of VLC are the plentiful bandwidth (400–800 THz, enabling higher data rates), robustness to electromagnetic interference, low-power consumption, and no license requirement [67]. Additionally, VLC networks provide appealing features from the perspective of confidentiality of the transmitted information as light can be confined and blocked by any opaque objects. However, in public areas, the emitted light can be received by possible eavesdroppers.

Several techniques have been proposed to improve the secrecy of communications in VLC networks, considering active or passive eavesdroppers. In general, zero-forcing precoding is applied to deal with an active eavesdropper so as to suppress the legitimate information at its receiver, whereas jamming-based techniques are used to hinder passive eavesdroppers. However, as pointed out in [68], designing an appropriate method to cope simultaneously with active and passive eavesdroppers is still an open problem. On the other hand, non-Lambertian beams are integrated into the design of the VLC transmitter, allowing a dynamic switching among multiple candidate non-Lambertian beams based on the CSI of legitimate receiver and eavesdropper, thus improving the probability of secure transmission [69].

Despite the advantages associated with VLC systems, several limitations exist that could result in a noticeable degradation of their performance, such as LoS blockages which lead to transmission failures, limited-coverage cell, bulky overhead due to frequent users' handovers, and inter-cell interference at edge users. To overcome these issues, some studies addressing hybrid solutions where VLC networks are complemented by RF communication have been proposed, as RF transmissions present high transmission range and capability of penetrating walls or other opaque barriers, thus complementing some of the shortcomings inherent to VLC [70–72]. For example, in [72], the problem of maximizing the sum secrecy rate for the downlink of a hybrid RF/VLC network with multiple legitimate users in the presence of an eavesdropper, as illustrated in Figure 20.6, is addressed. To do so, a joint precoding and user association strategy is proposed, which leads to significant gains in terms of secrecy performance regarding the solutions based on standalone RF or VLC.

20.3.5 Terahertz Communication

The THz frequency band (0.3–10 THz) has gained notable attention within the research community as a key enabler to provide Terabits-per-second data rates and microsecond latency, both of which are targets envisioned for 6G networks. THz systems will benefit from an extremely higher link directionality in comparison to mmWave, which allows to extend the transmission range or reduce the transmission power. In the context of PLS, another interesting feature due to the high directionality of THz beams is the lower eavesdropping probability, which entails that illegitimate users must be located within the transmitting beam to intercept messages.

However, even the use of very narrow beams may result in a considerably large area around the receiver (referred to as eavesdropping zone), where the

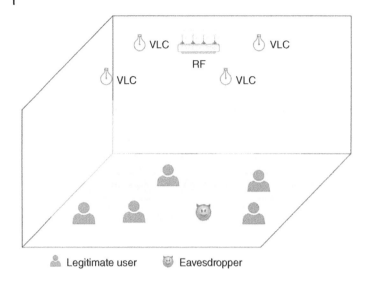

Legitimate user 　Eavesdropper

Figure 20.6 Hybrid RF/VLC network.

eavesdropper can capture all the sensitive data. Thus, PLS strategies can be a design for overcoming this issue. For instance, a multipath communication approach is proposed in [37], whereby the transmitter sequentially transmits different message shares of the sensitive data over different nonblocked propagation paths currently available toward the receiver. In that approach, a secure encoding is used so that the message can be decoded only if all the shares are successfully received.

20.3.6   Joint Communications and Sensing

JCS is touted as one of the key features in 6G systems, which consist in using the same spectrum and, possibly, the same hardware for both communications and radar sensing [73]. This integration of radar and communications services will allow a more efficient usage of radio resources, as well as enable a number of new services and applications, such as radar sensor fusion, interference coordination among radars, and improved beam searching.

Radar information can already be obtained by opportunistically using existing reference signals in communication systems, what was already demonstrated with Wi-Fi [74]. In the future, it is likely that wireless communications systems will also allow the generation of dedicated radar signals in order to achieve certain radar performance requirements. Also monostatic radar can be implemented, in which case, also data-bearing communications signals can be employed for radar sensing.

In all cases, radar data will be available to all nodes in a wireless system, providing real-time information about the environment, either as a dedicated service or as a by-product of communications. This information can be used for many purposes, and security is also one of those. It may help PLS methods by providing an awareness of the physical environment, which is part of a general context.

It is know that the performance of different PLS schemes is highly dependent on the statistics of the propagation channels, as well as on the threat model. Therefore, context-awareness is essential to allow the network to choose the most adequate security measures for each particular situation, including one of the different PLS flavors. For example, radar can be used to build a 3D image of the environment, even in the dark, which may help to identify possible external threats and to classify the propagation channel. Information about the physical environment can also be employed as an authentication method, e.g. by a challenge-response scheme to verify that all trusted communication nodes are sensing the same environment, thus being in the same room.

Research on applying sensing to improve security is still in its infancy, but it can be one of the enabling technologies to make PLS realizable in practice.

References

1 N. H. Mahmood, S. Böcker, I. Moerman, O. A. López, A. Munari, K. Mikhaylov, F. Clazzer, H. Bartz, O.-S. Park, E. Mercier et al., "Machine type communications: Key drivers and enablers towards the 6G era," *EURASIP Journal on Wireless Communications and Networking*, vol. 134, no. 1, p. 134, 2021.

2 A. Chorti, A. N. Barreto, S. Köpsell, M. Zoli, M. Chafii, P. Sehier, G. Fettweis, and H. V. Poor, "Context-aware security for 6G wireless: The role of physical layer security," *IEEE Communications Standards Magazine*, vol. 6, no. 1, pp. 102–108, 2022.

3 D. P. Osorio, J. D. V. Sánchez, and H. Alves, *Physical-Layer Security for 5G and Beyond*. John Wiley & Sons, 2019, Chapter 1, pp. 1–19. [Online]. Available: https://onlinelibrary.wiley.com/doi/abs/10.1002/9781119471509.w5GRef152.

4 C. E. Shannon, "Communication theory of secrecy systems," *Bell System Technical Journal*, vol. 28, no. 4, pp. 656–715, 1949.

5 A. D. Wyner, "The wire-tap channel," *Bell System Technical Journal*, vol. 54, no. 8, pp. 1355–1387, 1975.

6 U. Maurer and S. Wolf, "Information-theoretic key agreement: From weak to strong secrecy for free," in *Advances in Cryptology — EUROCRYPT 2000*, B. Preneel, Ed. Berlin, Heidelberg: Springer, 2000, pp. 351–368.

7 J. Barros and M. R. D. Rodrigues, "Secrecy capacity of wireless channels," in *2006 IEEE International Symposium on Information Theory*, July 2006, pp. 356–360.

8 I. Csiszar and J. Korner, "Broadcast channels with confidential messages," *IEEE Transactions on Information Theory*, vol. 24, no. 3, pp. 339–348, 1978.

9 S. Leung-Yan-Cheong and M. Hellman, "The Gaussian wire-tap channel," *IEEE Transactions on Information Theory*, vol. 24, no. 4, pp. 451–456, 1978.

10 J. D. Vega Sanchez, D. P. M. Osorio, E. E. B. Olivo, H. Alves, M. C. P. Paredes, and L. Urquiza-Aguiar, "On the statistics of the ratio of nonconstrained arbitrary α-μ random variables: A general framework and applications," *Transactions on Emerging Telecommunications Technologies*, vol. 31, no. 3, p. e3832, 2019.

11 E. N. Papasotiriou, A.-A.-A. Boulogeorgos, K. Haneda, M. F. de Guzman, and A. Alexiou, "An experimentally validated fading model for THz wireless systems," *Scientific Reports*, vol. 11, no. 1, pp. 1–14, 2021.

12 J. D. Vega-Sanchez, D. P. M. Osorio, F. J. López-Martínez, M. C. P. Paredes, and L. F. Urquiza-Aguiar, "Information-theoretic security of MIMO networks under κ-μ shadowed fading channels," *IEEE Transactions on Vehicular Technology*, vol. 70, no. 7, pp. 6302–6318, 2021.

13 W. Zeng, J. Zhang, S. Chen, K. P. Peppas, and B. Ai, "Physical layer security over fluctuating two-ray fading channels," *IEEE Transactions on Vehicular Technology*, vol. 67, no. 9, pp. 8949–8953, 2018.

14 J. D. Vega-Sanchez, D. P. M. Osorio, F. J. López-Martínez, M. C. P. Paredes, and L. F. Urquiza-Aguiar, "On the secrecy performance over N-wave with diffuse power fading channel," *IEEE Transactions on Vehicular Technology*, vol. 69, no. 12, pp. 15 137–15 148, 2020.

15 R. Negi and S. Goel, "Secret communication using artificial noise," in *VTC-2005-Fall. 2005 IEEE 62nd Vehicular Technology Conference, 2005*, vol. 3, 2005, pp. 1906–1910.

16 A. Sanenga, G. A. Mapunda, T. M. L. Jacob, L. Marata, B. Basutli, and J. M. Chuma, "An overview of key technologies in physical layer security," *Entropy*, vol. 22, no. 11, p. 1261, 2020. [Online]. Available: https://www.mdpi.com/1099-4300/22/11/1261.

17 H. Sharma, N. Kumar, and R. Tekchandani, "Physical layer security using beamforming techniques for 5G and beyond networks: A systematic review," *Physical Communication*, vol. 54, p. 101791, 2022. [Online]. Available: https://www.sciencedirect.com/science/article/pii/S1874490722001094.

18 A. I. Sulyman and C. Henggeler, "Physical layer security for military IoT links using MIMO-beamforming at 60 GHz," *Information*, vol. 13, no. 2, p. 100, 2022. [Online]. Available: https://www.mdpi.com/2078-2489/13/2/100.

19 I. W. G. Da Silva, J. D. V. Sanchez, E. E. B. Olivo, and D. P. M. Osorio, "Impact of self-energy recycling and cooperative jamming on SWIPT-based FD relay networks with secrecy constraints," *IEEE Access*, vol. 10, pp. 24 132–24 148, 2022.

20 D. P. M. Osorio, H. Alves, and E. E. B. Olivo, "On the secrecy performance and power allocation in relaying networks with untrusted relay in the partial secrecy regime," *IEEE Transactions on Information Forensics and Security*, vol. 15, pp. 2268–2281, 2020.

21 E. N. Egashira, D. P. M. Osorio, and E. E. B. Olivo, "Impact of wireless energy transfer strategies on the secrecy performance of untrustworthy relay networks," *IEEE Transactions on Vehicular Technology*, vol. 71, no. 6, pp. 6859–6863, 2022.

22 D. P. M. Osorio, E. E. B. Olivo, H. Alves, and M. Latva-Aho, "Safeguarding MTC at the physical layer: Potentials and challenges," *IEEE Access*, vol. 8, pp. 101 437–101 447, 2020.

23 P. L. Yu, J. S. Baras, and B. M. Sadler, "Physical-layer authentication," *IEEE Transactions on Information Forensics and Security*, vol. 3, no. 1, pp. 38–51, 2008.

24 N. Wang, W. Li, P. Wang, A. Alipour-Fanid, L. Jiao, and K. Zeng, "Physical layer authentication for 5G communications: Opportunities and road ahead," *IEEE Network*, vol. 34, no. 6, pp. 198–204, 2020.

25 H. Fang, X. Wang, and L. Hanzo, "Learning-aided physical layer authentication as an intelligent process," *IEEE Transactions on Communications*, vol. 67, no. 3, pp. 2260–2273, 2019.

26 H. Fang, X. Wang, and L. Xu, "Fuzzy learning for multi-dimensional adaptive physical layer authentication: A compact and robust approach," *IEEE Transactions on Wireless Communications*, vol. 19, no. 8, pp. 5420–5432, 2020.

27 H. Forssell, "Performance guarantees for physical layer authentication in mission-critical communications," Ph.D. dissertation, KTH Royal Institute of Technology, Malvinas väg 10, 100 44 Stockholm, Sweden, 2021.

28 U. Maurer, "Secret key agreement by public discussion from common information," *IEEE Transactions on Information Theory*, vol. 39, no. 3, pp. 733–742, 1993.

29 R. Ahlswede and I. Csiszar, "Common randomness in information theory and cryptographyPart I: Secret sharing," *IEEE Transactions on Information Theory*, vol. 39, no. 4, pp. 1121–1132, 1993.

30 S. Del Prete, F. Fuschini, M. Barbiroli, M. Zoli, and A. N. Barreto, "A study on physical layer security through ray tracing simulations," in *2022 16th European Conference on Antennas and Propagation (EuCAP)*, 2022, pp. 1–5.

322 | *20 Physical Layer Security*

31 C. Huth, R. Guillaume, T. Strohm, and T. G. P. Duplys, I. A. Samuel, "A study on physical layer security through ray tracing simulations," in *2022 16th European Conference on Antennas and Propagation (EuCAP)*, 2022, pp. 1–5.

32 M. Bloch, J. Barros, M. R. D. Rodrigues, and S. W. McLaughlin, "Wireless information-theoretic security," *IEEE Transactions on Information Theory*, vol. 54, no. 6, pp. 2515–2534, 2008.

33 Z. Li, J. Kang, R. Yu, D. Ye, Q. Deng, and Y. Zhang, "Consortium blockchain for secure energy trading in industrial internet of things," *IEEE Transactions on Industrial Informatics*, vol. 14, no. 8, pp. 3690–3700, 2018.

34 G. Li, H. Yang, J. Zhang, H. Liu, and A. Hu, "Fast and secure key generation with channel obfuscation in slowly varying environments," in *IEEE INFOCOM 2022 - IEEE Conference on Computer Communications*. IEEE, 2021.

35 M. Zoli, M. Mitev, A. N. Barreto, and G. Fettweis, "Estimation of the secret key rate in wideband wireless physical-layer-security," in *ISWCS*, 2021, pp. 1–6.

36 M. Zoli, A. N. Barreto, S. Köpsell, P. Sen, and G. Fettweis, "Physical-layer-security box: A concept for time-frequency channel-reciprocity key generation," *EURASIP Journal on Wireless Communications and Networking (JWCN)*, vol. 2020, p. 114, 2020.

37 V. Petrov, D. Moltchanov, J. M. Jornet, and Y. Koucheryavy, "Exploiting multipath terahertz communications for physical layer security in beyond 5G networks," in *IEEE INFOCOM 2019 - IEEE Conference on Computer Communications Workshops (INFOCOM WKSHPS)*, 2019, pp. 865–872.

38 Y. Liu, X. Liu, X. Mu, T. Hou, J. Xu, M. Di Renzo, and N. Al-Dhahir, "Reconfigurable intelligent surfaces: Principles and opportunities," *IEEE Communications Surveys & Tutorials*, vol. 23, no. 3, pp. 1546–1577, 2021.

39 M. Katz and I. Ahmed, "Opportunities and challenges for visible light communications in 6G," in *2020 2nd 6G Wireless Summit (6G SUMMIT)*, 2020, pp. 1–5.

40 Z. Wei, F. Liu, C. Masouros, N. Su, and A. P. Petropulu, "Toward multi-functional 6G wireless networks: Integrating sensing, communication, and security," *IEEE Communications Magazine*, vol. 60, no. 4, pp. 65–71, 2022.

41 R. Chen, C. Li, S. Yan, R. Malaney, and J. Yuan, "Physical layer security for ultra-reliable and low-latency communications," *IEEE Wireless Communications*, vol. 26, no. 5, pp. 6–11, 2019.

42 P. Porambage, G. Gür, D. P. M. Osorio, M. Liyanage, A. Gurtov, and M. Ylianttila, "The roadmap to 6G security and privacy," *IEEE Open Journal of the Communications Society*, vol. 2, pp. 1094–1122, 2021.

43 L. Mucchi, A. Martinelli, S. Jayousi, S. Caputo, and M. Pierobon, "Secrecy capacity and secure distance for diffusion-based molecular communication systems," *IEEE Access*, vol. 7, pp. 110 687–110 697, 2019.

44 L. Mucchi, S. Jayousi, S. Caputo, E. Panayirci, S. Shahabuddin, J. Bechtold, I. Morales, R.-A. Stoica, G. Abreu, and H. Haas, "Physical-layer security in 6G networks," *IEEE Open Journal of the Communications Society*, vol. 2, pp. 1901–1914, 2021.

45 D. P. M. Osorio, I. Ahmad, J. D. V. Sánchez, A. Gurtov, J. Scholliers, M. Kutila, and P. Porambage, "Towards 6G-enabled internet of vehicles: Security and privacy," *IEEE Open Journal of the Communications Society*, vol. 3, pp. 82–105, 2022.

46 S. Gong, X. Lu, D. T. Hoang, D. Niyato, L. Shu, D. I. Kim, and Y.-C. Liang, "Toward smart wireless communications via intelligent reflecting surfaces: A contemporary survey," *IEEE Communications Surveys & Tutorials*, vol. 22, no. 4, pp. 2283–2314, 2020.

47 Q. Wu, S. Zhang, B. Zheng, C. You, and R. Zhang, "Intelligent reflecting surface-aided wireless communications: A tutorial," *IEEE Transactions on Communications*, vol. 69, no. 5, pp. 3313–3351, 2021.

48 G. Zhou, C. Pan, H. Ren, K. Wang, and Z. Peng, "Secure wireless communication in RIS-aided MISO system with hardware impairments," *IEEE Wireless Communications Letters*, vol. 10, no. 6, pp. 1309–1313, 2021.

49 J. D. V. Sanchez, P. Ramírez-Espinosa, and F. J. López-Martínez, "Physical layer security of large reflecting surface aided communications with phase errors," *IEEE Wireless Communications Letters*, vol. 10, no. 2, pp. 325–329, 2021.

50 J. D. Vega-Sanchez, G. Kaddoum, and F. J. Lopez-Martinez, "Physical layer security of RIS-assisted communications under electromagnetic interference," *arXiv preprint arXiv:2203.08370*, 2022.

51 Z. Ji, P. L. Yeoh, D. Zhang, G. Chen, Y. Zhang, Z. He, H. Yin, and Y. Li, "Secret key generation for intelligent reflecting surface assisted wireless communication networks," *IEEE Transactions on Vehicular Technology*, vol. 70, no. 1, pp. 1030–1034, 2021.

52 S. Xu, J. Liu, and Y. Cao, "Intelligent reflecting surface empowered physical-layer security: Signal cancellation or jamming?" *IEEE Internet of Things Journal*, vol. 9, no. 2, pp. 1265–1275, 2022.

53 E. N. Egashira, D. P. Moya, N. T. Nguyen, and M. Juntti, "Secrecy capacity maximization for a hybrid relay-RIS scheme in mmWave MIMO networks," *arXiv preprint arXiv:2205.13904*, 2022.

54 Y. Zeng, Q. Wu, and R. Zhang, "Accessing from the sky: A tutorial on UAV communications for 5G and beyond," *Proceedings of the IEEE*, vol. 107, no. 12, pp. 2327–2375, 2019.

55 Q. Wu, W. Mei, and R. Zhang, "Safeguarding wireless network with UAVs: A physical layer security perspective," *IEEE Wireless Communications*, vol. 26, no. 5, pp. 12–18, 2019.

56 Y. Zhou, P. L. Yeoh, H. Chen, Y. Li, R. Schober, L. Zhuo, and B. Vucetic, "Improving physical layer security via a UAV friendly jammer for unknown eavesdropper location," *IEEE Transactions on Vehicular Technology*, vol. 67, no. 11, pp. 11 280–11 284, 2018.

57 X. A. F. Cabezas, D. P. M. Osorio, and M. Latva-Aho, "Distributed UAV-enabled zero-forcing cooperative jamming scheme for safeguarding future wireless networks," in *2021 IEEE 32nd Annual International Symposium on Personal, Indoor and Mobile Radio Communications (PIMRC)*, 2021, pp. 739–744.

58 X. A. F. Cabezas, D. P. M. Osorio, and M. Latva-aho, "Weighted secrecy coverage analysis and the impact of friendly jamming over UAV-enabled networks," in *2021 Joint European Conference on Networks and Communications and 6G Summit (EuCNC/6G Summit)*, 2021, pp. 124–129.

59 X. A. F. Cabezas, D. P. M. Osorio, and M. Latva-aho, "Positioning and power optimization for UAV-assisted networks in the presence of eavesdroppers: A multi-armed bandit approach," *PREPRINT (Version 1) available at Research Square*, 2022.

60 C. Wen, Y. Fang, and L. Qiu, "Securing UAV communication based on multi-agent deep reinforcement learning in the presence of smart UAV eavesdropper," in *2022 IEEE Wireless Communications and Networking Conference (WCNC)*, 2022, pp. 1164–1169.

61 M. T. Mamaghani and Y. Hong, "Terahertz meets untrusted UAV-relaying: Minimum secrecy energy efficiency maximization via trajectory and communication co-design," *IEEE Transactions on Vehicular Technology*, vol. 71, no. 5, pp. 4991–5006, 2022.

62 H. Q. Ngo, A. Ashikhmin, H. Yang, E. G. Larsson, and T. L. Marzetta, "Cell-free massive MIMO versus small cells," *IEEE Transactions on Wireless Communications*, vol. 16, no. 3, pp. 1834–1850, 2017.

63 H. Q. Ngo, A. Ashikhmin, H. Yang, E. G. Larsson, and T. L. Marzetta, "Cell-free massive MIMO: Uniformly great service for everyone," in *2015 IEEE 16th International Workshop on Signal Processing Advances in Wireless Communications (SPAWC)*, 2015, pp. 201–205.

64 X. Zhang, D. Guo, K. An, Z. Ding, and B. Zhang, "Secrecy analysis and active pilot spoofing attack detection for multigroup multicasting cell-free massive MIMO systems," *IEEE Access*, vol. 7, pp. 57 332–57 340, 2019.

65 X. Zhang, D. Guo, K. An, and B. Zhang, "Secure communications over cell-free massive mimo networks with hardware impairments," *IEEE Systems Journal*, vol. 14, no. 2, pp. 1909–1920, 2020.

66 J. Qiu, K. Xu, X. Xia, Z. Shen, W. Xie, D. Zhang, and M. Wang, "Secure transmission scheme based on fingerprint positioning in cell-free massive mimo systems," *IEEE Transactions on Signal and Information Processing over Networks*, vol. 8, pp. 92–105, 2022.

67 H. Abuella, M. Elamassie, M. Uysal, Z. Xu, E. Serpedin, K. A. Qaraqe, and S. Ekin, "Hybrid RF/VLC systems: A comprehensive survey on network topologies, performance analyses, applications, and future directions," *IEEE Access*, vol. 9, pp. 160 402–160 436, 2021.

68 M. Obeed, A. M. Salhab, M.-S. Alouini, and S. A. Zummo, "On optimizing vlc networks for downlink multi-user transmission: A survey," *IEEE Communications Surveys & Tutorials*, vol. 21, no. 3, pp. 2947–2976, 2019.

69 J. Ding, I. Chih-Lin, J. Wang, H. Yang, and L. Wang, "Multiple optical beam switching for physical layer security of visible light communications," *IEEE Photonics Journal*, vol. 14, no. 1, pp. 1–9, 2022.

70 I. W. G. da Silva, D. P. M. Osorio, E. E. B. Olivo, I. Ahmed, and M. Katz, "Secure hybrid RF/VLC under statistical queuing constraints," in *2021 17th International Symposium on Wireless Communication Systems (ISWCS)*, 2021, pp. 1–6.

71 J. Al-Khori, G. Nauryzbayev, M. M. Abdallah, and M. Hamdi, "Joint beamforming design and power minimization for friendly jamming relaying hybrid RF/VLC systems," *IEEE Photonics Journal*, vol. 11, no. 2, pp. 1–18, 2019.

72 I. W. G. da Silva, E. E. B. Olivo, M. Katz, and D. P. Moya Osorio, "Secure joint precoding and user association for multiuser hybrid RF/VLC systems," *TechRxiv preprint*, 2022.

73 C. De Lima, D. Belot, R. Berkvens, A. Bourdoux, D. Dardari, M. Guillaud, M. Isomursu, E.-S. Lohan, Y. Miao, A. N. Barreto et al., "Convergent communication, sensing and localization in 6G systems: An overview of technologies, opportunities and challenges," *IEEE Access*, vol. 9, pp. 26 902–26 925, 2021.

74 H. Jiang, C. Cai, X. Ma, Y. Yang, and J. Liu, "Smart home based on WiFi sensing: A survey," *IEEE Access*, vol. 6, pp. 13 317–13 325, 2018.

21

Quantum Security and Postquantum Cryptography*

* With additional contribution from Kimmo Halunen, University of Oulu, Finland, and Sara Nikula, VTT Technical Research Centre of Finland

This chapter focuses on the concepts of Quantum Security and postquantum cryptography (PQC). The core concepts of both are presented as well as recent developments in standardization of PQC that will have an impact in the 6G development. The main objectives of this chapter are listed as follows:

- Understand the impact of quantum computing to the security of modern cryptography.
- Gain an overview of Quantum Security and postquantum cryptography (PQC) that can provide security against quantum computers.
- Identify the major security and other considerations behind the PQC standardization effort.
- Gain an understanding of the issues in the future deployment of PQC that is relevant to 6G.
- Find relevant open research questions in the field of PQC for future study.

21.1 Overview of 6G and Quantum Computing

Within the next couple of years, it is expected that quantum computing will be commercially available and will impose a huge threat on the current cryptographic schemes. As stated in the current state-of-the-art, quantum computing is envisioned to be used in 6G communication networks for detection, mitigation, and prevention of security vulnerabilities. Quantum computing-assisted communication is a novel research area that investigates the possibilities of replacing quantum channels with noiseless classical communication channels to achieve extremely high reliability in 6G. Moreover, with the advancements of quantum computing,

Security and Privacy Vision in 6G: A Comprehensive Guide, First Edition.
Pawani Porambage and Madhusanka Liyanage.

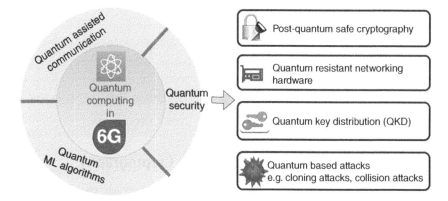

Figure 21.1 Role of quantum computing in 6G.

it is foreseen by the security researchers that quantum-safe cryptography should be introduced in the postquantum world. Figure 21.1 demonstrates the envisioned roles of quantum computing and quantum security in the 6G era.

21.2 Quantum Computing

In order to understand quantum security and the need for PQC, it is important to take a look at what quantum computing is and how it can change the way computers can solve problems. The basic unit of computation in traditional computers is a *bit* that can have at any time one of two distinct values 0 or 1. In quantum computers, the basic unit is a *qubit*. A qubit can be in a state of superposition, where it has the value of 0 or 1 with some well-defined probability. The actual measure is *the amplitude* of a qubit or a set of qubits, but it is beyond the scope of this chapter to go into more detail. This value of the qubit (0 or 1) does not need to be exactly determined in the intermediate steps of the computation and the computations (using different quantum gates to manipulate the different qubits in the computation) can be done on the qubits. Only at the end of the computation, the qubits' state and values need to be measured, and this then yields the result of the quantum computation. This is in contrast to traditional computing, where bits have at any given point of computation either the value 0 or 1.

Although the difference may seem small, it has been shown that there are many algorithms that can benefit from this type of computation and quantum computers can solve some problems much more efficiently than traditional computers. On the other hand, from a theoretical point of view, quantum computers cannot solve all the problems more efficiently. In particular, quantum computers cannot solve all problems in the complexity class NP in (quantum) polynomial time [1].

There are however algorithms, where the advantage is immense. For example with Simon's problem [2], a traditional computer runs in the complexity class $\mathcal{O}(2^{\frac{n}{2}})$, but a quantum computer can solve the problem in $\mathcal{O}(n)$! Although this is an impressive result, this does not have much implications toward the security of systems even if quantum computers would become commonplace.

The security implications of quantum computing come from two famous algorithms. The first is Grover's algorithm [3], which gives a quadratic speed-up in brute force searches. This means that for any key (for example in a symmetric cipher like advanced encryption standard [AES]), the effective security of that key length is halved (e.g. a 256 bit key gives only 128 bit of security). However, this threat can be mitigated in most of the cases by simply doubling the key lengths in algorithms.

Shor's algorithm [4] on the other hand has the potential to really break the protections afforded by modern cryptography. The algorithm can be used to solve efficiently (i.e. in polynomial time) all the mathematical problems (factoring, discrete logarithm in finite fields, and elliptic curves) that are used currently in public-key cryptography (PKC). This has prompted National Institute of Standards and Technology (NIST) to hold a competition for new standards in two of the most important application areas of PKC, namely key exchange (or key encapsulation) and digital signatures.[1]

At the moment, there is some research on how close we are toward having a quantum computer that could realistically apply Shor's algorithm to break encryption. Currently, for example IBM has demonstrated a 127-qubit quantum computer. Research on the applications of Shor's algorithm vary on the estimates of the number of qubits needed (from tens of millions to some thousands), with the low end also needing other special infrastructure such as quantum memory [5–7].

21.3 Quantum Security

In light of the above possible threats to the security of our digital systems, there is a need to find ways to secure cryptography against quantum computers. In this section, we will discuss further some possibilities to achieve this, namely quantum key distribution (QKD) and information-theoretic security. QKD is a method to distribute keys between two parties by using quantum entanglement. These keys are then used in a traditional, symmetric cryptosystems. Information-theoretic security is a very strong security level that is only attained by few cryptosystems. After these, we will present PQC as the most practical and versatile solution to the quantum threat.

---

1 https://csrc.nist.gov/Projects/post-quantum-cryptography/post-quantum-cryptography-standardization.

21.3.1 Quantum Key Distribution

Mathematically intractable problems, such as factoring large numbers, form the basis for the security of currently deployed asymmetric cryptography. This means that increases in the computing power of quantum computers gradually make these algorithms less secure. QKD offers a way to create and transmit cryptographical keys, which are needed to encrypt the communication in later phases, in a way which is secure even in the presence of powerful quantum computers. This is due to the fact that QKD is not dependent on mathematically complex problems, but rather on the laws of quantum physics: a quantum state cannot be cloned, and after measuring it, the original state is lost. By utilizing this phenomenon, we can create a key transmission scheme which cannot be broken by mathematical analysis, even though computing power of classical and quantum computers grows.

QKD protocols can be roughly divided into continuous-variable (CV) and discrete-variable (DV) protocols. Discrete-variable QKD protocols are based on encoding each bit of information in a single-quantum state, in practice usually in the polarization of a photon. The most known examples of these protocols are BB84 [8] and B92 [9]. Any attempt to eavesdrop the quantum channel, through which the information is transmitted, leads to an increased error rate in the received key. This is due to the fact that after measuring a photon, the potential eavesdropper is not able to return it to its original state. CV QKD protocols are based on encoding the key information in continuous properties, such as phase, of the signal. Whereas security of DV QKD protocols can be quite simply inferred, and the security proofs for CV QKD protocols are usually more complex and require more advanced analysis, as in [10]. One advantage of CV QKD protocols is that, as opposed to DV protocols, they can be implemented based on the existing communication infrastructure [11].

Before using the transmitted key material for encryption of messages, further processing is required in order to remove errors caused in the quantum transmission phase and to improve the security of the error-corrected key. This process is called key distillation, and it requires a classical channel for exchanging messages related to this phase. This classical channel must be authenticated, using for example hash functions [12] or postquantum digital signatures [13].

Key distillation process includes bit sifting, error rate estimation, error correction, and privacy amplification. Bit sifting is made according to the used protocol and it means selecting only a subset of the bits for further processing. For example, in [8] this means selecting only the bits which are prepared and sent in the same basis. Imperfections of the quantum channel usually cause some errors in the sifted key material. Because of error correction procedures, this will not prevent the parties from obtaining a shared key, but if the error rate is too high, the privacy of the key material can no longer be assured, because it is unknown if the errors are caused by an eavesdropper. Error estimation can be done by publishing a subset of the sifted bits and comparing the differences between these bits on the sender and the receiver side. Based on the found differences, an estimate of the total error rate can be done [8]. For error correction, there exist several different alternatives, e.g. [14], which is based on iterative parity checks. The last phase of the key distillation process is privacy amplification, which aims to mitigate the information leaked during the previous phases of key transmission and distillation. Privacy amplification is typically based on hashing the reconciled key, and the resulting string of bits is then considered as the final secure key [15].

By connecting QKD nodes to each other, for example using trusted nodes [16], QKD networks can be formed. This allows to not only send keys between two adjacent nodes but to share key materials with several different nodes included in the network.

In practice, QKD can be realized by different physical implementations. The quantum channel can be realized by sending the photons through a fiber [17, 18] or sending them in free space, possibly using satellites as the other endpoint [19]. Both of these techniques currently offer only limited transmission distances. Transmission through a fiber causes increasing amount of loss when the transmission distance grows [18]. In free space, environmental conditions, diffraction, and atmospheric turbulence decrease the efficiency of the communication [19]. The practical implementations of QKD are still being developed and improved. The main shortcomings of the currently available realizations of QKD tend to be issues like weak key rates and limited transmission distances between nodes [11].

21.3.2 Information-Theoretic Security

Another possibility to guard against quantum adversaries is to use *information-theoretic security* (also known as *unconditional security*). With this type of security guarantees, the computational resources of the adversary are basically irrelevant. The system is secure even if the adversary has access to quantum computers or any other computational technology that may come about in the future.

A well-known example of an unconditionally secure cryptosystems is the *one-time pad*. The one-time pad uses a key that is perfectly random and at least as long as the message is to be encrypted. The random key is then XORed with the message to produce the ciphertext. The results of Shannon show that this achieves unconditional security. However, there are many practical issues that make one-time pad highly impractical to use in modern digital systems.

First of all, the key length is directly proportional to the amount of data to be encrypted (the key needs to be at least as long as the message). In contrast to modern ciphers, where the security (although only computational) is achieved with a key of fixed length. Furthermore, key exchange and key management are very difficult to solve with such large keys.

Second, the system provides only encryption and not authentication or integrity checks on the data. Thus, it cannot achieve, for example digital signatures. All in all, there are only a limited number of cryptosystems that achieve unconditional security, and these cannot provide all the same security goals and functionalities than for example public key encryption. In addition, PKC (as it is currently understood) cannot be achieved with unconditional security [20–22].

In light of these practical issues, it is not possible to achieve security against quantum computers with the help of unconditionally secure solutions and to keep all the functionality that our modern digital society depends upon in an efficient and sustainable way. Luckily, there are solutions that can produce these in the field of PQC.

21.4 Postquantum Cryptography

As the Section (21.3) has shown, there is a need for more practical and versatile solutions toward cryptography than QKD and information-theoretic security. The answer to this call is PQC. PQC systems can be used with traditional computers and provide similar computational security guarantees as our current PKC, but are not known to be weak against quantum computers. In this section, we will detail PQC and present its effect on 6G development.

21.4.1 Background

To better understand the current developments in PQC, it is important to take a look at the PKC that is implemented today in our digital world. PKC is mainly based on three different types of mathematical problems, all of which are susceptible to Shor's algorithm.

The first mathematical problem is that of *integer factoring*. The problem is to find the prime factors of a given (very large) integer. The Rivest–Shamir–Adleman (RSA) cryptosystem [23] is based on this very problem. The best-known classical algorithm for solving the factoring problem is the General Number Field Sieve [24], which runs in superpolynomial complexity.

Another problem is the *discrete logarithm* problem (and different variants of the Diffie–Hellman problem). In the discrete logarithm problem, given $y = g^x$ and the generator element g of a finite cyclic group, one should find x, the discrete logarithm of y. Again, this problem is in general difficult and many systems are based on this such as Diffie–Hellman key exchange [20] and the ElGamal cryptosystem [25]. However, there are cases where this problem can be easy to solve, so care needs to be taken in the parameter selection.

The third problem is *elliptic curve discrete logarithm*, which is a variant of the discrete logarithm problem on the groups of points on an elliptic curve. Again, this is a difficult problem and cryptographic systems based on these are used extensively. One advantage of elliptic curve cryptography (ECC) is that the key lengths necessary for good security levels are smaller than with the previous two methods.

21.4.2 PQC Methods

As the currently used methods are susceptible to attacks by quantum computers, there is a need for cryptography that is resistant against these. Fortunately, there exist many mathematical problems that can be used for cryptography and for which there are no known quantum algorithms in existence. In general, the PQC methods are based on five different problems that are listed below

- Symmetric primitives which are based on known symmetric cryptography, e.g. hash functions.
- Error correcting codes (e.g. Goppa codes). The classic McEliece cryptosystem [26] is an example of this.
- Lattice-based constructions. There are several slightly different lattice problems like Learning with Errors that can be used to build cryptosystems.
- Multivariate polynomial based constructions. These are based on the difficulty on solving large systems of (quadratic) multivariate polynomials without trapdoor information.
- Supersingular isogenies on elliptic curves. This is a fairly new class of problems that have been used to develop PQC.

All of the above methods have been used to design proposals for the NIST standardization competition (see Section 21.4.3). Most of these are NP-hard problems, which give some assurances that these would not be susceptible to attacks from quantum computers. Because of the very different problems and approaches that are available for PQC, the different proposals exhibit a wide range of properties and the different parameters used in these (e.g. key size) vary greatly.

One key point to remember with PQC is that the proposed systems need to be resistant not only against quantum computers but also (and especially) against classical attacks with traditional computers. In many cases, the real attacks against PQC systems are made with traditional means and cryptanalysis. In the NIST standardization competition, none of the proposals was dismissed due to novel quantum attacks, but even one of the finalists and one of the chosen algorithms were found to be weak against attacks by traditional means [27, 28]. Also, other late-stage candidates were found susceptible to attacks (Tables 21.1 and 21.2) [29].

Table 21.1 The third round candidates of the NIST standardization competition for PQC KEMs (https://csrc.nist.gov/publications/detail/nistir/8309/final).

PQC KEM	Basis	Properties
BIKE	Code-based	Slightly larger keys and ciphertexts than lattice-based systems. Overall performance is good
Classic McEliece	Code-based	Underlying cryptographic problem studied since 1970s. Very large public keys can be difficult for many applications. On the other hand, small cipher texts can be an asset in some applications e.g. cellular networks
CRYSTALS-KYBER	Lattice-based	Very good overall performance, key size and ciphertext size. Based on the same mathematical problem as the CRYSTALS-DILITHIUM digital signature scheme. The only KEM chosen to the standard after Round 3 of the NIST competition
HQC	Code-based	OK overall performance. Keys and ciphertexts larger than with lattice-based alternatives
SIKE	Isogeny	Computationally intensive for both communicating parties. Transmission costs are good. Unfortunately an attack was published shortly after the announcement by NIST [27]

Table 21.2 Standardized digital signature schemes from the NIST PQC competition (https://csrc.nist.gov/publications/detail/nistir/8309/final).

PQC digital signature	Basis	Properties
CRYSTALS-DILITHIUM	Lattice-based	Very good overall performance, key size, and ciphertext size. Based on the same mathematical problem as the CRYSTALS-KYBER KEM. Easier to implement than FALCON
FALCON	Lattice-based	More complex to implement than CRYSTALS-DILITHIUM. Uses floating point arithmetic, which can be problematic in some applications. Fairly small key and signature size and good performance are the strong points of this algorithm
SPHINCS+	Stateless hash functions	Very complex to implement and much larger signatures than the other two. Unfortunately an attack was published shortly after the announcement by NIST [28]

21.4.3 PQC Standardization

In the last few years, the cryptographic community has been actively trying to find suitable algorithms for PQC standardization. NIST has run this competition like in many cases before with symmetric encryption and hash functions.

The NIST standards will produce algorithms for the two main use cases of PKC, key encapsulation mechanisms (for agreeing on symmetric keys for encryption and authentication) and digital signatures. These will be necessary in many use cases. However, these will not solve all possible cryptographic issues, where PQC is needed. Thus, there is need for further research in finding solutions to other cryptographic goals beyond these.

Table properties of the NIST chosen candidates (finalists, if not available).

NIST has also listed several alternate candidates that would need further research before they can be considered to be included in the standard. Because many of the proposed systems have not been under scrutiny for very long and even the underlying mathematics in some cases is relatively young, this is a warranted position. In time, some of the alternate candidates can be added to the standards.

21.4.4 Challenges with PQC

Transitioning from current public-key infrastructure (PKI) to public-key cryptography using PQC is not an easy task. Already in the early stages of the NIST standardization process there were results showing that some of the candidates pose serious issues for important protocols [30] such as transport layer security (TLS) [31] and secure shell (SSH) [32]. Thus, it is not simply the case of substituting the new algorithms for the old ones, but there may be a need to redesign the underlying protocols as well. This would then also require some new security analysis of the protocols in addition to the primitives defined in the upcoming standards.

Of course, there are many other standards and protocols that need to take the new standards into account. One of the major cryptography engineering challenges of the 2020s will be finding the best ways to implement PQC in the myriad of protocols and standards that are used in our digital society, e.g. smart vehicles.

Performance issues are also one possible challenge for the implementation of PQC. As there are increasingly many different applications that need cryptographic protection even in the very resource-constrained devices, the increased overhead that some of the PQC algorithms have can become an obstacle for their use in some cases. However, this is an active field of research and at least with earlier standards e.g. AES [33] optimizations have been developed over time, such as AES-NI (Native Instructions) [34].

One key area, where the implementations of PQC algorithms need to be carefully scrutinized is programming libraries [35]. Cryptographic libraries are available for most programming languages and are used to provide cryptographic protections in software written in those languages. Although it is beneficial to have fewer implementations of such complex algorithms as cryptographic primitives, it is not without perils. Both historically and recently, there have been vulnerabilities in many cryptographic programming libraries [36], e.g. the Heartbleed bug [37].

The PQC algorithms are fairly challenging to implement correctly as they involve advanced mathematics. Thus, it is more likely that some vulnerabilities will emerge in these implementations. In order to avoid this, it would be good to have more thorough testing, reviewing and researching on the libraries and PQC implementations. A good starting point would be to direct some special attention to the open-source cryptographic libraries, for example through special bug bounty programs.

An important challenge for the adoption of the PQC standards is trust. The recent attacks that came just after the announcement of the winners [27, 28] have not increased the trust in the security of the standards. This can be contrasted with for example AES, which has withstood over two decades of cryptanalysis without major theoretical breakthroughs. Also, other issues such as patents have undermined the consensus of the cryptographic community on the benefits and weaknesses of PQC algorithms. If there will be a lack of trust for the PQC standards, it will be difficult to get them adopted widely and this will result in protracted vulnerability against quantum computing attacks.

21.4.5 Future Directions of PQC

After the standardization and the possible amendments to it from alternate candidates, it is important to address the challenges mentioned in Section 21.4.4. However, it is also important to look beyond those issues.

The new standards provide solutions to two important areas of cryptography (key exchange and digital signatures), but these are not the only areas where such solutions are needed. For example, in many applications, there is a need to use zero-knowledge proofs (ZKP). Details of ZKPs are beyond the scope of this chapter, but it suffices to know that these are different from the two earlier primitives. As there are applications that need ZKPs, it would be necessary to have also postquantum secure ZKPs. Currently, some of the applied systems are quantum-safe and some are not [38].

In addition to ZKPs, there are many more advanced cryptographic systems that are used in various applications such as homomorphic encryption, multi-party computation, attribute- and identity-based encryption, and more. Although some algorithms are "inherently" quantum-safe (meaning that they are based on problems other than the ones susceptible to Shor's algorithm), not all enjoy such properties.

Furthermore, it is also important to study how the new standards affect the security proofs of more complex protocols that apply for example cryptographic signatures. Similarly as in the case of implementations, it is not straightforward to replace one primitive with a PQC primitive and claim that the proof of security immediately holds. Thus, there is a great need for research on this field as well.

21.4.6 6G and PQC

With the development of 6G, it is important to take the PQC standards into account from the beginning. These will be the cryptographic primitives that the 6G systems will need to use when they are built. In addition, it is essential to look into the alternate candidates as well. Because the properties of different PQC methods are so different, it is also necessary to see, which of the algorithms and with what parameters are best suited for different 6G applications.

One consideration is also whether to have some short-lived and not necessarily critical and/or widely distributed data that only have traditional cryptographic methods applied to them. This could be an option, when none of the PQC standards can be used due to resource constraints, but previous systems such as ECC could be applied. This can be a better option than leaving the data without any cryptographic protections. However, this will also add to the complexity of the overall system design, which may be disadvantageous. Also, deciding which data are "not-critical" in case there is a possibility for a long-term storage of the data and then decryption attacks in the long distance future that may be severe is a problem that is not easy and with an "all-or-nothing" approach to PQC would not need consideration.

References

1 C. Zalka, "Grover's quantum searching algorithm is optimal," *Physical Review A*, vol. 60, pp. 2746–2751, 1999. [Online]. Available: https://link.aps .org/doi/10.1103/PhysRevA.60.2746.

2 D. R. Simon, "On the power of quantum computation," *SIAM Journal on Computing*, vol. 26, no. 5, pp. 1474–1483, 1997. [Online]. Available: https://doi .org/10.1137/S0097539796298637.

3 L. K. Grover, "A fast quantum mechanical algorithm for database search," in *Proceedings of the 28th Annual ACM Symposium on Theory of Computing*, 1996, pp. 212–219.

4 P. W. Shor, "Polynomial-time algorithms for prime factorization and discrete logarithms on a quantum computer," *SIAM Review*, vol. 41, no. 2, pp. 303–332, 1999.

5 M. Webber, V. Elfving, S. Weidt, and W. K. Hensinger, "The impact of hardware specifications on reaching quantum advantage in the fault tolerant regime," *AVS Quantum Science*, vol. 4, no. 1, p. 013801, 2022. [Online]. Available: https://doi.org/10.1116/5.0073075.

6 C. Gidney and M. Ekerå, "How to factor 2048 bit RSA integers in 8 hours using 20 million noisy qubits," *Quantum*, vol. 5, p. 433, 2021. [Online]. Available: https://doi.org/10.22331/q-2021-04-15-433.

7 E. Gouzien and N. Sangouard, "Factoring 2048-bit RSA integers in 177 days with 13 436 qubits and a multimode memory," *Physical Review Letters*, vol. 127, p. 140503, 2021. [Online]. Available: https://link.aps.org/doi/10.1103/PhysRevLett.127.140503.

8 C. H. Bennett and G. Brassard, "Quantum cryptography: Public key distribution and coin tossing," *Theoretical Computer Science*, vol. 560, pp. 7–11, 2014, theoretical Aspects of Quantum Cryptography celebrating 30 years of BB84. [Online]. Available: https://www.sciencedirect.com/science/article/pii/S0304397514004241.

9 C. H. Bennett, "Quantum cryptography using any two nonorthogonal states," *Physical Review Letters*, vol. 68, pp. 3121–3124, 1992. [Online]. Available: https://link.aps.org/doi/10.1103/PhysRevLett.68.3121.

10 F. Grosshans and P. Grangier, "Continuous variable quantum cryptography using coherent states," *Physical Review Letters*, vol. 88, p. 057902, 2002.

11 D. Huang, P. Huang, D. Lin, and G. Zeng, "Long-distance continuous-variable quantum key distribution by controlling excess noise," *Scientific Reports*, vol. 6, p. 19201, 2016.

12 E. O. Kiktenko, A. O. Malyshev, M. A. Gavreev, A. A. Bozhedarov, N. O. Pozhar, M. N. Anufriev, and A. K. Fedorov, "Lightweight authentication for quantum key distribution," *IEEE Transactions on Information Theory*, vol. 66, no. 10, pp. 6354–6368, 2020.

13 L.-J. Wang, K.-Y. Zhang, J.-Y. Wang, J. Cheng, Y.-H. Yang, S.-B. Tang, D. Yan, Y.-L. Tang, Z. Liu, Y. Yu, Q. Zhang, and J.-W. Pan, "Experimental authentication of quantum key distribution with post-quantum cryptography," *npj Quantum Information*, vol. 7, p. 67, 2021.

14 G. Brassard and L. Salvail, "Secret-key reconciliation by public discussion," in *Advances in Cryptology — EUROCRYPT '93*, T. Helleseth, Ed. Berlin, Heidelberg: Springer-Verlag, 1994, pp. 410–423.

15 B.-Y. Tang, B. Liu, Y.-P. Zhai, C.-Q. Wu, and W.-R. Yu, "High-speed and large-scale privacy amplification scheme for quantum key distribution," *Scientific Reports*, vol. 9, p. 15733, 2019.

16 M. Grillo, A. Dowhuszko, M.-A. Khalighi, and J. Hämäläinen, "Resource allocation in a Quantum Key Distribution Network with LEO and GEO trusted-repeaters," in *17th International Symposium on Wireless Communication Systems (ISWCS)*, 2021, pp. 1–6.

17 Y. Tian, P. Wang, J. Liu, S. Du, W. Liu, Z. Lu, X. Wang, and Y. Li, "Experimental demonstration of continuous-variable measurement-device-independent quantum key distribution over optical fiber," *Optica*, vol. 9, no. 5, pp. 492–500, 2022. [Online]. Available: http://opg.optica.org/optica/abstract.cfm?URI=optica-9-5-492.

18 R. J. Hughes, G. L. Morgan, and C. G. Peterson, "Quantum key distribution over a 48 km optical fibre network," *Journal of Modern Optics*, vol. 47, no. 2–3, pp. 533–547, 2000. [Online]. Available: https://doi.org/10.1080/09500340008244058.

19 R. Bedington, J. M. Arrazola, and A. Ling, "Progress in satellite quantum key distribution," *npj Quantum Information*, vol. 3, p. 30, 2017.

20 W. Diffie and M. E. Hellman, "New directions in cryptography," *IEEE Transactions on Information Theory*, vol. 22, no. 6, pp. 644–654, 1976.

21 L. Panny, "Guess what?! On the impossibility of unconditionally secure public-key encryption," Cryptology ePrint Archive, Paper 2019/1228, 2019, https://eprint.iacr.org/2019/1228. [Online]. Available: https://eprint.iacr.org/2019/1228.

22 U. Maurer, "Information-theoretic cryptography," in *Annual International Cryptology Conference*. Springer, 1999, pp. 47–65.

23 R. L. Rivest, A. Shamir, and L. Adleman, "A method for obtaining digital signatures and public-key cryptosystems," *Communications of the ACM*, vol. 21, no. 2, pp. 120–126, 1978.

24 A. K. Lenstra, H. W. Lenstra, M. S. Manasse, and J. M. Pollard, "The number field sieve," in *The Development of the Number Field Sieve*. Springer, 1993, pp. 11–42.

25 T. ElGamal, "A public key cryptosystem and a signature scheme based on discrete logarithms," *IEEE Transactions on Information Theory*, vol. 31, no. 4, pp. 469–472, 1985.

26 R. J. McEliece, "A public-key cryptosystem based on algebraic," *Coding Thv*, vol. 4244, pp. 114–116, 1978.

27 W. Castryck and T. Decru, "An efficient key recovery attack on SIDH (preliminary version)," *Cryptology ePrint Archive*, 2022.

28 R. Perlner, J. Kelsey, and D. Cooper, "Breaking category five SPHINCS$^+$ with SHA-256," *Cryptology ePrint Archive*, 2022.

29 W. Beullens, "Improved cryptanalysis of UOV and rainbow," in *Annual International Conference on the Theory and Applications of Cryptographic Techniques*. Springer, 2021, pp. 348–373.

30 E. Crockett, C. Paquin, and D. Stebila, "Prototyping post-quantum and hybrid key exchange and authentication in TLS and SSH," Cryptology ePrint Archive, Paper 2019/858, 2019. [Online]. Available: https://eprint.iacr.org/2019/858.

31 E. Rescorla, "The Transport Layer Security (TLS) Protocol Version 1.3," RFC 8446, August 2018. [Online]. Available: https://www.rfc-editor.org/info/rfc8446.

32 C. M. Lonvick and T. Ylonen, "The Secure Shell (SSH) Transport Layer Protocol," RFC 4253, January 2006. [Online]. Available: https://www.rfc-editor.org/info/rfc4253.

33 M. J. Dworkin, E. B. Barker, J. R. Nechvatal, J. Foti, L. E. Bassham, E. Roback, and J. F. Dray Jr., "Advanced encryption standard (AES)," *Federal Inf. Process. Stds. (NIST FIPS)*, National Institute of Standards and Technology, Gaithersburg, MD, 2001.

34 G. Hofemeier and R. Chesebrough, "Introduction to intel AES-NI and intel secure key instructions," *Intel, White Paper*, vol. 62, 2012.

35 J. Hekkala, K. Halunen, and V. A. Vallivaara, "Implementing post-quantum cryptography for developers." in *ICISSP*, 2022, pp. 73–83.

36 J. Blessing, M. A. Specter, and D. J. Weitzner, "You really shouldn't roll your own crypto: An empirical study of vulnerabilities in cryptographic libraries," *CoRR*, vol. abs/2107.04940, 2021. [Online]. Available: https://arxiv.org/abs/2107.04940.

37 M. Carvalho, J. DeMott, R. Ford, and D. A. Wheeler, "Heartbleed 101," *IEEE Security & Privacy*, vol. 12, no. 4, pp. 63–67, 2014.

38 J. Partala, T. H. Nguyen, and S. Pirttikangas, "Non-interactive zero-knowledge for blockchain: A survey," *IEEE Access*, vol. 8, pp. 227 945–227 961, 2020.

Part VI

Concluding Remarks

22

Concluding Remarks

There is obviously a long journey to get to 6G, while current 5G will continue to evolve over the next few years. Every new generation brings a big leap with respect to previous generation. However, in the long run, 6G will be a revolution rather than an evolution due to the self-managing networks and will drive toward a more sustainable and trustworthy society.

The goal of 6G networks is to fulfill the connectivity requirements of the 2030s and beyond human society. 6G will be the key communication infrastructure to satisfy the demands of future needs of hyper-connected human society in the 2030 and beyond. The development of new technologies, such as smart surfaces, zero-energy Internet of Things (IoT) devices, advance artificial intelligence (AI) techniques, possible quantum computing systems, AI-powered automated devices, AI-driven air interfaces, humanoid robots, self-sustained networks, and future trends of digital societies' such as massive availability of small data, increasing elderly population, convergence of communication, sensing, and computing, gadget-free communication, will demand new applications. Thus, 6G will support new applications such as unmanned aerial vehicle (UAV)-based mobility, connected autonomous vehicles (CAV), Smart Grid 2.0, Collaborative Robots, Hyper-Intelligent Healthcare, Industry 5.0, Digital Twin, and Extended Reality.

In this book, we have identified mainly four key technological domains which may bring the highest impact on 6G security and privacy. In a way it is possible to summarize the benefits and challenges with using blockchain/distributed ledger technologies (DLT) for security, quantum security, distributed AI/machine learning (ML) security, Explainable AI, and physical layer security (PLS). The security, surveillance, accountability, and governance of the network can be implemented through blockchain and DLT in general. As DLT allows to store immutable and transparent logs for each event which can be utilized in the auditing of events, it may introduce trust among unknown entities in the system. However, DLTs may introduce lots of issues with the user and data privacy and extra computation and storage overhead when they try to achieve this trust level. With quantum security

Security and Privacy Vision in 6G: A Comprehensive Guide, First Edition.
Pawani Porambage and Madhusanka Liyanage.
© 2023 The Institute of Electrical and Electronics Engineers, Inc. Published 2023 by John Wiley & Sons, Inc.

algorithms and their implications in network protocols and related security procedures, such as postquantum cryptography and quantum key distribution, should be considered in the design of next-generation networks. AI/ML has two aspects regarding security: it can enable security as well as suffer from threats and vulnerabilities as a founding element of 6G networks. In 6G, AI/ML will be pushed closer to the source of data for ultralow latency while distributing ML functions over the network to attain performance gains due to optimized models and ensemble decision-making. However, overcoming practical constraints of some network elements (e.g. IoT) will be challenging with AI security. PLS mechanisms are expected to advocate and develop relying on the unique characteristics and properties of wireless channels to secure wireless communication. This may include the list of security operations such as authentication, encryption, and key exchange.

Index

Security and Privacy Vision in 6G: A Comprehensive Guide, First Edition.
Pawani Porambage and Madhusanka Liyanage.

Printed and bound by CPI Group (UK) Ltd, Croydon, CR0 4YY

27/10/2024

14580270-0002